# Springer Earth System Sciences

The Springer Earth System Sciences series focuses on interdisciplinary research linking the lithosphere (geosphere), atmosphere, biosphere, cryosphere, and hydrosphere that build the system earth. The series seeks to publish a broad portfolio of scientific books, aiming at researchers, students, and everyone interested in this extremely interdisciplinary field. It covers the entire research area of earth system sciences including, but not limited to, Earth System Modeling, Glaciology, Climatology, and Human-Environment/Earth interactions. Springer Earth System Sciences includes peer-reviewed monographs, edited volumes, textbooks, and conference proceedings.

More information about this series at http://www.springer.com/series/10178

Mirian M. Collantes · Laura Perucca ·
Adriana Niz · Jorge Rabassa
Editors

# Advances in Geomorphology and Quaternary Studies in Argentina

## Special Symposium from the Argentine Association of Geomorphology and Quaternary Studies, October 2017

Springer

*Editors*
Mirian M. Collantes
Instituto de Geociencias y Medio Ambiente
National University of Tucumán
Tucumán, Argentina

Laura Perucca
Instituto de Geología, CIGEOBIO
National University of San Juan
San Juan, Argentina

Adriana Niz
Departamento de Geología
National University of Catamarca
San Fernando del Valle de Catamarca,
Catamarca, Argentina

Jorge Rabassa
Laboratorio de Geomorfologia y Cuaternario
CADIC-CONICET
Ushuaia, Tierra del Fuego, Argentina

ISSN 2197-9596 ISSN 2197-960X (electronic)
Springer Earth System Sciences
ISBN 978-3-030-22620-6 ISBN 978-3-030-22621-3 (eBook)
https://doi.org/10.1007/978-3-030-22621-3



This Springer imprint is published by the registered company Springer Nature Switzerland AG
The registered company address is: Gewerbestrasse 11, 6330 Cham, Switzerland

# Preface

Most of the papers that are included in this volume were presented during a special Symposium on Geomorphology, Quaternary and Climatic Change studies, within the framework of the 20th Argentine Geological Congress, that took place in 2017 in the province of Tucumán, northern Argentina.

This Symposium had as a main objective to offer a discussion space, about the climatic and/or anthropogenic mechanisms that generate geomorphological processes, from a space–time perspective and from an integrated and multidisciplinary viewpoint. The characteristics of such processes in different regions of Argentina were discussed, both in relationship to present climatic and hydrological variability, as well as from palaeoclimatic studies, as an expression of past environmental conditions. Several aspects related to the landscape genesis and evolution were taken into consideration, attending to the different morphogenetic and morphoclimatic environments. The scientific approach on the effects of climatic variability, both seasonal, annual or pluriannual, was taken into account, as well as the impact of global climatic change on the surface dynamics and the anthropogenic activity.

Besides, the integration of the palaeoclimatic and neotectonic influence on the reconstruction of the present landscape evolution was considered, by means of papers related to the study and analysis of the crustal deformation during the Quaternary and the evidence of palaeoseismicity in the sedimentary record, oriented to unveil the potentiality of an area as a potential seismogenic source. Diverse contributions were presented focused on a neotectonic evaluation which would allow a better knowledge of the Quaternary activity of the regional faults, their co-seismic displacement, the fault slip rate, the earthquake magnitude and recurrence interval, the seismotectonic association of the historical earthquakes and the morphological evolution of the landscape due to tectonics.

Thus, this volume gets together important scientific contributions, and it becomes an appropriate ambient to communicate and discuss the advances of those investigations devoted to the Quaternary palaeoclimatic and palaeoenvironmental reconstruction and evolution and the basic and applied geomorphology of Argentina. This volume also displays the complex framework of the Argentine landscapes, as exposed by contributions from different regions of this country, which extends from tropical to sub-Antarctic environments.

This volume starts with a challenging contribution of Rosa Compagnucci et al. on the "**Relationships among a supernova, a transition of polarity of the geomagnetic field and the Plio-Pleistocene boundary**", which considers that there is an increase of Galactic Cosmic Rays (GCR) in space due to the birth of a supernova near our solar system. This event would have started around 2.8 to 2.6 Ma and finished between 1.7 and 1.5 Ma with a peak around 2.2 Ma; a consequent increase of radiation input to Earth occurred. This increase was favoured by the weakening and even suspension of the Earth's magnetic field during the reversion between the Gauss and Matuyama Chrons. All these factors together generated a relative maximum of the GCR flux on the Earth, which could act as trigger of the beginning of the first Quaternary glaciation and the appearance of the oscillations between warmer and colder stages that characterized the Pleistocene.

The following contribution is by SC Grill et al., "**Neogene continental deposits of Bahía Blanca, Argentina. Palaeoenvironmental and palaeoclimatics inferences**", which involves deposits of continental Neogene that outcrop in the Bahía Blanca region, Argentina. The studies that were carried out in order to contribute to the palaeoenvironmental and chronostratigraphical knowledge, involve sedimentological, palaeomagnetic and magnetic susceptibility analyses. One of the outcrops studied (Cueva de Los Leones) is mainly constituted of facies of fine to thick sandstones, largely pedogenized and with important amounts of calcrete, whereas the other (Loma Sarmiento) has a predominance of loess-like silt facies, reddish brown in colour, an abundance of rhizoconcretions and evidence of pedogenic processes. Both outcrops culminate in an important layer of calcrete of regional scope, with particular morphologies at each of the sites. The palaeomagnetic analysis leads to the conclusion that the sedimentation of the deposits underlying the calcareous crust at the two sites was not synchronous. Reverse polarity was recorded in the Cueva de Los Leones in contrast to Loma Sarmiento, where normal polarity is predominant.

"**Perched dunes in the Fuegian steppe, southern Argentina: archaeological reservoirs of Holocene information**" is the title of the contribution by A. Coronato et al., comprising the description of the genesis and the geomorphological characteristics of three perched dune fields, located in shallow lakes named as O´Connor, Amalia and Arturo. Perched dunes develop over the cliffs of most of the ephemeral shallow lakes of the Fuegian steppe (lat. 53° S). They are formed by the alternation of massive aeolian deposits and soils, which are distinguished from each other by depth, colour, edaphic structures and organic matter content. The edaphized levels are interpreted as palaeosoils formed during wetter periods when the aeolian sedimentation was less intense. A set of radiocarbon dating, obtained from organic matter in most of them and in different dunes, shows that they were formed since the Late Glacial to the Late Holocene. This fact is considered as evidence of humid–arid periods cyclicity in the region.

**"Morphotectonic and gravimetric analysis of the La Burra valley, central Precordillera. San Juan, Argentina"** is the contribution by LP Perucca et al. This work describes new evidence of Quaternary tectonic activity in the La Burra intermountain valley in the central Precordillera, San Juan Province, Argentina. Quaternary structures are located in the eastern piedmont of the Sierra de la Crucecita, and western piedmont of the Sierra de Talacasto (30° 54'–31° S and 68° 47'–68° 55' W), about 70 km north-west of San Juan. By means of morphotectonic analysis, the main deformation features have been identified and located in both piedmonts, named from north to south: Las Crucecitas and Vertientes faults (in the eastern piedmont of the Sierra de La Crucecita) and the Western Talacasto fault system (in the western piedmont of the homonymous range). The main geological structures were identified by the analysis and interpretation of gravimetric anomalies, employing data obtained from World Gravity Map 2012 (WGM12), which includes earth and satellite gravity data derived from the EGM2008 model. The gravimetrical response of the crustal blocks that compose the area under study shows a correlation with the most superficial crustal structure and could be associated with the different morphotectonic domains recognized in the region.

Likewise, the chapter on **"Morphotectonic analysis in the eastern piedmont of the Sierra de la Huerta, Sierras Pampeanas Occidentales, San Juan, Argentina"** presents a similar contribution by LM Rothis et al., about a geomorphological and morphotectonic analysis that was carried out in the eastern flank of the Sierra de la Huerta to characterize the available evidence of tectonic activity. The Andean back arc between 27° S and 33° S is characterized by an intense seismicity, being the epicentre of the most damaging earthquakes in Argentina. Those events produced deformation, causing changes on the land surface morphology. Three morphostructural units (mountain, transition and depression) were recognized. The transition unit shows Quaternary active faulting where eight fault traces with a NW trend and associated scarps were identified. These evidences of Quaternary faulting were assigned to the Marayes Fault System. Detailed topographic profiles perpendicular to the strike of fault scarps allowed to measure main deformation to the north of the study area. Neotectonic deformation in the Arroyo Los Papagayos (arroyo=creek, sometimes an ephemeral one) produced anomalies in its pattern and longitudinal profile. Thus, this study allowed the consideration of the Marayes Fault System as a potential seismogenic source.

MY Esper Angellieri et al. presented the paper "**Morphometric analysis of river basins using GIS as a basis for peak discharge calculation. Case of study: Pre-Andean section of National Route 150, Argentina**", which analysed various morphometric characteristics of basins in order to evaluate debris flood or flash flood hazards. This study, when properly combined with geomorphology and geology analyses, helps to elaborate a primary hydrological diagnosis to predict the approximate behaviour of the river basin at the time of heavy rainstorms. The chosen area includes a stretch of National Road No. 150 up to the International Border Pass Agua Negra between Argentina and Chile in a sector of the western and central Pre-Andean. A total of 122 river basins were identified and delineated, but only 3 (those of larger surface area) were selected for morphometric analysis and as a suitable case to evaluate the potential magnitude of flash floods in relation to their associated hazard level in the region.

The contribution by P Santi Malnis et al., **"Distributary drainage systems in La Huerta Range, western Pampean Ranges, San Juan, Argentina"**, comprises a geomorphological and sedimentological survey of the western piedmont of the La Huerta range, San Juan (Argentina), showing that there is a north–south trend change of drainage basin morphometry and facies architecture. The study showed facies distribution on the systems under study. On Las Chacras river, in the northern portion of the piedmont, fits with the classical alluvial fan model, with a divergent net of channels from the fan apex. In the middle portion of the piedmont (the Arenal and Puquial rivers), alluvial fan facies are also present with aeolian interaction, triggering disconnection of middle fan facies from distal facies. In the southern portion of the piedmont, drainage system flows parallel to axial drainage system (Bermejo River). In the Cañada River, flows keep channelized until its distal portion in middle to high sinuosity channels that deposit alternate and scroll bars. In the distal portion, main channel diverge in a distributary net of channels that develops a fan-shaped deposit characterized by terminal splays and aeolian dunes.

The paper by CE Ginestra Torcivia and NN Ríos López on **"Preliminary morphometric analysis. Río Talacasto basin, Central Precordillera of San Juan, Argentina"** analyses various morphometric characteristics of the Río Talacasto basin, situated approximately 80 km north-west of the San Juan city, in the central Precordillera geological province. Various morphometric parameters were analysed in order to evaluate its flash flood hazard. For this purpose, some measurements (area, perimeter, length, width, maximum and minimum height and Strahler order, slope, aspect, curvature and convergence index, topographic position index, and roughness index) were calculated. These measurements allows future analysis to predict approximately the behaviour of the basin in the presence of a series of theoretical rainstorms that may generate unusual runoff volumes. The importance of this evaluation is that the basin includes the National Route 149 and the Provincial Route 436, both being the main accesses to the northern and western portions of the province. The results obtained conclude that the basin has a significant susceptibility to the occurrence of flash floods and other landslides.

The paper by EV Ortiz et al., **"Risk analysis due to the variation of chemical parameters induced by lithology in Fiambalá: Chaschuil River, Catamarca, Argentina"**, discussed that, in predominantly semiarid environments, such as the town of Fiambalá, Catamarca, the water resource directly impacts on the socio-economic development of the region. The demand of the resource increases the vulnerability and can produce a deterioration in quality and quantity due to different environmental and anthropic factors. Therefore, it is necessary to carry out studies that improve the knowledge of water resources and help a more efficient planning in the management of these resources for the development and the general welfare of the population. The Abaucán River is mainly supplied by the Fiambalá and Chaschuil rivers, the latter being the most important tributary. Lithology affects the location and exploitation of aquifers; therefore, it is necessary to analyse the fluvial contributions of the area, the Quaternary deposits, and the physical–chemical quality of the water that flows over them. From the analysis carried out, it is concluded that the geochemical alteration due to the lithological influence is almost negligible, which is why it presents strong performance in the agricultural development of the region.

Boretto et al., on the "**Identification of knickpoints in littoral basins of Argentine Patagonia: geomorphic markers in a passive margin".** These authors analysed the knickpoints that are an expression of disequilibrium in a fluvial system, and they occur in both alluvial and bedrock rivers and provide a method for using knickpoints as geomorphic markers in coastal basins associated with the Atlantic Ocean coast. However, less attention is paid to the fluvial systems in well-developed environments along the passive coastal margin as the Atlantic Ocean coast in Patagonia. For the first time in this region, knickpoint analysis together with hypsometric curves and longitudinal stream profiles studies was conducted to understand the landform evolution in coastal basins along the passive margin in Bustamante and Solano Bays (Chubut province, Patagonia, Argentina) during the recent past.

E. García Aráoz et al. have contributed a paper on "**Granitic caves from the Achala Batholith in the Province of Córdoba, Argentina: A study on three particular stories with a common past"**.

This paper discussed three caves developed especially in granitic rocks composition, which are analysed in order to specify their particular origins and classify it according to its morphological characteristics and the genetic processes involved. These caves are located on the same scarp of the rock massif and, therefore, share a geological and geomorphological common history, which is shown in their hypothesis of evolution. In addition, based on the type of cavity and certain observations of their environments, it is possible to establish a relative chronological order in their ages of formation. In this context, it is proven that these underground environments (as the other geographical accidents and granitic landforms) are not chaotic but—on the contrary—they are structured in geological and geomorphical processes over time and they respond to a common evolutionary framework to the entire rock massif.

The paper by JP Velázquez et al., **"Scenic and geotouristic potential of volcanic landscapes in southern Patagonia: necks, buttes and vents, Santa Cruz Province, Argentina",** analyses the spatial distribution of volcanic landforms that compose natural elements, which occur as positive features in the Southern Patagonia tablelands. In the province of Santa Cruz (Southern Patagonia, Argentina), an inventory of 106 relevant landforms has been completed: 52 necks, 28 buttes and 26 isolated eruptive centres, which provide a high degree of variability in the visual components of the landscape. A database has been prepared, containing information about their geographical position and localization with respect to the main circulation routes, cities and towns, elevation, relative relief and geological age. Likewise, a database of satellite imagery and aerial photographs has been compiled, and thematic cartography was prepared. This information allowed the identification of the most appropriate sectors in which detailed case studies could be performed, with the aim of evaluating their scenic possibilities and their potential to develop activities associated with geodiversity and geotourism.

The paper on "**Early human occupations in the valleys of northwestern Argentina: contributions to dating by the varnish micro-laminations technique",** by JP Carbonelli et al., presents the results of the micro-laminations of rock varnish (VML), made on artefacts from a quarry-workshop located in the Santa María valley, Argentina. The data obtained indicate that the minimum age of exposure of the surface of these artefacts is 9400 radiocarbon years AP, which allows us to begin discussing the settlement of the region of the intermontane valleys of north-western Argentina, throughout the Holocene. The results obtained through this methodology, mainly the detection of ten humid events throughout the studied sequence, contribute to the knowledge of the palaeoclimatic variability that occurred during the Holocene in the study region.

Mirian Collantes
Laura Perucca
Adriana Niz
Jorge Rabassa

# Acknowledgements

The editors acknowledge the authorities of the 20th Argentine Geological Congress, in which framework the Symposium on Geomorphology, Quaternary and Climatic Change studies took place. We also want to thank the authorities of the Argentine Association of Quaternary Studies and Geomorphology (AACYG) for the support to the cited Symposium and for sponsoring this volume. We would also like to thank Springer Nature for accepting our proposal of this publication.

We are greatly indebted to those colleagues and specialists who generously reviewed the contributions published in this volume: E Aguilera, P Bouza, E Brunetto, M Cioccale, C Colombi, S Degiovanni, N Doffo, MY Esper Angilieri, F Hongn, A Mehl, G Ojeda, JL Palacio Prieto, A Rapalini, M Salemme and J Suriano.

Finally, we want to express our gratitude to the corresponding authors without whose contribution this volume would not have been possible.

# Obituary

**Professor Dr. José Manuel Sayago (December 8th, 1940–June 8th, 2018)**

Alfred Zinck
(former Professor at the International Institute for Aerospace Survey and Earth Sciences, The Netherlands)

Jorge Sanabria
(National University of Córdoba, Argentina)

Mario Caria, Julián Gómez Augier, Mirian Collantes
(National University of Tucumán, Argentina)
Mario Toledo (Miguel Lillo Foundation)

José Manuel Sayago was born in the province of Córdoba, Argentina, on December 8th, 1940. After finishing his studies in geology at the National University of Córdoba, he started his activities as a lecturer of Physical Geography in a prestigious high school belonging to the local university, the Colegio Monserrat, a condition that he kept for three years. During the same period, he worked as a technician in the Cartography Department of the province of Córdoba.

In 1973, he joined the Instituto Nacional de Tecnología Agropecuaria of Argentina (National Institute for Agricultural Technology, INTA), as Principal Investigator at the Estación Experimental Regional Agropecuaria Roque Sáenz Peña (province of Chaco) and the Estación Experimental Regional Agropecuaria La Banda (province of Santiago del Estero). In such organization, he worked essentially in the development of soil cartography with a geomorphological base, participating in the preparation of the regional soil map of the Gran Chaco. According to his own original words, during this period he learned about the intricate relationship between geomorphology and soils, together with the most important specialists on soil mapping and taxonomy of INTA. He also took part of the team responsible of the Soil Mapping Plan of the Pampean region, specifically the soil map of the province of Córdoba, under the responsibility of the Dirección Provincial de Asuntos Agrarios of the cited province.

In 1975, he obtained his doctoral degree in geology at the National University of Córdoba, with a dissertation on "Applied Geology to the hydrology and soils in the San Alberto valley (province of Córdoba)". In 1978, after having been sent off of INTA for political reasons during the military dictatorship that had taken up the government in Argentina, he decided to move to the city of San Miguel de Tucumán, where he started his university career on geomorphology and Quaternary geology, at the Facultad de Ciencias Naturales and the Instituto Miguel Lillo of the National University of Tucumán, where he stayed for the rest of his life.

He was one of the first geologists in becoming aware of the environmental problems of the northernmost region of Argentina, and thus, he created the Instituto de Geociencias y Medio Ambiente (Institute of Geosciences and Environment) of the National University of Tucumán. In the framework of this Institute, his research became oriented in the field of geo-environmental hazards. Likewise, he was a pioneer in the study of the impact of climatic change and global warming in this region and developed predictive models in these disciplines. Among other tasks, he edited a book on "Geomorphology and Climatic Change", which includes contributions from the most important specialists of Argentina on these areas.

His permanent interest in acquiring knowledge and becoming acquainted with new technologies and cartographic technique moved him in 1994 to obtain a master of science in the International Institute for Aerospace Survey and Earth Sciences (ITC, today the Faculty of Geo-Information Science and Earth Observation, University of Twente, the Netherlands). His thesis was on the subject of "Methodological aspects of water erosion inventory using remote sensing techniques in the subtropical region of northwestern Argentina".

In 2003, he promoted, together with other colleagues, the foundation of the Argentine Association of Quaternary Studies and Geomorphology (AACYG), in the framework of the II Argentine Congress of Quaternary Studies and Geomorphology that he had organized together with his research team in the province of Tucumán.

He was Principal Investigator of CONICET (the National Research Council of Argentina), and he directed many research projects financed by such organization, as well as by the Secretary of Science and Technology of the National University of Tucumán (SCAIT) and the National Agency of Scientific and Technological Promotion (FONCYT), among other institutions. His projects were always focused in the study of the Quaternary and the basic and applied geomorphology of north-eastern Argentina, the environment and climatic change. Among his most successful projects, the following should be noted: "Hydrological, agronomical and social sustainability in selected areas of north-western Argentina facing the present and future climatic changes", "Prevention and mitigation of the impact of climatic change in the rural and urban areas of the province of Tucumán" and "Climatic change and landscape resilience in northwestern Argentina", a topic on which he was working until his last days.

José Manuel Sayago was an excellent pedologist, integrating with great success the disciplines of geomorphology and pedology. Those of us who were so fortunate to share with him some time in the field were able to appreciate the precision of his soil descriptions and the relevancy of his landscape observations. His capacity of synthesis enabled him to properly interpret the geo-pedological information with the aim of reconstructing the landscape evolution and detecting climatic changes in environments that are quite different as the lowlands of Burruyacú and the highlands of Tafí del Valle. In addition to his scientific interest and his passion for the soil–landscape relationships, José Manuel was a person very worried for the increasing degradation of the land in the subtropical region of north-western Argentina. His regional vision allowed him to estimate, by means of predictive scenarios, the loss of soil in the province of Tucumán under estimated, future climate changes and to propose techniques of protection and recovery of the degraded lands. To put the scientific information at the reach of local and regional users was one of the permanent objectives of his professional activities.

Sayago also constructed, during his long academic and scientific careers, a very fruitful connection with foreign research centres and colleagues, mainly from the Netherlands, France (Bordeaux), London, Spain and Brazil, thus permitting that researchers of those countries conducted investigation and/or postgraduate formation in the University of Tucumán. Additionally, he promoted and coordinated research projects of international cooperation between the University of Tucumán and foreign research centres. Among them, it should be mentioned the research projects with the ITC (the Netherlands) and the Institut de Géographie et Aménagement, of the Université Michel de Montaigne-Bordeaux. With this Dutch institute, the scientific activities lasted for over four years and their main objective was to study the relationship between the land degradation and the deforestation process during the last 40 years in the Tucumán lowlands. The achieved results were published in the book edited by ITC entitled “Land use change and land degradation in the western Chaco plain (Tucumán province, north-west Argentina, Burruyacú region)”. With the French institute, the results of the research projects performed in Tucumán during two years were included in the book “Le piédmont andin argentin: environnement, risques et enjeux”, edited by the Laboratoire de Géographie Physique Appliquée, of the cited institute.

In addition to his distinguished career as a researcher, Sayago was a brilliant lecturer, as Full Professor of geomorphology (as part of the geology degree) and of Quaternary geology (within the degree on archaeology) of the University of Tucumán. Through his lectures, the geology students learned to interpret the geomorphological evolution of the landscape and the impact of anthropogenic activities on the development of morphodynamic processes. He taught his students the importance of the integrated geomorphological cartography, including morphogenetic, morphostructural, morphodynamic and morphochronological aspects. Similarly, the archaeology students applied methodologies developed by Sayago, aimed to the study of Native American, pre-Hispanic settlements with geomorphological bases, oriented to understand the perception and management of the landscape that these human groups had developed. He always exposed the need of recovering the soil and water management practices that the pre-Hispanic dwellers had improved in the mountain valleys of north-western Argentina, to prevent and/or mitigate the soil erosion and degradation processes.

We must mention that José Manuel Sayago was a pioneer geologist promoting a substantial approach of the Quaternary studies to the archaeological problems of north-western Argentina. His permanent relationship with the archaeology lecturers and researchers of the University of Tucumán provided a fundamental contribution in the development of research projects that had, as a core objective, the study of the palaeoenvironments in pre-Hispanic times, as well as on the study of the use of space and the past social relationships. His contributions in geoarchaeology, together with those of archaeologists trained in his laboratory, implied the starting point in this discipline concerning the archaeology of north-western Argentina, a methodology that his disciples have followed until today.

During his docent and researcher life, Sayago published more than one hundred scientific papers in international journals, chapters of books and proceedings of scientific meetings and symposia. He always defended that the results of scientific research should have as a priority the resolution of regional problems without ignoring the social needs. In such sense, he privileged the publication of his results at the regional and national levels. He supervised many graduation projects and doctoral theses in geology, geography, archaeology, biology and agricultural sciences. He motivated and helped his students to have the opportunity of completing their formation in prestigious centres abroad and to actively participate in international events.

Finally, it is important to highlight his firm attitudes, as a neat sign of his robust personality. Those who were his disciples and colleagues remember him, not only as a distinguished scientist but as a man of steady and well-founded principles and generous character, a gentleman always available for enjoying good food and quality wines, particularly those made from grapes grown in his beloved part of the world. He loved to tell stories of his numerous trips to certain parts of the world, which he greatly appreciated and visited frequently. In the field, his straw hat and his neck handkerchief were a classical characteristic of his figure. Either under the merciless sun of the semiarid mountain landscape or under the shade of a centennial "algarrobo" tree, his enthusiasm for science never diminished. We may say that he was a fearless, impulsive and sentimental person. He lived and died as he wanted to, taking up to the last drop of the green sap of life and trying to be loyal with himself and with his history until the very end.

José Manuel Sayago leaves behind an extensive knowledge of the basic and applied geomorphology and the Quaternary studies of north-western Argentina. His legacy will undoubtedly survive through those who had the opportunity and the privilege to meet and work with him, particularly his disciples and his students who shared his passion for fieldwork in the most remote corners of the Andean mountains and the endless piedmont areas of north-western Argentina.

## Some of His Most Relevant Publications

### *Book Chapters*


Sayago JM, 1983. Geomorfología. In: Geología de la Sierra de Ancasti. **Munster Forsch. Geol. Palaont.** 59, 265–284.

Sayago JM, 1983. Suelos. In: Geología de la Sierra de Ancasti. **Munster Forsch. Geol. Palaont.** 59, 285-305.

Sayago JM, 1984. Suelos de la provincia de Tucumán. In: **Geología de Tucumán**, Aceñolaza G, Toselli A, Bossi G (editors), 15, 133–141. Tucumán.

Sayago JM, Ratto L, Collantes MM, 1984. Geomorfología. In: **Geología de Tucumán**, Aceñolaza G, Toselli A, Bossi G (editors), 16, 143–154. Tucumán.

Meijerink AMJ, Sayago JM, Emaruchi B, Ziadi K, 1986. A spatial assemblage model for the estimation of gross erosion and sediment yield using remote sensing and geodata base operations. In: **Land Evaluation for Land Use Planning and Conservation in slopping areas**. Public. International Institute for Aerospace Survey and Earth Sc. (ITC). Enschede, The Netherlands, 174–195.

Sayago JM, 1987. Grandes unidades geomorfológicas de la Argentina. In: **Territorio (Centro Editor de América Latina),** 108–109.

Collantes MM, Sayago JM, 1989. Interpretabilidad de las imágenes "MOMS" en el mapeo geológico y geomorfológico de un sector de la Puna argentina. In: **La Teledetección en las Ciencias de la Tierra, Public. Espec. INGEMA (San Miguel de Tucumán)**: 57–70.

Sayago JM, Powell J, Collantes MM, Neder L, 1998. El Cuaternario. In: **Geología de Tucumán**, Gianfrancisco M, Puchulu ME, de Cabrera JD, Aceñolaza G (editors), 11, 111–128. Tucumán.

Sayago JM, Collantes MM, Toledo MA, 1998. Geomorfología. In: **Geología de Tucumán**, Gianfrancisco M, Puchulu ME, de Cabrera JD, Aceñolaza G (editors), 21, 241–257. Tucumán.

Sayago JM, Puchulu ME, Neder L, 1998. Suelos. In: **Geología de Tucumán**, Gianfrancisco M, Puchulu ME, de Cabrera JD, Aceñolaza G (editors), 24, 275–286. Tucumán.

Sayago JM, 1999. El Cuaternario de la Provincia de Tucumán. **Relatorio XIV Congreso Geológico Argentino**, Salta, 1, 451–462.

Puchulu ME, Sayago JM, 1999. Los Suelos de la Provincia de Tucumán. **Relatorio XIV Congreso Geológico Argentino**, Salta, 2, 161–171.

Sayago JM, 1999. El Cuaternario de la región pre-puneña del noroeste Argentino. In: **Geología Argentina,** 23, Cuaternario. Anales, 29, 688–691. Servicio Geológico Minero Argentino.

Sayago JM, Toledo M, 2000. Rainfall increase, land use and morphodynamic changes in northwestern Argentina as indicators of the effects of future climatic changes. In: **Geomorphology and human activity: Their role in Global Environmental Changes**, Olaf Slaymaker (editor), 8, 185–195. John Willey & Sons, Ltd.

Sayago JM, Sampietro MM, Caria M, Collantes MM, 2003. Paleoclimatic changes and human crisis in northwest Argentina during the European medieval warm period. In: **Quaternary Climatic Changes and Environmental Crises in the Mediterranean Region**, Ruiz Zapata MB, Dorado Valiño M, Valdeolmillos Rodríguez A, Gil García MJ, Bardajalí T, Azcárate I, De Bustamante Gutiérrez, Martínez Mendizábal I (editors), 10, 81–87. Madrid.

Sayago JM, Zinck A, Collantes MM, Toledo M, 2003. Evolución ambiental de los valles pre-andinos y la llanura chaqueña occidental argentina durante el Pleistoceno tardío y el Holoceno. In: **Cuaternario y Geomorfología**, Collantes MM, Sayago JM, Neder L (editors), 185–200. San Miguel de Tucumán.

Sampietro Vattuone MM, Sayago JM, Caria M, Collantes MM, 2003. Cambio climático y dinámica poblacional en el noroeste argentino durante los periodos "Formativo" y "Desarrollos Regionales". In: **Cuaternario y Geomorfología**, Collantes MM, Sayago JM, Neder L (editors), 463–473. San Miguel de Tucumán.

Sayago JM, Collantes MM, Neder L, Toledo M, 2006. Cambio climático y riesgos ambientales en el Gran San Miguel de Tucumán. In: **Desafíos Ambientales del Gran San Miguel de Tucumán**, Aquino AL, Collantes MM, Mon R (editors), 24, 405–420, MAGNA Publicaciones, Tucumán.

Prat MC, Sayago JM., 2006. Riesgos ambientales y desafíos en el Gran San Miguel de Tucumán, Noroeste de Argentina. In: **Desafíos Ambientales del Gran San Miguel de Tucumán**, Aquino AL, Collantes MM, Mon R (editors), 21, 345–369. MAGNA Publicaciones, Tucumán.

Sayago JM, Zinck JA, Collantes MM, 2006. Paleoenvironmental evolution. In: **Land use change and land degradation in the western Chaco plain (Tucumán province, Northwest Argentina, Burruyacú region)**, Zinck A (editor), Special Public ITC, 3, 19–32, The Netherlands.

Van Zuidam R, Collantes MM, Sayago JM, 2006. Geomorphology. In: **Land use change and land degradation in the western Chaco plain (Tucumán province, Northwest Argentina, Burruyacú region)**, Zinck A (editor), Special Public ITC N° 84, 4, 33–52, The Netherlands.

Zinck JA, Flores E, Sayago JM, 2006. Soil compaction and fertility depletion. In: **Land use change and land degradation in the western Chaco plain (Tucumán province, Northwest Argentina, Burruyacú region)**, Zinck A (editor), Special Public ITC N° 84, 10: 169–233. The Netherlands.

Sayago JM, Collantes MM, Sayago F, 2007. El deterioro de las tierras en el valle de Tafí. Propuesta para la preservación del patrimonio natural y cultural. In: **Paisajes y procesos sociales en Tafí del Valle. Una mirada interdisciplinaria desde el Valle (Tucumán, Argentina)**, Arenas P, Manasse B, Noli E (editors), Instituto de Arqueología y Museo, Universidad Nacional de Tucumán, 335–368. Tucumán.

Sayago JM, Collantes MM, 2009. 'Cambio climático o cambio geomorfodinámico' In: Geomorfología y Cambio Climático, Sayago JM, Collantes MM (editors), INGEMA (Inst. de Geociencias y Medio Ambiente), 1, 9–24. MAGNA Ediciones, San Miguel de Tucumán.

Busnelli J, Sayago JM, Collantes MM, 2009. Riesgo erosivo ante diferentes escenarios de cambio climático en la provincia de Tucumán (Argentina). In: Geomorfología y Cambio Climático, Sayago JM, Collantes MM (editors), INGEMA (Inst. de Geociencias y Medio Ambiente), 7, 97–118. MAGNA Ediciones, San Miguel de Tucumán.

Sayago JM, Collantes MM, 2016. Geopedology and land degradation in northwest Argentina. In: **Geopedology−An Integration of Geomorphology and Pedology for Soil and Landscape Studies, Zinck A, Metternicht G, Bocco G, Del Valle HF** (editors), Springer International Publishing Switzerland 2016, 27, 441–455.

Angueira C, Cruzate G, Zamora EM, Olmedo GF, Sayago JM, Castillejo I, 2016. Soil mapping base on landscape classification in the semiarid Chaco, Argentina. In: **Geopedology−An Integration of Geomorphology and Pedology for Soil and Landscape Studies, Zinck A, Metternicht G, Bocco G, Del Valle HF** (editors), Springer International Publishing Switzerland 2016, 17, 285–303.

Collantes MM, Sayago JM, Carbonelli JP, Isuani A, 2017. Evolución paleoambiental del Pleistoceno y Holoceno de la provincia de Tucumán. In: **Relatorio XX Congreso Geológico Argentino**, Muruaga C, Grosse P (editors), Ciencias de la Tierra y Recursos Naturales del NOA, 15, 291–306. San Miguel de Tucumán.

Collantes MM, Sayago JM, González L, Isuani A, 2017. Amenazas geomorfológicas y cambio climático en la provincia de Tucumán. In: **Relatorio XX Congreso Geológico Argentino**, Muruaga C, Grosse P (editors), Ciencias de la Tierra y Recursos Naturales del NOA, 38, 783–803. San Miguel de Tucumán.

## *Scientific Journals*

Sayago JM, Tenchini A, 1973. Estudio integrado de los recursos naturales en el chaco-santiagueño. **Revista Asociación Geológica de Córdoba**, 1, 3.

Sayago JM, 1976. Utilidad de las imágenes orbitales (LANDSAT) en el inventario de los recursos naturales. **Revista IDIA (INTA)**, 337–342, 46–51.

Sayago JM, 1980. Estudio de la fisiografía y los suelos en el sureste de Santiago del Estero (Argentina), mediante el empleo de imágenes Landsat. **Acta Geológica Lilloana, 15,** 2, 5–20.

Sayago JM, 1980. Geomorfología aplicada a hidrología y suelos en el valle de San Alberto (Provincia de Córdoba-Argentina). **Acta Geológica Lilloana** 15, 2, 120–138.

Aceñolaza FG, Sayago JM, 1980. Análisis preliminar sobre la estratigrafía, morfodinámica y morfogénesis en la región de Villa Urquiza, Provincia de Entre Ríos. **Acta Geológica Lilloana** 15, 2, 139–154.

Sayago JM, 1981. Las grandes unidades geomorfológicas de la República Argentina. **Revista "Atlas Total".**

Sayago JM, 1981. Morfogénesis de los "barreales" y su relación con el deterioro del paisaje en el valle de Catamarca. **Acta Geológica Lilloana** 15, 3, 75–86.

Sayago JM, 1982. Las unidades geomorfológicas como base para la evaluación integrada del paisaje natural. **Acta Geológica Lilloana** 16, 1, 169–180.

Sayago JM, 1982. La formación Hernandarias y su influencia en los procesos de deterioro de las tierras del centro oeste de la Provincia de Entre Ríos. **Acta Geológica Lilloana** 16, 1, 141–152.

Sayago JM, 1983. Introducción al conocimiento de la Geomorfología del Chaco. **Revista de la Asociación de Ciencias Naturales del Litoral (Argentina)** 8, 125–127.

Sayago JM, 1986. Morfoclimas y paleoformas en la evolución geomorfológica de la Sierra de Ancasti (Cuaternario Superior) Catamarca. **Revista Asociación Geológica Argentina (RAGA)** 41, 1–2, 155–164.

Sayago JM, Collantes MM, 1991. Evolución paleogeomorfológica del valle de Tafí (Tucumán, Argentina) durante el Cuaternario Superior. **Bamberger Geographische Schriften, 11,** 109–124, Bamberg.

Sayago JM, Collantes MM, Arcuri C, 1991. El glaciarismo Finipleistoceno-Holoceno y su relación con los depósitos clásticos pedemontanos en la región montañosa de Tucumán (Argentina). **Bamberger Geographische Schriften, 11,** 155–168, Bamberg.

Sayago JM, 1992. El deterioro del ambiente en el noroeste Argentino. **Revista "Estudios Geográficos"**, Spain, 53, 208, 543–567.

Sayago JM, 1995. The Argentine neotropical loess: an overview. **Quaternary Science Reviews** 14, 755–766. Pergamon.

Zinck A, Sayago JM, 1999. Loess-paleosol sequence of La Mesada in Tucumán Province. Characterization and paleoenvironmental evolution. **Journal of South American Earth Sciences** 12, 293–310.

Sayago JM, Collantes MM, Karlsson A, Sanabria J, 2001. Genesis and distribution of late Pleistocene and Holocene loess of Argentina: a regional approximation. **Journal Quaternary International, 76–77, 1,** 247–257.

Zinck A, Sayago JM, 2001. Climatic periodicity during the late Pleistocene from a loess-paleosol sequence in northwest Argentina. **Quaternary International** 78, 1, 11–16. Pergamon.

Kemp RA, Toms P, Sayago JM, Derbyshire E, King M, Wagoner L, 2003. Micromorphology and OSL dating of the basal part of the loess-paleosol sequence at La Mesada, Tucumán province, Northwest Argentina. Quaternary International 106/107, 111–117. Pergamon.

Sayago JM, Zinck A, Collantes MM, Toledo M, Neder L, 2003. "Environmental evolution of western Chaco plain and pre-Andean valleys (northwest Argentina) during the Late Pleistocene and Holocene". **Terra Nostra 2, 2003, 18. Lateinamerika Kolloquium, TU Bergakadecmic Freiberg**, 67–68.

Collantes MM, Sayago JM, 2003. Los flujos de detritos como indicador de cambios climáticos durante el Cuaternario tardío en los piedemontes de Cumbres Calchaquíes y Aconquija, Tucumán, Argentina. **Terra Nostra 2, 2003, 18. Lateinamerika Kolloquium, TU Bergakadecmic Freiberg**, 21.

Prat MC, Sayago JM, 2003. Risques Naturles et enjeux a San Miguel de Tucumán (north-ouest argentin), **"Cahiers d'Outre Mer"** 223, 301–326. France.

Kemp RA, King GM, Toms P, Derbyshire E, Sayago JM, Collantes MM, 2004. Pedosedimentary development in part of a Late Quaternary loess-palaeosol sequence in Northwest Argentina. Journal of Quaternary Science 19, 6, 567–576. Willey & Sons.

Sayago JM, Zinck A, Collantes MM, Toledo M, 2005. Environmental changes in the pre-Andean valleys and Chaco plain (Northwest Argentina) during the Late Pleistocene and Holocene. **Neues Jahrbuch für Geologie und Paläontologie, Abh.** 236, 1–2, 245–265. Stuttgart.

Busnelli J, Collantes MM, Sayago JM, 2008. Evaluación del grado de torrencialidad en la cuenca del río Gastona y su relación con hidrogramas de crecientes, Tucumán, Noroeste Argentino. **Revista de Geología Aplicada a la Ingeniería y al Ambiente** 22, 49–59.

Sayago JM, 2007. Cambio climático: 'Mitigación de emisiones o prevención de desastres' **Serie Monográfica y Didáctica.** Facultad de Ciencias Naturales e Instituto Miguel Lillo, 46, 15–17. San Miguel de Tucumán.

Caria MA, Sayago JM, 2007. Arqueología y ambiente en un valle intermontano del piedemonte oriental de las Cumbres Calchaquíes (Tucumán, Argentina). **Revista RUNA** 29, 11–28.

Collantes MM, Sayago JM, Busnelli J, 2009. Paleofluvial systems in the western Chaco plain, Tucumán Province, Argentina. 9th. International Conference on Fluvial Sedimentology, **Acta Geológica Lilloana**, 21, 25–26. San Miguel de Tucumán.

Sayago JM, Collantes MM, Neder L, Busnelli J, 2010. Cambio climático y amenazas ambientales en el área metropolitana de Tucumán (Argentina). **Revista de la Asociación Geológica Argentina** 66, 4, 546–556.

Busnelli J, Pérez H, Sayago JM, 2010.Impactos geoambientales de la planta de residuos urbanos de "Pacará Pintado", Area metropolitana San Miguel de Tucumán, provincia de Tucumán (Argentina). **Revista del Departamento de Geografía de la Universidad Nacional de Tucumán 11,** 12, 9–26.

Karlin MS, Bachmeier O, Dalmasso A, Sayago JM, Sereno R, 2011. Environmental dynamics in Salinas Grandes, Catamarca (Argentina). **Arid Land Research and Management** 25, 4.

Sayago JM, Collantes MM, Niz A, 2012. El umbral de resiliencia del paisaje en el proceso de desertificación de los valles preandinos de Catamarca (Argentina). **Revista Acta Geológica Lilloana 24,** 1–2, 62–79.

Sayago F, Sayago JM, 2012. Las predicciones sobre los riesgos geoambientales del cambio climático en Tucumán y su aplicación en las acciones preventivas de adaptación. **Revista Acta Geológica Lilloana 24,** 1–2, 98–107.

**Fig. 1** Prof. Dr. José Manuel Sayago

**Fig. 2** José Manuel Sayago during fieldwork in the Chaco Forest, north-east of the province of Tucumán, Argentina.

**Fig. 3** José Manuel Sayago at the Archaeological site Yasyamayo, Santa María valley, province of Tucumán.

**Fig. 4** José Manuel Sayago, together with aboriginal students of the Yasyamayo school, Santa María valley, province of Tucumán, Argentina.

**Fig. 5** José Manuel Sayago in the shade of a pluri-centennial *"algarrobo",* Santa María valley, province of Tucumán, Argentina.

**Fig. 6** José Manuel Sayago at the Archaeological Site "Sacred City of the Quilmes", together with members of the Quilmes aboriginal community, province of Tucumán, Argentina.

**Fig. 7** José Manuel Sayago during fieldwork together with a doctoral student, in the Río Acequión valley, province of San Juan, Argentina.

# Contents

# Relationship Among a Supernova, a Transition of Polarity of the Geomagnetic Field and the Pliocene-Pleistocene Boundary

R. H. Compagnucci[1], M. J. Orgeira[2,3(✉)], A. M. Sinito[4], L. Cappellotto[3], and S. Plastani[5]

[1] Universidad de Buenos Aires-CONICET, Buenos Aires, Argentina
rhcompagnucci@gmail.com
[2] Facultad de Ciencias Exactas y Naturales, Universidad de Buenos Aires, Buenos Aires, Argentina
orgeira@gl.fcen.uba.ar
[3] IGEBA, Universidad de Buenos Aires-CONICET, Buenos Aires, Argentina
[4] Universidad del Centro de la Provincia de Buenos Aires, Tandil, Argentina
[5] IDEAN, Universidad de Buenos Aires-CONICET, Buenos Aires, Argentina

**Abstract.** After the Middle Miocene, two important climatic changes took place, consisting mainly of cooling in both hemispheres. One occurred between 7.0 and 5.4 Ma and another at the end of the Pliocene, which marked the beginning of the Pleistocene in approximately 2.58 Ma. The proposal of this presentation is to analyze diverse forcings of these climatic changes, such as the influence of the joint occurrence of reversions of the geomagnetic field and explosions of a supernova. These events occurred coincidentally with the cooling of Earth. Also, biological changes in those time intervals are analyzed, especially the evolution of the Hominins since the oldest hominin fossils. The characteristics of the Galactic Cosmic Rays, its influence on the climate and its potential mutogenetic effect were taken into account.

Briefly, according to our analysis, it seems to be evident that together with other factors, the joint occurrence of the explosion of a supernova at less than 100 pc from the Earth and the weakening and/or reversion of the Geomagnetic Field was an important factor that promoted these two climatic and ecosystem changes.

**Keywords:** Geomagnetic field reversion · Supernovae · Pliocene-Pleistocene boundary · Paleoclimatology · Global changes · Mutations · Human evolution

## 1 Introduction

The scientific community has been interested in detecting and understanding the forcings that have acted and are acting on climate. There are multiple natural causes. Orgeira et al. (2016) presented a brief review of the different hypotheses for climate change, such as solar variability, changes of the Total Solar Irradiance (TSI) received due to terrestrial orbital changes, length of day variations and volcanic activity.


M. M. Collantes et al. (Eds.): *Advances in Geomorphology and Quaternary Studies in Argentina*, SPRINGEREARTH, pp. 1–39, 2020.
https://doi.org/10.1007/978-3-030-22621-3_1

Additionally, Orgeira et al. (2016) and Kitaba et al. (2017), among other authors, proposed a connection between the variations of the Geomagnetic Field (GF) during the reversal of its polarity and climatic changes that occurred in the Late Cenozoic. The mechanism would be linked to the influence of GF over the entrance in the atmosphere of Galactic Cosmic Rays (GCRs). In turn, the GCRs favor the formation of a low cloud cover, which would promote the cooling of the planet (the "Umbrella" effect).

## 2 Paleoclimatology

After the Middle Miocene Climatic Optimum, a Late Miocene global cooling took place synchronously in both hemispheres (Herbert et al. 2016) (Fig. 1). The Antarctic glaciations began around 34 Ma under $CO_2$ levels above 600 ppm (Galeotti et al. 2016). Approximately between 7.0 and 5.4 Ma, the cooling culminated with ocean temperatures dipping to near-modern values (Fig. 1). The Late Miocene shifts from an equable climate towards the modern world conditions of strong equator-pole temperature gradients in both hemispheres. The changes in climate, temperature and ecosystems were a direct consequence of a decline in atmospheric $CO_2$ (Bartoli et al. 2011, Herbert et al. 2016).

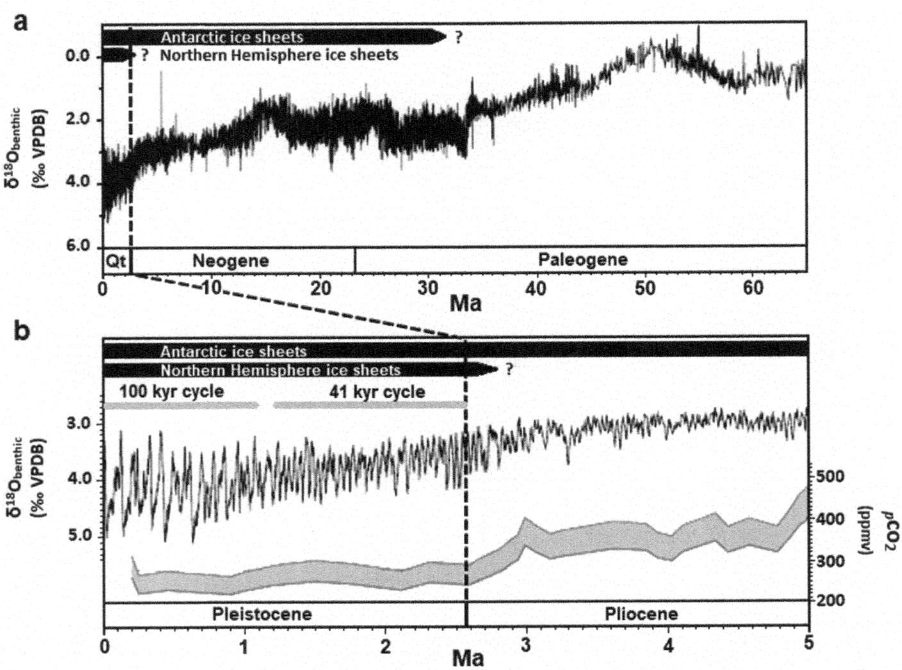


**Fig. 1.** Global deep ocean temperature in the Cenozoic Era; (a) the past 65 Ma, using $\delta^{18}O$ data of Zachos et al. (2008), (b) the past 5 Ma (Pliocene and Pleistocene) using Lisiecki and Raymo (2005) and Atmospheric $pCO_2$ range estimates (Seki et al. 2010) (Adapted from Jakob 2017).

During the Pliocene (5.3 Ma to 2.58 Ma), the Earth experienced a transition from relatively warm climates to the prevailing cooler climates of the Pleistocene. As it was mentioned before, Lisiecki and Raymo (2005) presented climate data through an average of 57 globally distributed benthic $\delta^{18}O$ records for the last 5.3 Ma, which measured global ice volume and deep ocean temperature.

Although the external factors that determine climate were essentially the same, the Pliocene prior to 3.0 Ma was generally 2–3 °C warmer than today. This period shows smaller terrestrial and sea ice extent (Haywood and Valdés 2004, Jiang et al. 2005), a partially deglaciated West Antarctic Ice Sheet (Naish et al. 2009) and global sea level 10–40 m higher than today (Raymo et al. 2011). This epoch is characterized by pronounced climatic oscillations and then, there was a cooling during the Quaternary. In Fig. 1, it is possible to observe the cooling that occurred gradually and in a stepped way. Warmer and cooler periods occurred throughout the Pliocene, but with an overall cooling trend.

Climatically, the Pliocene could be divided into three periods: (1) an Early Pliocene warm period, (2) the Middle Pliocene, a period of sustained global warming, and (3) a climatic deterioration during the Late Pliocene leading to the high-magnitude climate variability associated with Pleistocene glacial/interglacial cycles.

The early warm period had climate conditions that were forced by $CO_2$ concentrations. Different evidence determined a gradual decline from $CO_2$ values of around 410 ppm during the Early Pleistocene to $\sim$300 ppm around 3.0 Ma, coinciding with the start of transient glaciations on Greenland (Bartoli et al. 2011).

Conditions in the equatorial Pacific Ocean were characterized by sea surface temperature gradient considerably lower than today, and mean sea surface temperature in the east similar to that in the west. These conditions have been described as a permanent "El Niño state" or "El Padre" (Fedorov et al. 2006).

This period was marked by a number of significant tectonic events which created the present landscape. One of these events was the configuration of a continuous land bridge between the North America and South America tectonic plates. This was the product of the interplay of several plates, mainly the Caribbean Plate and resulted in a land bridge, i.e. the Isthmus of Panama, ending the surface water exchange between the Pacific and Atlantic oceans, known as the Central American Seaway. These changes had impacts on the evolution of climate through large-scale changes in ocean circulation and could have contributed to the initial intensification of the Atlantic Meridian Overturning Circulation (AMOC) and the North Atlantic warming prior to 3.6 Ma. Furthermore, it led to enhance flow of warm and salty surface water into the North Atlantic via the Gulf Stream (Bell et al. 2015).

However, Hoorn and Flantua (2015), Montes et al. (2015) and Erkens (2015), among others, claimed that rather than the generally accepted date of $\sim$3.0 Ma, the Panama Isthmus was effectively completed by the Middle Miocene, around 13.0 Ma. This was disputed by others researchers as Herbert et al. (2016) and O'Dea et al. (2016) who proposed a younger formation at approximately 2.8 Ma, and a strong criticism against the uncritical acceptance of the old isthmus hypothesis. It is crucial to determine the time of closure, mostly because it has been linked to three major events in the history of Earth: the onset of the Thermohaline Circulation, the onset of Northern Hemisphere glaciations and the birth of the Caribbean Sea. Recently, Jaramillo et al.

(2017) rejected both age hypothesis, and claimed that new data are needed to understand the rise of the Panama Isthmus.

The Northern Hemisphere glaciation (NHG) started as early as 3.6 Ma (Mudelsee and Raymo 2005). The global average temperature in the Middle Pliocene was 2–3 °C higher than today (Raymo et al. 2006, Lisiecki and Raymo 2007, Robinson et al. 2008).

During the Late Pliocene, from 3.5 to 2.58 Ma, the deterioration of the Earth's climate state led to an increase in Northern Hemisphere glaciations (Fig. 1). During this period, the temperature reached values comparable to those of present-day and Holocene-like climate conditions took place.

The increasing glaciation of Greenland may have been driven indirectly through variations in atmospheric aerosol concentration or volcanic dust (Prueher and Rea 1998) and/or directly through variations in precession, obliquity and eccentricity orbital parameters, which produced a decrease in the incoming solar radiation (Milankovitch 1941).

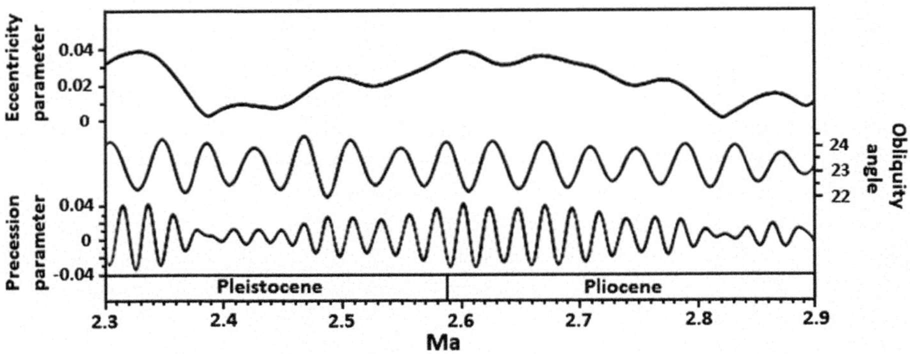


**Fig. 2.** Orbital parameters (eccentricity, obliquity and precession) (with information from Laskar et al. 2004, adapted from Jakob 2017).

Although Milankovitch forcing has been constant over the past several million years, the amplitude of the climatic response underwent remarkable changes as the long-term global cooling introduced different climate feedbacks.

At the Middle Pliocene, around 3 Ma an amplification of the response of climate to orbital forcing started due to the permanent ice sheets over high latitudes of both hemispheres (McGehee and Lehman 2012) and the decreasing $CO_2$ (Willeit et al. 2015).

Development of glaciation required a low insolation during the summer to avoid full melting of the accumulations of winter snow, as it normally happens in conditions of great amplitude of obliquity.

The amplitude of obliquity cycles increased for ~3.0–2.5 Ma and a strong rise in the amplitude of precession affected insolation at 60° N, at ~2.8–2.55 Ma (Fig. 2). Both processes may have forced a rapid ice-sheet buildup in the Northern Hemisphere (Maslin et al. 1998) during the Pliocene-Pleistocene (Neogene-Quaternary) boundary. The glacial amplitudes of the Northern Hemisphere ice sheets started to present substantial differences between glacial and interglacial periods. This represents a characteristic of the Quaternary climate (Raymo et al. 2006). The beginning of a marked cycle of obliquity can be seen (Fig. 2) during the Early Pleistocene in the proxy temperature series, which is replaced later during the last 1.0 Ma by the cycle of eccentricity of 100 ka.

Hennissen et al. (2015) listed proxy records evidence of the Northern Hemisphere major climatic changes around the Pliocene-Pleistocene boundary. The compilation includes a wide range of data from Hennissen et al. (2014), Kleiven et al. (2002), Bailey et al. (2013), Naafs et al. (2010, 2013), Friedrich et al. (2013), Versteegh et al. (1996), Thierens et al. (2013), Haug et al. (2005), Martínez-García et al. (2010), Jansen and Sjøholm (1991), Froese et al. (2000), Duk-Rodkin and Barendregt (2011), Pross and Klotz (2002), Herbert et al. (2015), Klotz et al. (2006), Ding et al. (2005), Sun et al. (2006, 2010a, 2010b), Demske et al. (2002), Melles et al. (2012), Brigham-Grette et al. (2013) and Andreev et al. (2014). Furthermore, Hennissen et al. (2015) analyzed SSTs proxy data of the central North Atlantic Subtropical Gyre, and noted that seasonality increased significantly over the 2.8–2.5 Ma time period. This process would be caused mainly by the drop in the early spring temperatures while summer temperatures remained relatively stable. The seasonality tended to peak during glacial intervals and the maximum occurred at ~2.6 Ma, a time when the North Atlantic Current and the Arctic Front shifted to more southerly positions. According to the present authors, it seems unlikely that the increase in the seasonality could be explained only by changes in orbital parameters.

Other causes should be involved in the changes of the Pliocene to the Pleistocene, as the changes in ocean gateways:

(i) the Central American Seaway that was equivalent to a ~5 cm sea-level difference between the "closed" versus "open" CAS scenario (Lunt et al. 2008),
(ii) the Indonesian Seaway ~4.0–3.0 Ma that may have altered atmospheric and oceanic circulation pattern over the Pacific Ocean (Cane and Molnar 2001, Karas et al. 2009) causing a cooling and shoaling of the thermocline in the tropical Indian Ocean. This seaway may have also contributed to the global thermocline shoaling recorded for the Pliocene (Philander and Fedorov 2003, Wara et al. 2005, Steph et al. 2006) strengthening the tropical Pacific Walker Circulation and reducing poleward heat transport. Thereby, favoring high-latitude cooling and the buildup of extensive Northern Hemisphere ice sheets (Cane and Molnar 2001).

A consequence of these changes was the termination of the "permanent El Niño-like conditions" during the transition from the warm Pliocene climate state to a climate state that approximates modern conditions with strong upwelling in the east of the basin and more developed zonal temperature and primary productivity gradients (Bolton et al. 2010). During the Late Pliocene and the Early Pleistocene, in the equatorial Pacific Ocean a climatic shift occurred from the "El Niño-like" to the "La Niña-like state". According to Philander and Fedorov (2003) "in the tropics and subtropics the response to obliquity variations is in phase with, and corresponds to, El Niño conditions when tilt is large and La Niña conditions when tilt is small".

Additional forcing mechanism was the decrease in carbon dioxide concentrations (De Schepper et al. 2014). The reconstructed $CO_2$ record shows that glaciation in the northern hemisphere begins once the average long-term $CO_2$ concentration falls below 265 ppm after a period of strong $CO_2$ decrease. Different simulations show that Greenland glaciation was controlled mainly by a decrease in atmospheric $CO_2$ during the Pliocene (Lunt et al. 2008, Willeit et al. 2015).

## 3 Pliocene-Pleistocene Boundary

Lisiecki and Raymo (2005) presented a 5.3 Ma stack of benthic $\delta^{18}O$ records from a global distribution. Their age model situates the Pliocene-Pleistocene boundary around 2.6 Ma (Marine Isotope Stage –MIS- 104).

According to Bintanja and Van de Wal (2008), the onset of major glaciations in the Northern Hemisphere, around 2.7 Ma, was most probably induced by climate cooling during the Late Pliocene. The reconstructed marine isotope signal increased progressively since 3.0 Ma, showing climatic cooling and increase of ice sheets. Its amplitude also increased with time, with noticeable changes occurring around 2.7 Ma. According to calculations, the inception of the Northern Hemisphere ice sheets at about 2.7 Ma may be linked to air temperatures dropping below –5 °C in relation with present time (Bintanja and Van de Wal, 2008).

Concurrent results are presented by Venti and Billups (2012) who, based on the North Pacific Ocean first orbital-scale benthic-foraminiferal $\delta^{18}O$ and $\delta^{13}C$ time series that span the Pliocene-Pleistocene climate transition, found that glacial-interglacial climate cycles began to impact terrestrial plant habitat in middle latitude North Pacific Ocean at 2.7 Ma.

It has long been known that global cooling began in the Late Tertiary/Neogene, with multiple major cooling phases between 2.8 and 2.4 Ma, which vary regionally: North Atlantic Ocean ice-rafted debris at 2.72 Ma; loess–palaeosol accumulation in China at 2.6 Ma; severe cooling in northwestern Europe at 2.52 Ma; arrival of sub-Antarctic molluscs in New Zealand at 2.4 Ma (Gibbard et al. 2010).

In 1996, IUGS ratified a new Pliocene stage, the Gelasian, between the underlying Piacenzian Stage and the overlying Pleistocene Series. The base of the Gelasian was defined by a Global Stratotype Section and Point (GSSP) at Monte San Nicola in southern Sicily and dated by astronomical tuning to 2.58 Ma. This stratotype lies just 1.0 m above the Gauss–Matuyama palaeomagnetic transition (Gibbard et al. 2010, Ogg et al. 2016).

This transition at the base of the Gelasian was determined by Gibbard and Head (2009) at 2.58 Ma or 2.60 Ma, based on revised $^{40}Ar/^{39}Ar$ dating of the underlying Gauss–Matuyama polarity boundary as 2.61 Ma (Singer 2014).

The base of the Gelasian stage in Sicily is near above the base of reversed-polarity Chron C2r (Gauss–Matuyama polarity boundary) (Rio et al. 1998, Gibbard et al. 2010). This level is slightly younger than the 2.6–2.7 Ma (MIS 104 and 110) onsets of the first major influx of ice-rafted debris into middle latitudes of the North Atlantic Ocean, deposition of glacial till into middle continent North America, and the major glacial-caused global sea-level sequence boundary and lowstand (Ogg et al. 2016). Then, the Gelasian stage and its coincidence with the GF transition Matuyama-Gauss enabled an unambiguous and precise global marker between continental and oceanic deposits. The

Quaternary System was defined by the ICS/IUGS at this established Gelasian GSSP. This level has a recommended age from astronomical calibration of 2.588 Ma (Rio et al. 1998); but other estimates are 2.58 Ma, based on its relative position within that precession cycle.

## 4 Geomagnetic Field Reversions

The Geomagnetic Field (GF) has reversed its polarity many times in the geological past. Particularly, during the Late Miocene and the Early Pleistocene several reversions of the GF have been recorded (Fig. 3). For the last 6.0 Ma, four Magnetic Chrons are accepted: the Bruhnes, Matuyama, Gauss and Gilbert chrones, respectively. In addition to Gauss-Matuyama, Matuyama-Brunhes, and Gilbert-Gauss polarity transitions, there are eight other reversals boundaries, which may represent instability in the geodynamo that is associated with a precipitous drop in paleointensity (Singer 2014).

The causes of GF reversals are still not clearly known, although some hypotheses and models have been presented (among others, Gallet and Pavlov 2016). GF is accepted to be generated by fluid motions in the liquid, outer part of the Earth's core, which is mainly composed of iron. The fluid motions are driven by buoyancy forces that develop at the base of the outer core as the Earth slowly cools and iron condenses onto the solid, inner solid core below. The rotation of the Earth causes the buoyant fluid to rise in curved trajectories, which generate new magnetic field by twisting and shearing the existing magnetic field. Over 99% of the Earth's magnetic energy remains confined entirely within the core. The temporal patterns of dynamo behavior may reflect physical interactions between the molten outer core and the solid inner core or lowermost mantle (Singer 2014). These interactions may control reversal frequency and shape the weak magnetic fields that arise during successive dynamo instabilities.

Particularly, Deino et al. (2006) estimated a calibrated age of 2.610 Ma for the Gauss-Matuyama reversal, using $^{40}Ar/^{39}Ar$ ages of two tuffs in normally and reversely magnetized lake sediments in Kenya.

After the transition, a conspicuous characteristic defined the GF. Ahn et al. (2016) estimated a weak virtual axial dipole moment of 3.66 ($\pm$1.85) $\times$ $10^{22}$ $Am^2$ during the Early Matuyama Chron (inferred time period of 2.34 to 1.96 Ma). It means that the early stage of the Matuyama Chron can be characterized by a long period of low geomagnetic field intensity.

Regarding variations of the GF in more recent times, Korte et al. (2018) recently reconstructed past geomagnetic field morphology and variations based on paleo- and archeomagnetic data. They improved the Late Pleistocene and Holocene geomagnetic field models and provided high-quality data.

According to the information compiled and provided for these authors, it is evident that the intensity of the GF was substantially lower in the Late Pleistocene than during the Holocene. However, the trend appears to be reversing in the last hundred years. Measurements of the Earth's GF have been done practically continuously since about 1840 (Fig. 4). The recorded downward trend of the strength of the dipole moment is noticeable.

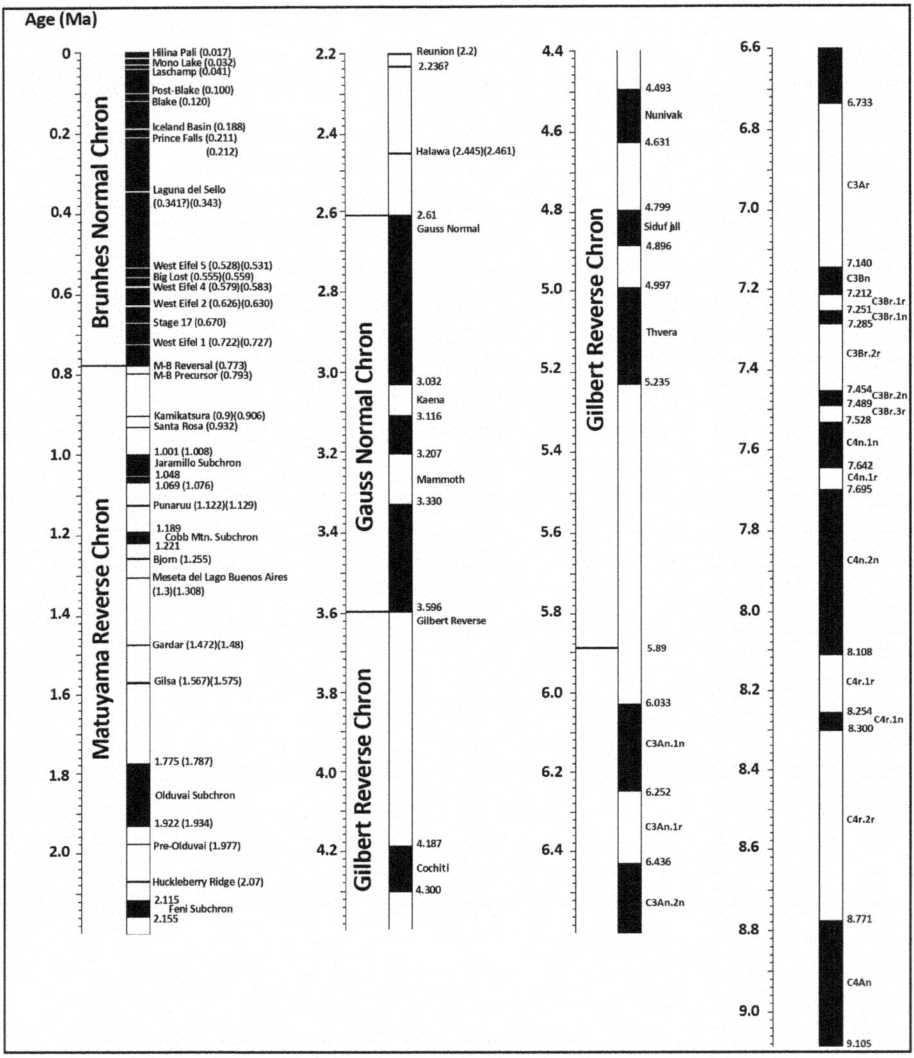


**Fig. 3.** GF reversion time table, following Cande and Kent (1995), Langereis et al. (2010), Singer (2014) and Ogg et al. (2016).

If this trend continues, the dipole moment may reach near 0 in about 1,500 years (British Geological Survey, http://www.geomag.bgs.ac.uk). However, this is only one of the possibilities taking into account the oscillations of the GF in the past. Up to now, it is not possible to know if the geomagnetic field is beginning a reversal, an excursion, or simply a variation in GF dipole intensity.

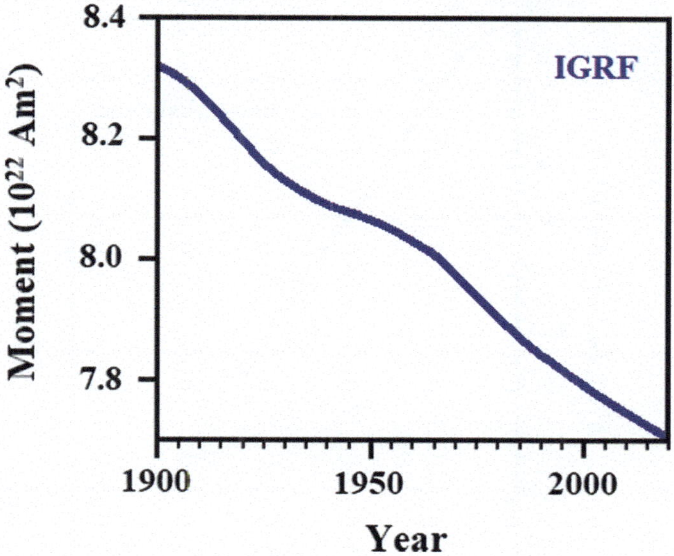


**Fig. 4.** Moment of the geomagnetic field from 1900 to present times (modified from http://www.geomag.bgs.ac.uk). IGRF: International Geomagnetic Reference Field.

## 5 Galactic Cosmic Rays Origin and Effects

Particles that bombard the Earth from anywhere beyond its atmosphere are known as cosmic rays (CRs). They are energetic, subatomic particles that arrive from outside the Earth's atmosphere. The basic unit of measurement used in discussing the energy of CRs is the electron-volt, or eV.

The range of energies encompassed by CRs is truly enormous, starting at about $10^7$ eV and reaching $10^{20}$ eV for the most energetic CRs ever detected.

The CRs spectrum in Fig. 5 clearly shows that the number of CRs (the CRs flux) detected drops off dramatically when going to higher energies. The spectrum exhibits a 'knee' and an 'ankle', both of which deviate from the standard exponential decline (dotted line).

In 1912, Viktor Hess discovered the CRs. After 100 years of mystery, the CRs origin finally seems to be resolved. Current suggestions are that the CRs with energies less than about $10^{10}$ eV are primarily solar CRs or solar energy particles (SEPs). They have a composition similar to that of the Sun, and are produced in the corona by shock acceleration, or when part of the solar magnetic field reconfigures itself (Miroshnichenko 2001, Reames 2013).

On the other hand, Galactic Cosmic Rays (GCRs) come from outside the Solar system but generally from within our Milky Way galaxy. They are charged particles with energies between $\sim 10^{10}$ and $10^{15}$ eV. The GCRs helical motion around the magnetic field lines of the Galaxy causes their distribution to appear isotropic (they are detected equally in all directions); even though astronomers believe that they originate in the shocks of supernova remnants (IceCube Collaboration 2013). For the particles of

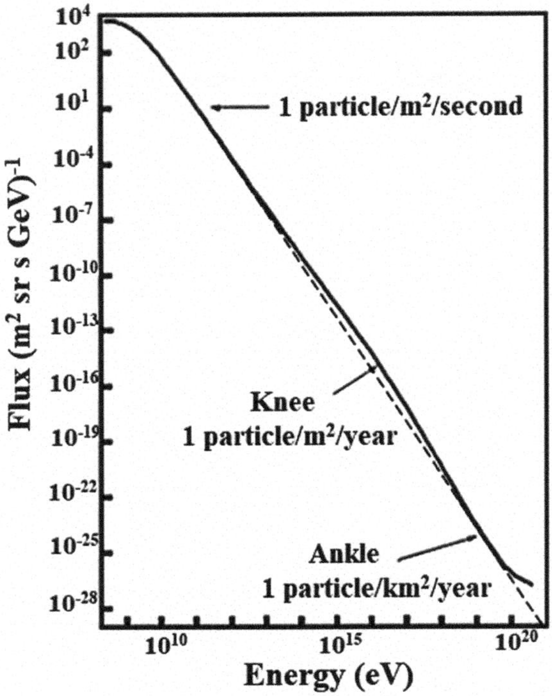


**Fig. 5.** The flux of cosmic ray particles as a function of their energy. The fluxes for the lowest energies are mainly attributed to solar CRs, intermediate energies to galactic CRs, and highest energies to extragalactic CRs. Adapted from Helder et al. (2012)

galactic origin, then, the energy comes mainly (but not exclusively) from supernova explosions which explains the cosmic ray particles at all energies below the 'ankle' at $3 \times 10^{18}$ eV (Fig. 5, Helder et al. 2012, Drury 2012, Cardillo et al. 2014).

During periods of intense solar activity - not just during the maxima of the solar cycles - SEPs may be released in large numbers. Energetic particles from the Sun, as well as its magnetic field, can scatter GCRs, leading to a much smaller number reaching Earth, phenomenon known as a Forbush decrease (Cane 2000). On a larger scale, there is an anti-correlation between the solar cycle and GCRs flux given variations connected to the ~11 yr sunspot cycle and ~22 yr solar magnetic cycle (Ahluwalia 1997, Cane et al. 1999, Mangeard et al. 2018). Variation of GCRs flux is therefore the product of the solar wind speed and the interplanetary magnetic field strength (Sabbah 2000), with higher cosmic ray fluxes reaching the troposphere during the solar minima (Carslaw et al. 2002). The composition of GCRs is slightly different from that of SEPs and anomalous cosmic rays (ACR) insofar as they are slightly enriched in heavy elements as well as in lithium, beryllium and boron (Mewaldt et al. 1994). These latter elements are thought to result from the fragmentation of heavy nuclei during collisions with interstellar matter, and observations of their composition allow to estimate their age (3–10 million years) and the amount of material traversed on the way to the Solar System.

Cosmic ray research has recently discovered new sources of ultra-high energy cosmic rays. The Pierre Auger Collaboration (2017) reported the first observational evidence that cosmic rays originate from much further away than our galaxy. Eichmann et al. (2018) discussed radio galaxies as the source of GCRs with energies higher than 6 Joules ($39 \times 10^{18}$ eV), namely ultra-high-energy cosmic rays (UHECRs).

Pudovkin and Veretenenko (1995) were the first to mention the possible causal relationship between the decrease of cloud cover and the Forbush decrease of the GCRs that occurred following a coronal mass ejection.

However, the correlation between the GCRs flux and the global cloud cover of the Earth was first postulated by Svensmark and Friis-Christensen (1997). They proposed the link between the solar variation and climate by changes in the global cloud coverage produced by the variations in the GCRs flux. This was followed by many articles either rejecting or questioning the link between GCRs flux and cloud formation. Kuang et al. (1998) using data for 1983–1993 period analyzed variations of cloud optical thickness and concluded that is not clear if low level cloudiness variations were caused by the solar cycle or by the El Niño/Southern Oscillation (ENSO) cycle. Sun and Bradley (2002) suggested that there is not a solid relationship between low cloud covering and GCRs flux.

Different doubts were expressed by Kernthaler et al. (1999), Kristjánsson et al. (2000), Wagner et al. (2001), Sloan and Wolfendale (2008), among others. The main argument against the GCRs link with low cloud formations was the lack of an established physical mechanism.

The SKY Experiment to investigate the role of cosmic rays in cloud formation in the low Earth's atmosphere, at the Danish National Space Science Center and the European Organization for Nuclear Research (CERN), produced a comprehensive verification through the Cosmics Leaving Outdoor Droplets (CLOUD) Project (see https://home.cern/about/experiments/cloud).

Since the first publication (Duplissy et al. 2010) and up to 2017, there were thirty publications and ten technical reports on the subject. The most relevant are those from Kirkby et al. (2011) and Kirkby et al. (2016), both widely cited (741 and 236 citations respectively). The GCRs effect over cloud formation is now established, despite rigorous attempts to disprove it, i.e. the proper scientific method.

Recently, Svensmark et al. (2017) determined the effect of ionization on the growth of aerosols into cloud condensation nuclei, both theoretically and experimentally.

The hypothesis (Fig. 6) is explained in short by Svensmark in the web page of the Technical University of Denmark, (http://www.space.dtu.dk/english/news/2017/12/missing-link-found-between-exploding-stars-clouds-and-earths-climate) as follows (Fig. 7):

- "Cosmic rays, high-energy particles raining down from exploded stars, knock electrons out of air molecules. This produces ions, that is, positive and negative molecules in the atmosphere.
- The ions help aerosols - clusters of mainly sulfuric acid and water molecules - to form and become stable against evaporation. This process is called nucleation. The

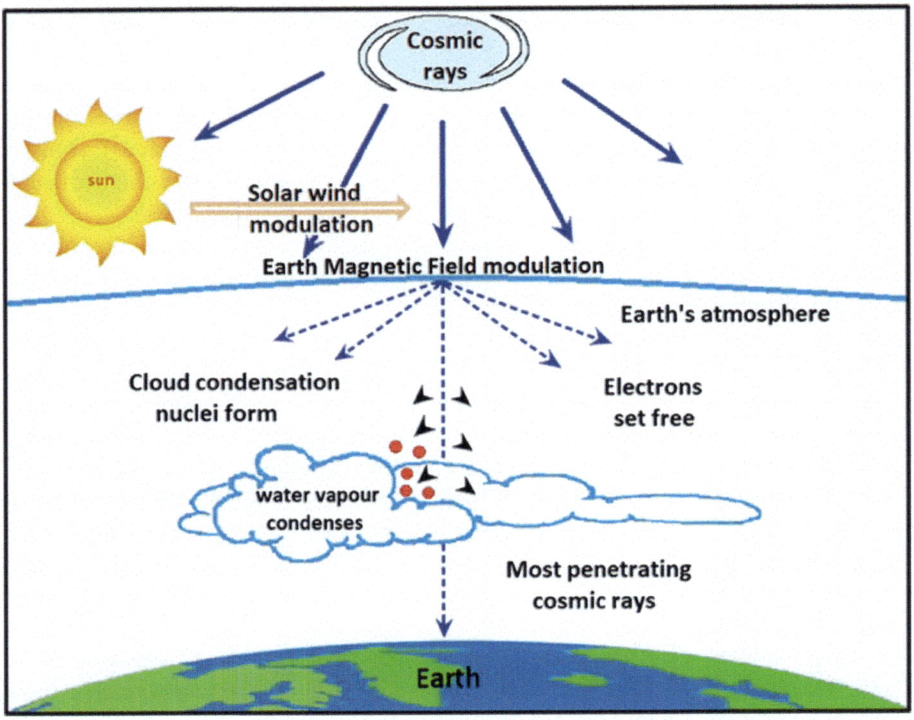


**Fig. 6.** Henrik Svensmark's hypothesis proposed a link between solar wind, cosmic rays, and cloud cover. When cosmic rays hit the atmosphere, they form ions, which give rise to cloud condensation nuclei and thereby spur the growth of clouds. An active sun partially shields the Earth from the normal barrage of cosmic rays, leading to fewer low-altitude clouds. Changes in cosmic-ray influx do not affect high clouds, as there are always plenty of cosmic rays at higher altitudes, but fewer cosmic rays penetrate to low altitudes, so increases or decreases due to changes in the solar wind have more noticeable effects there.

small aerosols need to grow nearly a million times in mass in order to have an effect on clouds.

- The second role of ions is that they accelerate the growth of the small aerosols into cloud condensation nuclei - seeds on which liquid water droplets form to make clouds. The more ions the more aerosols become cloud condensation nuclei. It is this second property of ions which is the new result published in Nature Communications (Svensmark et al. 2017)
- Low clouds made with liquid water droplets cool the Earth's surface.
- Variations in the Sun's magnetic activity alter the influx of cosmic rays to the Earth.
- When the Sun is lazy, magnetically speaking, there are more cosmic rays and more low clouds, and the world is cooler. When the Sun is active fewer cosmic rays reach the Earth and with fewer low clouds the world warms up."

In the long-term scale, the geomagnetic intensity is controlling the GCR flux on the Earth, being more intense during the reversals with its weakening of the magnetic field

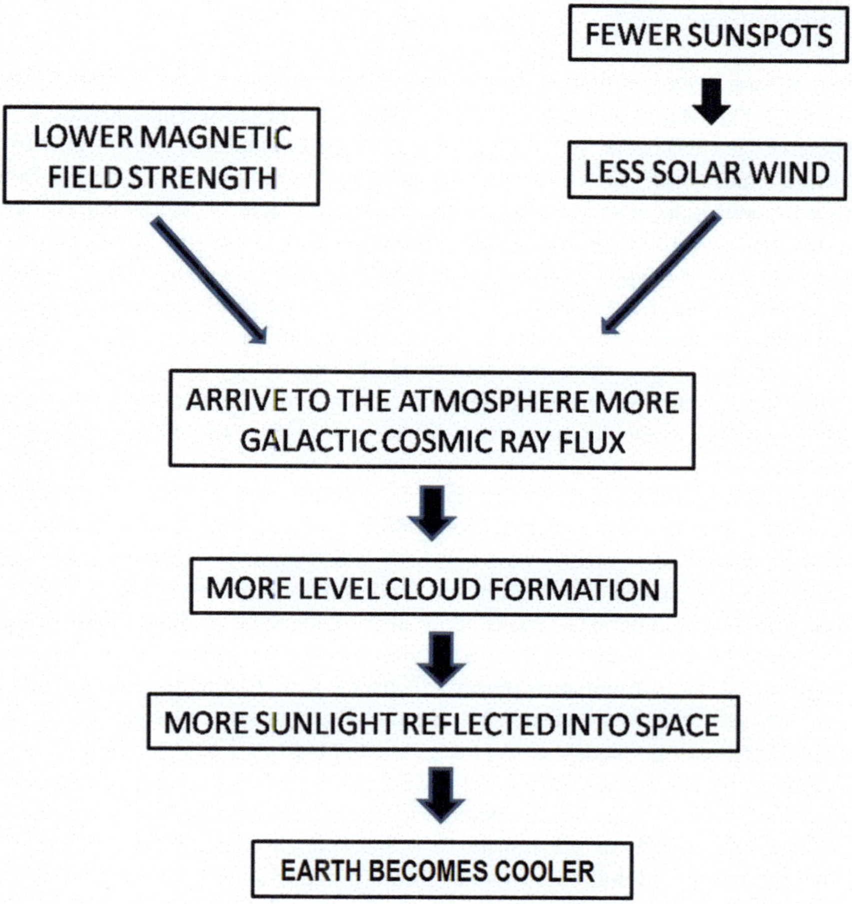


**Fig. 7.** Schematic relationship representing the effect of the GCRs in the Earth climate. This is specially connected with increased GCR flux related with supernova explosion, if it is coeval with weak magnetic field.

(Frank 2000, Kitaba et al. 2013). Therefore, the cooling that occurred during or immediately after a reversal would be associated to the increased of the GCRs (Orgeira et al. 2016).

Furthermore, Svensmark (2015) suggests that the atomic particles coming from supernova remnants left by exploded stars appear to play a major part in the transition to a cool/glacial climate. "It gives an understanding of how changes caused by solar activity or by super nova activity can affect climate."

## 6 Supernovae

There has long been speculation (Schindewolf 1954) that terrestrial mass extinctions (Bambach 2006, Melott and Bambach 2014) might be related to supernovae.

Supernovae (SN) occur in two main types, named as I and II. Type I is primarily a result of a small object, such as a white dwarf star, gravitationally robbing gas from a companion until it becomes too massive to support itself against gravity and collapses. Type II usually results at the end of the lifetime of a very massive star (much larger than the Sun). As this type exhausts its fuel, it will swell up and then suddenly collapse. The rebound results in a blast of light, radiation and a shell of ejected material. The typical released kinetic energy for the two types is about the same.

Including both types, there are about three SN per century in our galaxy.

There have been a few recorded within historical times (Stephenson 2016), and in the cases of most of them, a remnant can still be seen, in the form of an expanding shell of hot gas.

Recent massive-star and SN activity in Earth's vicinity may be traced by radionuclides with half-lives of up to 100 million years, if trapped in interstellar dust grains that penetrate the Solar System. One such radionuclide is $^{60}Fe$ (with a half-life of 2.6 Ma), which is ejected in SN explosions and winds from massive stars.

The density and temperature distribution of the interstellar medium (ISM) is highly variable, superbubbles with typical substructures of about 50–150 pc (pc: parsec is a unit of length used in astronomical objects, 1 parsec = 3.2616 light-years = $3.0857 \times 10^{16}$ m) having lifetimes of some ten Ma. Several SN explosions over the last 14 Ma or so shaped the present structure of the local superbubble (the Local Bubble).

Although no mass extinction can be definitively linked to a SN, circumstantial evidence in the form of the agreement of geographical extinction patterns, with expectations of atmospheric UV transmission, has been presented for one possible link to a radiation event (Melott and Thomas 2009). Using available information, Gehrels et al. (2003) modeled the effects of a hypothetical nearby SN and found that a very close event (within $\sim$8 pc) is necessary for a major mass extinction.

More recently, Breitschwerdt et al. (2016) looked at the $^{60}Fe$ deposition history and trajectories of stars in the Local Bubble. They found that the nearest explosions were at $\sim$100 pc, $\sim$2 Ma, showing a consistent picture of the evolution of the Local Bubble and the $^{60}Fe$ deposition in the crust.

Wallner et al. (2016) presented a greatly expanded set of data from three different deep-sea archives, including four sediment cores, two iron–manganese crusts and two iron–manganese nodules. This dramatically expanded the amount of available $^{60}Fe$ data and confirmed the same broad picture: two SN events within 100 pc (1.7–3.2 Ma and 6.5–8.7 Ma, Fig. 8), or isotope transport from multiple SN via the moving interstellar medium. The same parameters are indicated by the closely following publication of the detection of $^{60}Fe$ in lunar samples (Fimiani et al. 2016) and the general chronology supported by recent direct detection in space (Binns et al. 2016). Schulreich et al. (2018) showed also a peak actually broader (1.5–3.2 Ma) than indicated by the previous measurements (Fig. 9). Furthermore, noteworthy is the occurrence of an additional smaller peak at 6.5–8.7 Ma, whose origin is still unknown.

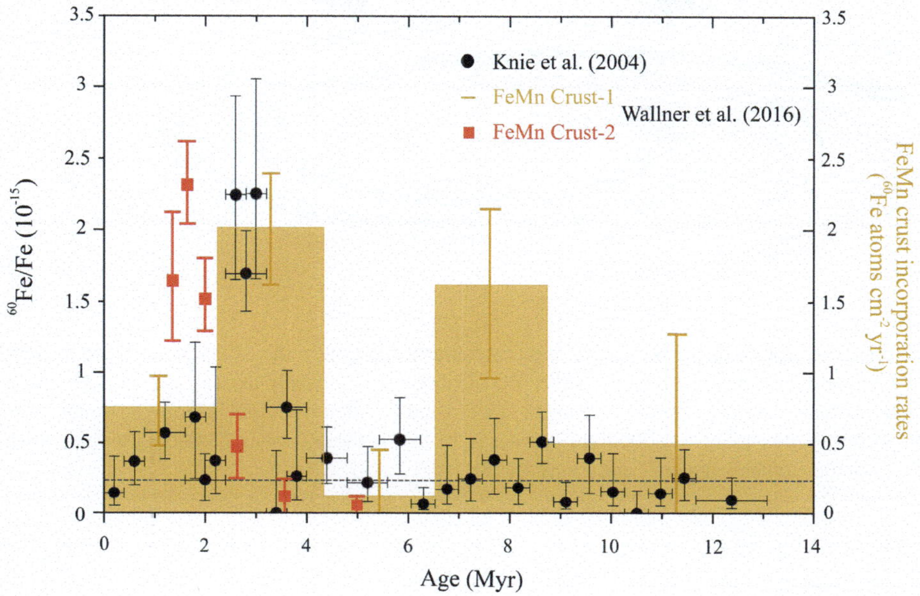


**Fig. 8.** Results from ferromanganese crust measurement showing a peak in $^{60}$Fe/Fe and ferromanganese (FeMn) crusts incorporation rates adapted from Knie et al. (2004) and Wallner et al. (2016) respectively.

## 7 Galactic Interaction Over the Solar System

The Solar System revolves around the galaxy with one orbit ~240 Ma (Svensmark 2006) and carries out an oscillation in the direction perpendicular to the galactic disk, with a ~60 to 70 million years period (Schwartz and James 1984, Rampino and Stothers 1984). These two movements would produce the galactic interaction with the Solar System carrying out consequences in the Earth.

One forcing mechanism acting over the Earth climate would take place due to the Solar System's oscillation, when the Sun pass through the galactic middle plane about every ~30 to 42 million years (Rampino 2015). Taking into account the last 250 Ma, several studies have reported a ~26–30 million years cycle in mass extinctions and terrestrial impact crater may exhibit a similar cycle of 31 (Rampino and Stothers 1984, Lieberman and Melott 2007, 2012, Rampino 2015).

In fact, the number of cosmic rays that hit the Earth can increase when the Solar System is closer to the galactic plane than when it is at a maximum distance (≈100 pc) from it. Therefore, the expectation is that climate on Earth would be cooler when the Solar System is at the galactic middle plane (Svensmark 2006). The impact cratering record initially turned up evidence for a similar cycle of 28–31 million years (Chang and Moon 2005, Wickramasinghe and Napier 2008).

The position of the Sun relative to the galactic plane may be inferred from the distribution of populations of disk stars. According to Bahcall and Bahcall (1985) the last Sun passage through the galactic plane occurred in the past 3 Ma, if the present

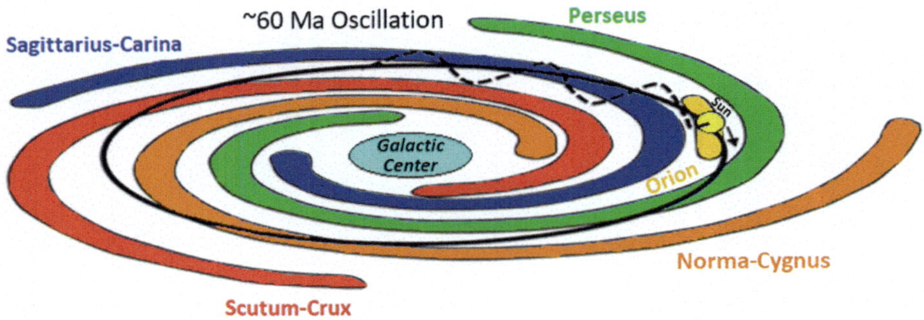


**Fig. 9.** The motion of the Sun in our outer spiral-arm location in the Milky Way. Solar System traveling through spiral arms in the Milky Way Galaxy. It is currently located near the Orion arm (yellow), between two major arms: Perseus (green) and Sagittarius-Carina (blue). The diameter of the Milky Way is about 100,000 light-years and the Sun is located about 28,000 light-years from the Galactic Center (adapted from Brink 2015).

position of the Sun is between 0 and 20 pc above the plane. Rampino and Haggerty (1996) evaluated the pulses of asteroids or comets impacts in the last 2–3 Ma. Humphreys and Larsen (1995) suggested, using star count information, a distance of 20.5 $\pm$ 3.5 pc above the galactic plane. More recently, Weissman (2014) proposed 52 to 74 million years of period and $\pm$49 to 73 pc of amplitude for the harmonic perpendicular movement of the Solar System through the galactic plane, and the occurrence of the last passage about 2–3 Ma. However, he indicated the uncertainty in the values that could be due to the uncertainty in the content of dark matter in the galaxy. A new analysis by Karim and Mamajek (2016) obtained values consistent with the true median of 55 previous estimates published over the past century of the Sun's height above the galactic middle plane ($\approx$17 $\pm$ 2 pc). Reed (2005) results of 19.5 + 2.2 pc, agrees well with various other determinations cited in his paper. According to the mentioned contributions, the Solar System could have crossed the plane of the galaxy around or before the Pliocene-Pleistocene boundary, which could explain, at least in part, the cooling that occurred through the increasing flow of GCRs. The speed of the Sun in the vertical direction is well measured, but the distance from the Sun to the galactic plane is not.

The other potential forcing over the Earth climate would be the Solar System passage through one spiral arm; its potential importance was noted in the late 1970s and early 1980s (Napier and Clube 1979, Clube and Napier 1984, Raup and Sepkoski 1984). The Milky Way has four spiral-arms (Englmaier et al. 2009). The lapse between spiral arms was considered $\approx$140 million years (Svensmark 2012). Presently, as shown in Fig. 9, the Solar System is located in the small arm Orion-Spur, between the Sagittarius-Carina and Perseus spiral-arms (Gies and Helsel 2005).

Shaviv (2002, 2003) and Shaviv and Viezer (2003) argued that the exposure of the Earth to the increased flow of cosmic rays during the incursion through the spiral arm of the galaxy correlates with the extended cold periods on Earth. Spiral arms are regions of stellar formation (Foyle et al. 2010). Therefore, they are regions where stars that end their lives enlarge the probability of interaction and molecular-dust clouds and supernova remnants are more likely to appear.

The amount of matter inside the galactic arms is more than on the outside. The gravitation influence of this matter attracts the inflow of comets to the Solar System. It results in an increase in concentration of interplanetary dust in the zodiac cloud (location of cosmic dust particles which occur along the Ecliptic plane), and a cooling of the Earth's climate (Brink 2015). Consequently, the increase of cosmic rays flux reach a maximum after each encounter with a spiral arm, and then went to a minimum in the dark spaces between the arms.

During the last 520 Ma, the Solar System crossed the galaxy arms approximately four times. Therefore, there were four alternating warming and cooling periods with temperature changes of more than 5.0 °C. Figure 10 shows the $\delta\,^{18}O$ proxy data of Earth's temperature from the Phanerozoic database covering the last 500 Ma. The abundance of $^{18}O$ in fossils is linked to the ocean water isotopic ratio. The $^{18}O$ in the ocean is enriched when $^{16}O$ water preferentially evaporates, and it is locked up in ice.

The violet curve is a 60 million years low pass filtered data. The light blue curve is a low passed filtered, 1/20 million years. These proxy data reflect changes in temperature of the Oceans ($\sim$2.0 °C). The $\sim$140 million year period had already been connected to the passing through the spiral arms of the Milky Way (Shaviv 2002, Shaviv and Veizer 2003, Benjamin 2008, Shaviv et al. 2014).

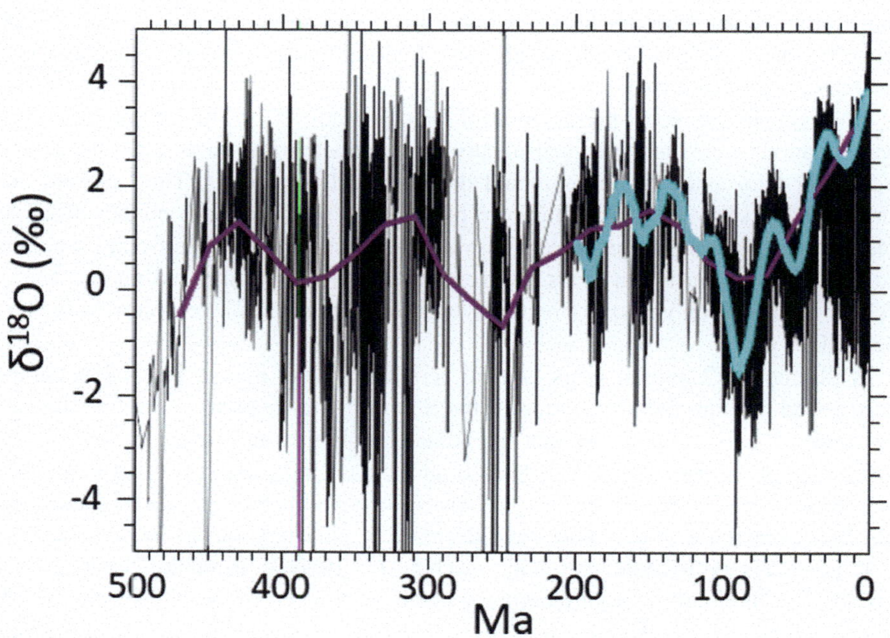


**Fig. 10.** $\delta^{18}O$ proxy data from the Phanerozoic database according to Veizer et al. (1999) (adapted from Svensmark 2006).

In Fig. 10, a ~30 million year period is visible. It had already been demonstrated by Prokoph and Veizer (1999) that there is a ~32 million years variation in the geological proxy data that could be related to the crossing of the galactic plane.

From paleoclimate data, Svensmark (2006) determined several features of two most recent spiral arm passages. He restricted to modulation of time scales between 20–60 Myr, much shorter than the characteristic time for spiral passage ≈140 Myr. His results were in agreement with the preceding reported properties of the Milky Way given additional confidence in the significance of CRs variations and their influence in climate. There were authors such as Overholt et al. (2009) that, based on new data of the structure of the galaxy, conclude that there is no evidence to suggest any correlation between the transit of our Solar System through the spiral arms of our galaxy and the terrestrial climate. In contrast to other published studies, Overholt et al. (2009) model does not force azimuthal symmetry into the spiral-arm structure. The asymmetry of the arms near the solar circle erases any correlation to the 140 million years cycle and any periodic trend less than the orbital period of our Solar System relative to the spiral pattern as a whole. Bailer-Jones (2009) reviewed the passage of the Sun through the galactic midplane or the spiral arms with respect to the evidence for and against the relevance of these mechanisms for climate change and evolution. The cited author noted the need to include in the on-going research a critical evaluation of the time series analysis techniques and hypothesis testing. The analysis suggests that galactic midplane and spiral arm crossings have little impact on biological or climate variation above background level. Non-periodic impacts and terrestrial mechanisms (volcanism, plate tectonics, sea level changes), possibly occurring simultaneously, remain likely causes of many environmental catastrophes. Internal dynamics of the biosphere may also play a significant role.

However, in the recent presentation of Boulila et al. (2018), long-term cyclic sea-level variations were quantified statistically from a Phanerozoic dataset that provided a rare longer-term and continuous record of geological parameters. They showed that some detected sea-level periodicities are of the same order as those predicted for the Solar System motions in the galaxy. They showed a statistically significant prominent and persistent ~36 million years sedimentary cyclicity superimposed on two mega-cycles (~250 million years) in a relatively well-constrained sea-level record of the past 542 million years.

While the ~72 million years vertical period was very robust, the ~250 million years period through radial variation of the Sun orbit still requires some fine tuning.

Briefly, it is possible to say that the climate cooling would be caused by the two movements of the Sun, since both lead to an increase in cosmic rays that may affect cloud formation. The bigger intensity of cosmic rays and higher density of space dust, both lead to increase of cloudiness and decrease of planetary surface temperature.

It is necessary to note that both mechanisms derived from the Solar System movement around and through the galaxy plane are theoretical models. These models are only constrained by the measurements of present observational velocity of the Sun, with respect to an inertial frame, as proposed by Schönrich and Binney (2009). Bovy (2017) presented a detailed new star inventory of the galaxy in the vicinity of the Solar System; it agrees with and extends the previous studies. This is a good beginning to map the galaxy with the full set of Gaia data.

## 8 Galactic Cosmic Rays Mutogenetic Effect

Initially, Muller (1927) suggested the relation between the incidence of energetic particles and the mutations. The question asked by Babcock and Collins (1929) "Does cosmic radiation have any influence on rate of mutation?", however, received little attention.

Only speculative hypothesis was produced by Krassovskij and Šklovskij (1958) with respect to periods of many hundred years when the flux of cosmic rays on the Earth will be hundred times larger than the mean common conditions. They argued that it is necessary to recognize that the increase of cosmic rays due to supernovas at certain stages of evolution could have important, if not decisive, significance. They also supposed that a significant increase in cosmic rays flux that could be prolonged for millennia may involve catastrophic consequences for many specialized animal species with comparatively small population.

Curtis and Smith (1963) observed a light increase in chromosomal deletions in corn seeds affected by heavy CR particles. Most of CRs energy arrives to the ground in the form of kinetic energy of muons. Micke et al. (1964) tested the mouns mutation action over dormant seeds.

Sagan (1973) remarked how important the effect of cosmic rays on Earth's atmosphere is, and that these rays can produce significant changes as well, such as mutations in the hereditary material.

More recent researches have shown that a strong increase in the flow of CR can produce mutations in living organisms and may affect biodiversity by direct or indirect mechanisms (Medvedev and Melott 2007, Melott and Thomas 2009). These mutations are the effects of: (i) direct radiation mainly through the muons (McNulty et al. 1974, Atri and Melott 2011), (ii) the significant climatic change induced by the increase in cloudiness through the seeding of ionization that produce cooling (Hadly et al. 2004, Kozma et al. 2016; see Sect. (4), (iii) the effects on atmospheric NOx chemicals and their precipitation as nitric acid (e.g. Nguyen et al. 1992), and (iv) the DNA (deoxyribonucleic acid) changes due to the increase in solar UVB (ultra-violet radiation at 280-315 nm), schematically shown in Fig. 11, as a consequence of ozone depletion due to ionizing radiation in the atmosphere (Gehrels et al. 2003).

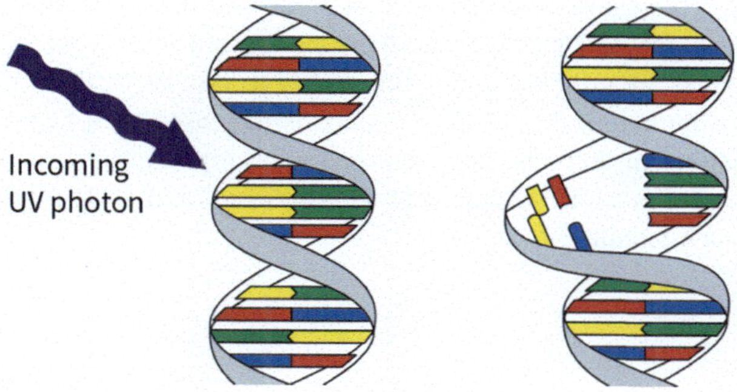


**Fig. 11.** Changes in the DNA of the genome due to UV light from the sun.

Despite previous researches, the question continues in the 21st century: "Is there a connection between enhanced rate of cosmic rays and the origin of life? Have the variations in cosmic ray flux affected the evolution of life on Earth?"(Atri and Melott 2014).

However, some advances may start to produce an answer. Briefly, SN remnants are considered the major source of the increased CR flux into the Earth. The radiation that arrive to the land consists primarily of neutrons and muons which would have a substantial effect on cancer and mutation rates and the larger and longer-lived organisms would experience a greater relative increase in radiation dose.

The terrestrial effects of both photon and charged particle radiation from supernovae 50–100 pc from Earth have been modeled by Thomas et al. (2016) and Melott et al. (2017a).

Thomas et al. (2016) ask whether such SNs are expected to have had substantial effects on the terrestrial atmosphere and biota. Combining photon and CR effects, they find that a SN at less than 100 pc can have substantial effect on terrestrial organisms. Troposphere ionization will increase by nearly an order of magnitude for thousands of years. The mouns irradiation will increase, twenty times which will approximately triple the overall radiation load on terrestrial organisms.

In the case of a supernova at 50 pc, as they were at about 2.7 to 1.7 Ma (Pliocene-Pleistocene boundary) and at 6.5 to 8.7 Ma at the Late Miocene, muon irradiation on the ground may have increased by a factor of about 150 to 2000 (Melott et al. 2017a, 2017b).

The SN irradiation of muons in the soil and hundreds of meters to the ocean would increase cancer and the rate of mutation. Especially large and long-lived organisms are susceptible to developing cancer (Melott and Thomas 2018).

Ancient pathogenic DNA was evidenced by the record of abnormal growth in a 1.6–million- to 1.8-million-year-old toe bone produced by a malignant tumor. Oncogenesis has been in existence in the hominin lineage for at least 2 million years (Odes et al. 2016, Randolph-Quinney et al. 2016). New research associated a long-term exposition to GCR with the present time cancer mortality in São Pablo (Vieira et al. 2018).

The potential biological impacts at Earth's surface of stratospheric ozone depletion caused by SN nearby at 50 pc, as it occurred about 2.5 and 8.0 Ma ago, was investigated by Thomas (2018). He used ionization profiles for three-time frames following arrival of SN CR. The results of the simulation of the CR ionization were compared to a control run without SN CR input. Since the runs are not connected to any particular dates, all time values are arbitrary and given simply to provide scale of the time frames involved. In the simulation, the UVB irradiance is increased by a factor of 1.1 to 2.8, with large variation in latitude and seasonally at high-latitude regions (Fig. 12). Changes in UVA (ultra-violet radiation at 315–400 nm) and PAR (400–700 nm, visible light) are much smaller.

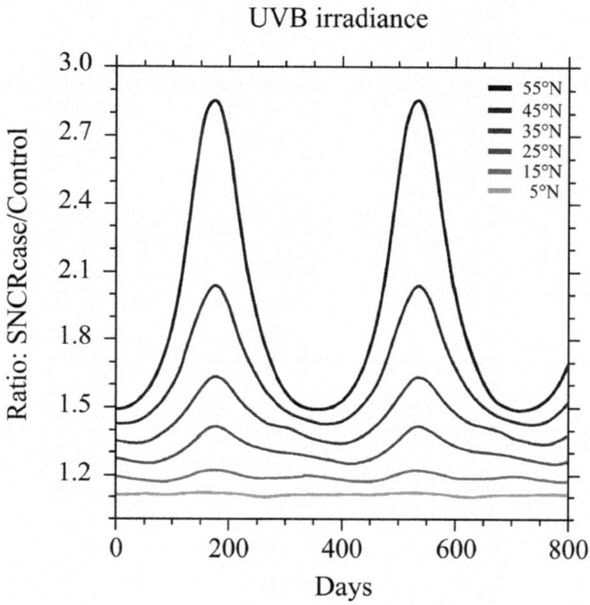


**Fig. 12.** The increase in UVB (as a ratio of the SN case to the control run) as a function of time and latitude (adapted from Thomas 2018).

A ratio plot for DNA damage (from Setlow 1974) versus ratio of the SN CR case to the control run is shown in Fig. 13. The DNA damage (in vitro) is increased by factors similar to UVB, while other biological impacts (erythema, skin cancer, cataracts, marine phytoplankton photosynthesis inhibition and plant damage) are increased by smaller amounts. However, the effects on organisms from changes in irradiance can be quite complicated and in most cases are strongly wavelength-dependent. He concludes that biological impacts due to increased UV irradiance in this SN case are not at a mass-extinction level but might be expected to contribute to changes in certain species abundance. This result fits well with species turnover observed around the Pliocene-Pleistocene boundary.

Although, as it was already mentioned, ozone depletion and the increase in hazardous UVB have an important biological impact, new effects come to the foreground (Mellot and Thomas 2018). Muon irradiation on the ground and hundreds of meters down into the ocean will increase cancer and mutation rates, the differences being most notable in terrestrial megafauna and benthic organisms. Typically, larger organisms live long enough to develop cancer; in microorganisms, the primary effects would be associated with mutation rates. Atmospheric ionization in the troposphere will greatly increase lightning rates, with a concomitant increase in the rate of wildfires.

With respect to the enhanced mutations, it should be noted that the ~8.0 Ma SN was roughly contemporaneous with the oldest Graecopithecus, dated at around 7.2 Ma. The oldest remains attributed to Hominini are coeval with about 2.7 to 1.7 Ma SN. Contemporaneously hominin brain size began to increase (Homo erectus), as it can be seen in Sect. 8.2.

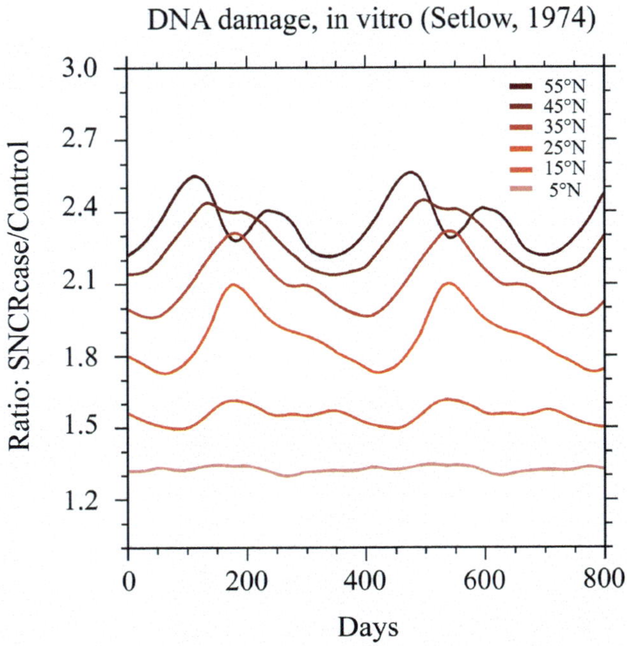


**Fig. 13.** Ratio of surface-level DNA damage (in vitro) in the 300 yr SN case compared to a control run, as a function of time, at latitudes between 5° and 55° N (adapted from Thomas 2018).

## 8.1 Fauna Extinctions

Traditionally, there was a consensus in placing the Pliocene-Pleistocene boundary at the base of the Calabrian stage, due to the first appearance of cold fauna from the North Atlantic Ocean in the Mediterranean Sea. This was interpreted as the first indication of climatic deterioration in the Northern Hemisphere. However, it is now known through stable isotopes analyses that a major cooling occurred earlier, around 2.55 Ma (Cita 2008), and even earlier cooling events are known (Cohen and Gibbard 2010).

From a biostratigraphic point of view, the Lowest Occurrence (LO) of some taxa occurs close to the Piacenzian-Gelasian boundary (ICS 2017). It could mention that during the Pliocene-Pleistocene turnover, 18 coral genera were lost from the Caribbean region. As an example, large colonies of Pocillopora and Stylophora highly vulnerable to the climate shifts in the Pliocene-Pleistocene were extinct (Klaus and Budd 2003).

Furthermore, nannofossils that are one of the best biostratigraphic tools in open oceans, also showed extinctions rate. In this case, there are the LO of Discoaster pentaradiatus Tan (2.39 Ma) and the LO of D. surculus that took place at 2.53 Ma (Agnini et al. 2017, Young et al. 2017, Rio et al. 1998). Regarding foraminifera, a marker of this boundary is Globoconella puncticulata (Deshayes 1832) (= Globorotalia puncticulata) which reached its extinction level at 2.35 Ma (Wei 1994).

Pimiento et al. (2017) recognized extinction events among marine megafauna during the Pliocene-Pleistocene transition. The extinction rates were three times higher than in the rest of the Cenozoic, and with the highest rates occurring during the Late Pliocene (between 3.8 and 2.4 Ma). The only exception for these times were seabirds that presented their maximum extinction during the Paleocene. The fossil megafauna record revels that 36% of genera of this fauna became extinct during the Pliocene, being the marine mammals the most affected, losing 55% of their genetic diversity. Meanwhile, seabirds lost 35%, sea turtles 43% and sharks 9%. Pimiento et al. (2017) found that the marine regression caused a sharp drop in neritic areas during the Late Pliocene (in coincidence with the highest extinction rates found). This reduction in coastal habitat, probably acting alongside oceanographic alterations such as changes in productivity and ocean circulation, was the most likely extinction driver for the Pliocene marine fauna.

The ecological preyscapes that energetically support gigantism only arose in the Pliocene-Pleistocene, with the onset of seasonally intensified upwelling regimes (Slater et al. 2017).

## 8.2 Hominin Evolution

Graecopithecus would be the oldest known hominin and the oldest known crown hominine, as the evidence for the gorillin status of Chororapithecus is much weaker than the hominin status of Graecopithecus (Fuss et al. 2017). The fossils were dated at around 7.2 Ma, which would make them the oldest remains attributed to Hominini, even older than Sahelanthropus tchadensis dated at around 6.0–7.0 Ma (Benoit and Thackeray 2017, Fuss et al. 2017). This age agrees with genetic estimates calibrated by mutations per generation, which place the human-chimpanzee split as early as 8.0–7.0 Ma (Langergraber et al. 2012), but slightly precedes genetic estimates calibrated using the fossil record, which suggest the split occurred after 6.3 Ma (Patterson et al. 2006).

From at least 6.0 to 3.0 Ma, early humans combined apelike and humanlike ways of moving around. Fossil bones record a gradual transition from climbing trees to walking upright on a regular basis. Australopithecus africanus is the first of an early ape-form species to be classified as hominin (Dart 1925) which apparently evolved in eastern Africa around 4.0 Ma before spreading throughout the continent and eventually becoming extinct at 2.0 Ma. The oldest direct evidence of stone tool manufacture comes from Gona (Ethiopia) and dates between 2.6 and 2.5 Ma, being attributed to Australopithecus afarensis (McPherron et al. 2010). The curved spine is uniquely human. Lower back absorbs shock when humans walk and there was a similar curve in the spine of Australopithecus africanus, who walked upright in a way very similar to modern humans (Willams et al. 2013, Tardieu et al. 2017). However, the brains of most species of Australopithecus were roughly 20% larger than the average cranial capacity in modern chimpanzees (DeSilva 2011).

It is now clear that key events in human evolution occurred within the time interval that also encompassed the Pliocene–Pleistocene boundary (i.e., within the limits of the Matuyama paleomagnetic chron, from 2.6 to 0.8 Ma) (Fig. 14). It was during that time that the genus Homo first appeared in Africa. Two different Homo species appeared. Homo habilis, who lived at the beginning of the Pleistocene (Leakey et al. 1964), had a cranial capacity slightly less than half of the size of modern humans. Homo habilis

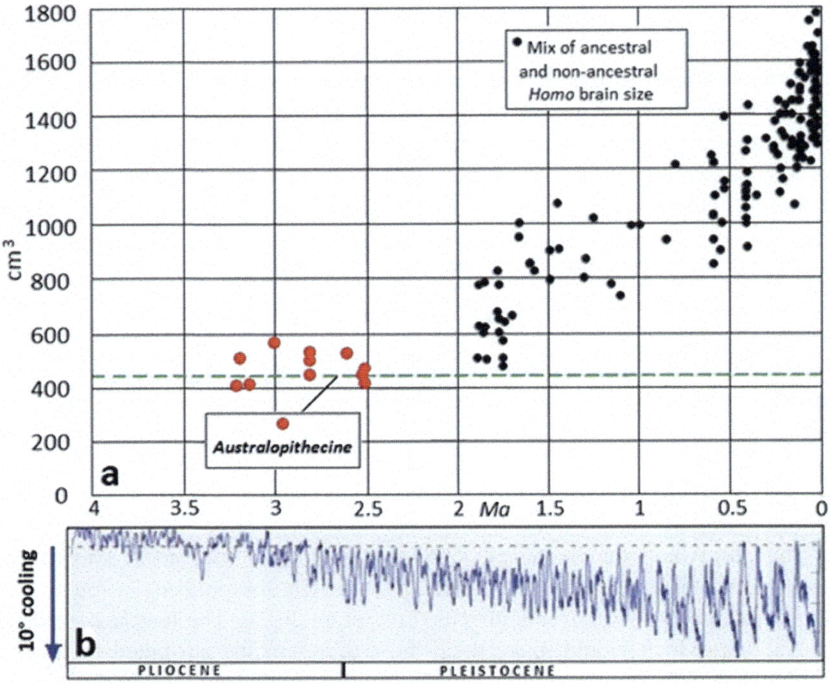


**Fig. 14.** (a) Evolution of the Homo brain size through time (Adapted from Calvin 2016); (b) Proxy of ocean temperature showing ice age fluctuations in climate, averaged over 57 sediment-core sites (adapted from Lisiecki and Raymo 2005).

coexisted with Homo erectus (meaning "upright man") for a substantial period after 2.0 Ma. Although it is commonly assumed that this species dispersed out of Africa, it is possible that it first evolved in Asia. Dennell (2008) explain that it is possible that H. erectus originated in Southwest Asia from a population that left Africa before 2.0 Ma, and then dispersed back into East Africa as well as eastwards to Java and China. From around 1.9 Ma Homo erectus started to spread in Africa, Europe, South and South East Asia. Depending upon which dates are preferred, early Homo was at Dmanisi by 1.85 Ma, and in Central China (Lantian Gongwangling) at 1.63 Ma, in Java by 1.5 Ma, and in the Iberian peninsula by 1.2 Ma.

Spoor et al. (2007) confirmed the distinctiveness of H. habilis and H. erectus, independently of overall cranial size, and suggested that these two early taxa co-existed in eastern Africa for nearly half a million years in the same lake basin.

The size and broad shape of the hipbones of Homo erectus are similar to those of modern humans, showing that this early human species had given up climbing for walking (Ward et al. 2015, Tardieu et al. 2017). Pelvic and proximal femoral morphology in early Homo (namely H. erectus) was uniquely derived (Churchill and Vansickle 2017). Therefore, Homo erectus shows adaptations of long distance running. During the evolution of Homo erectus, there was a significant increase in their body size compared to earlier hominids.

One of the most distinct features of human evolution is the trend towards increasingly large brains. The rapid three-fold enlargement of the hominin brain began about 2.3 Ma (Calvin 2016). Early hominin Australopithecines had a cranial capacity (CC) slightly larger than that of extant apes (Robson and Wood 2008).

As the Pleistocene cooling began 2.5 Ma, the hominin brain size began to increase and inexplicably doubled over the next million years. The scatter plot of hominin brain size over time (Fig. 14) shows major features: firstly a long period of near-stasis in the Pliocene where brain size in the Australopithecines remained stable; secondly, a fast 3.3 times increase during the Pleistocene up to present times. The tendency towards the increase started near 2.0 Ma or before.

Most explanations for hominin brain size increase have been focused on micro evolutionary mechanisms. These hypotheses can explain anagenetic patterns but may not be relevant for patterns caused by origination and extinction (Gould 2002, Simpson 2016). Patterson et al. (2017) also related the Hominid change, primarily to differences in dietary ecology.

The results of Schroeder and Ackermann (2017) indicate that for 95% of taxon comparisons, throughout the cranium (face, maxilla, neurocranium, temporal, mandible), showed that most of the phenotypic cranial and mandibular diversity within Homo, from ~2.8 Ma, is consistent with random genetic drift. These authors also proposed that it is necessary to reconsider the traditional view that selection was the main evolutionary process driving changes in the neurocranium, and most other cranial regions, within Homo, and to consider the implications of that for our understanding of how and why our lineage evolved.

Campisano et al. (2017) carried out the Hominin Sites and Paleolakes Drilling Project (HSPDP) to explicitly explore key hypotheses linking environmental history and mammalian (including hominin) evolution and potentially develop new testable hypotheses. The results improved the environmental history of eastern Africa during much of the Late Neogene and Quaternary even if they cannot refute particular hypotheses of environmental-evolutionary linkages. They demonstrated that temporal correlation does not prove causation in evolutionary history, but they note that a lack of correlation in multiple, well-constrained records is prima facie evidence against any particular hypothesis. There is also the possibility that environmental changes cannot be linked with observed events in faunal evolution. Competitive environmental hypotheses would be false in some cases. Non-ecological factors (for example, sexual selection, and competition, among others) may have been paramount in the evolution of hominids.

Du et al. (2018) emphasize that origination and extinction were also important in shaping endocranial volume (ECV) patterns at the clade level, and both micro- and macro-evolutionary change influenced hominin brain size to different extents at different times. They noted that almost all the first and last appearances are associated with an increase in average clade-level brain size, and the importance of each is staggered in time. The changes, as occurred in the appearance of large-brained species, were mostly from ca 2.3 to 1.7 Ma. According to Du et al. (2018) the complicated multi-causal nature of hominin endocranial volume or ECV evolution need future hypotheses and models to recognize and incorporate this hierarchical complexity.

Therefore, new, comprehensive theories to explain potential influences on hominin brain size evolution must be constructed.

As it can be seen in the final discussions, it is herein proposed that the conjunction of the explosion of the supernova and the weakening of the Earth's magnetic field would have produced a strong global change with climate-ecological consequences, plus an increase in the flux of GCR and neutrinos, both with probably mutogenetic effects.

## 9 Discussion and Conclusions

In a succinct summary, the analysis of this contribution considers the influence of GF weakening or/and reversions together with SNs explosions close than 100 pc. These SNs occurrence is likely to happen because of the Solar System location in one arm of the galaxy (Sect. 7). These forcing could promote climate changes, biotic variations, mutations and human evolution that occurred in the Earth history since the Middle Miocene.

The major cooling event during the Pliocene took place at 3.3 Ma, in Marine Isotope Stage (MIS) M2 (during the Mammoth reversed polarity subchron). During the Late Pliocene, from 3.5 to 2.58 Ma, deterioration in Earth's climate state led to an increase in Northern Hemisphere glaciations (Sect. 2). Furthermore, ecosystems changes (Sect. 8.1) occurred together with the earliest Homo erectus found in sediments assigned to this period in Africa (Sect. 8.2). The changes in ocean gateways led to the climatic shift during the Pliocene and the Pleistocene (Sect. 2). However, it is necessary to keep in mind that these important and relatively sudden change could also be explained by the conjunction of two haphazard factors, which are the explosion of a SN at ~2.7 to 1.7 Ma (Sect. 6) and the geomagnetic weakening due to the Gauss-Matuyama reversal (Sect. 4). The SN provided the strong increase of the GCR flow, and a weak geomagnetic field during the Early Matuyama Chron, allowed the entry of a greater flow of the GCR into the atmosphere. Another hypothesis proposed that the Solar System could have crossed the plane of the galaxy around or before the Pliocene-Pleistocene boundary (Weissman 2014 and Sect. 7), which would explain, at least partly, the cooling that occurred through the increase flow of GCRs. As explained in Sect. 5, low cloudiness increased in suchs way, causing the start of a greater cooling that continued increasing by a process of feedback with the ice. The enhanced cloud formation created an increase of the reflection of solar energy back into space, which was thus connected to a subsequent cooling of the Earth's atmosphere. This cooling promoted the formation of continental glaciers and therefore a significant drop in sea level with the interplay of further influences like orbital parameters of the Earth. The increased ice surface over the Earth also amplified the Milankovitch signal on the climate, characteristic of the Pleistocene.

The same mechanism could have occurred during the Late Miocene-Early Pliocene. Approximately between 7.0 and 5.4 Ma, the Miocene cooling culminated with ocean temperatures dropping to near-modern values (Sect. 2). The changes in climate, temperature and ecosystems were a direct consequence of a decline in atmospheric $CO_2$ (Sect. 2). However, other factors need to be considered because frequent reversals of

GF (Sect. 4) together with SN explosion (Sect. 6) took place concurrently. Therefore, the enhanced GCRs due to the SN together with the weakening of the GF that lead the considerably increased of the GCRs input on the Earth atmosphere producing a cooling effect on the climate (Sect. 2).

The SN events occurred around 6.5–8.7 Ma and 1.7–3.2 Ma, which coexisted with the collapse of the GM field, implied that a large GCR flow entered the Earth atmosphere. Then, the mutagenic effects (Sect. 8) of the GCRs, in conjunction with other factors due to climatic and environmental changes, could have led to an important step of human evolution. During the first mentioned event, the divergence between the chimpanzee and hominini (coeval with the started cooling at 7–6 Ma; Lebatard et al. 2008, Simpson et al. 2015) took place. Through the second, the hominini showed a big number of mutations being the most relevant the rapid enlargement of the brain that began about 2.3 Ma (Calvin 2016) (for more detail see Sect. 8.2.).

Therefore, the two main Earth´s climate changes occurred during the Late Miocene and the Pliocene-Pleistocene boundary. This lead us to hypothesize that the occurrence of the explosion of a SN at a distance of 100 pc or less might have been responsible for the increase in GCRs together with the reversal and significant weakening of the GF, for the occurrence of climate cooling, mutations, extinctions and evolution of the species.

The first SN recorded by humans occurred ~5000 years ago. Iqball et al. (2009) presented evidence in two sites that the ancient inhabitants of Kashmir definitely noted and registered, at a date between 7,000 and 3,000 BP, different celestial events that could be related to a SN explosion. According to the authors, the SNs candidates are HB9 and G182.4 + 4.3 are shell-type SNs remains and as such, they lack pulsars. But this SN HB9 is estimated to have exploded at around 1.0 kpc (Araya 2014), that is 10 times farther than was postulated to be necessary for the biological and climatic effects depicted previously. Furthermore, the intensity of the GF was substantively lower in the Late Pleistocene than during the Holocene, when the possible SN event was taking place. Therefore, the GCRs flux could had the "umbrella" effect of the GF that prevented the increased the input in the atmosphere as in the two previously situations described above.

For the above mentioned SNs which occurred during the Holocene, they are lacking of at least two conditions necessary for the explosions of SNs to affect the Earth as in the past. One that they must occur nearby, it means 100 pc or less, and another that the GF must be weakened and / or near zero due to the occurrence of a reversion.

**Acknowledgements.** The authors wish to express their deepest gratitude to the institutions that financed the present contribution. the National Council of Scientific and Technical Research of Argentina (CONICET), the University of Buenos Aires (UBA) and the National University of the Center of Buenos Aires Province (UNCPBA), Argentina. The present authors also appreciate the critical reading and the relevant contributions of Dr Augusto Rapalini (UBA), who reviewed a previous draft of this paper.

## References

Agnini, C., Monechi, S., Raffi, I.: Calcareous nannofossil biostratigraphy: historical background and application in Cenozoic chronostratigraphy. Lethaia **50**(3), 447–463 (2017)

Ahluwalia, H.S.: Galactic cosmic ray intensity variations at a high latitude sea level site 1937–1994. J. Geophys. Res. Space Phys. **102**(A11), 24229–24236 (1997)

Ahn, H.S., Kidane, T., Yamamoto, Y., Otofuji, Y.I.: Low geomagnetic field intensity in the Matuyama Chron: palaeomagnetic study of a lava sequence from Afar depression. East Afr. Geophys. J. Int. **204**(1), 127–146 (2016)

Andreev, A.A., Tarasov, P.E., Wennrich, V., Raschke, E., Herzschuh, U., Nowaczyk, N.R., Brigham-Grette, J., Melles, M.: Late Pliocene and Early Pleistocene vegetation history of northeastern Russian Arctic inferred from the Lake El'gygytgyn pollen record. Clim. Past. **10**, 1017e1039 (2014). https://doi.org/10.5194/cp-10-1017-2014

Araya, M.: Fermi LAT observation of supernova remnant HB9. Mon. Not. R. Astron. Soc. **444** (1), 860–865 (2014)

Atri, D., Melott, A.L.: Biological implications of high-energy cosmic ray induced muon flux in the extragalactic shock model. Geophysical Research Letters **38**(19) (2011)

Atri, D., Melott, A.L.: Cosmic rays and terrestrial life: a brief review. Astropart. Phys. **53**, 186–190 (2014)

Babcock, E.B., Collins, J.L.: Does natural ionizing radiation control rate of mutation? Proc. Natl. Acad. Sci. **15**(8), 623–628 (1929)

Bahcall, J.N., Bahcall, S.: The Sun's motion perpendicular to the galactic plane. Nature **316** (6030), 706 (1985)

Bailer-Jones, C.A.: The evidence for and against astronomical impacts on climate change and mass extinctions: a review. Int. J. Astrobiol. **8**(3), 213–219 (2009)

Bailey, I., Hole, G.M., Foster, G.L., Wilson, P.A., Storey, C.D., Trueman, C.N., Raymo, M.E.: An alternative suggestion for the Pliocene onset of major northern hemisphere glaciation based on the geochemical provenance of North Atlantic Ocean ice-rafted debris. Quat. Sci. Rev. **75**, 181–194 (2013)

Bambach, R.K.: Phanerozoic biodiversity mass extinctions. Annu. Rev. Earth Planet. Sci. **34**, 127–155 (2006)

Bartoli, G., Hönisch, B., Zeeba, R.E.: Atmospheric $CO_2$ decline during the Pliocene intensification of Northern Hemisphere glaciations. Paleoceanography **26**(4) (2011)

Bell, D.B., Jung, S.J., Kroon, D., Hodell, D.A., Lourens, L.J., Raymo, M.E.: Atlantic deep-water response to the early Pliocene Shoaling of the Central American seaway. Sci. Rep. **5**, 12252 (2015)

Benjamin, R.A.: The Spiral Structure of the Galaxy: Something Old, Something New. In: Beuther, H., Linz, H., Henning, T. (eds.) Massive Star Formation: Observations Confront Theory, Astronomical Society of the Pacific Conference Series 387, p. 375. Astronomical Society of the Pacific, San Francisco (2008)

Benoit, J., Thackeray, F.J.: A cladistic analysis of Graecopithecus. S. Afr. J. Sci. **113**(11–12), 1–2 (2017)

Binns, W.R., Israel, M.H., Christian, E.R., Cummings, A.C., de Nolfo, G.A., Lave, K.A., Leske, R.A., Mewaldt, R.A., Stone, E.C., von Rosenvinge, T.T., Wiedenbeck, M.E.: Observation of the $^{60}$Fe nucleosynthesis-clock isotope in galactic cosmic rays. Science **352**, 677–680 (2016)

Bintanja, R., Van de Wal, R.S.W.: North American ice-sheet dynamics and the onset of 100,000-year glacial cycles. Nature **454**(7206), 869 (2008)

Bolton, C.T., Wilson, P.A., Bailey, I., Friedrich, O., Beer, C.J., Becker, J., Baranwal, S., Schiebel, R : Millennial scale climate variability in the subpolar North Atlantic Ocean during the late Pliocene. Paleoceanography **25**(4) (2010)

Boulila, S., Laskar, J., Haq, B.U., Galbrun, B., Hara, N.: Long-term cyclicities in Phanerozoic sea-level sedimentary record and their potential drivers. Global Planet. Change **165**, 128–136 (2018)

Bovy, J.: Stellar inventory of the solar neighborhood using Gaia DR1. Mon. Not. R. Astron. Soc. **470**(2), 1360–1387 (2017)

Breitschwerdt, D., Feige, J., Schulreich, M.M., de Avillez, M.A., Dettbarn, C., Fuchs, B.: The locations of recent supernovae near the Sun from modelling $^{60}$Fe transport. Nature **532**(7597), 73 (2016)

Brigham-Grette, J., Melles, M., Minyuk, P., Andreev, A., Tarasov, P., DeConto, R., Koenig, S., Nowaczyk, N., Wennrich, V., Rosen, P., Haltia, E., Cook, T., Gebhardt, C., Meyer-Jacob, C., Snyder, J., Herzschuh, U.: Pliocene warmth, polar amplification, and stepped Pleistocene cooling recorded in NE Arctic Russia. Science **340**, 1421e1427 (2013). https://doi.org/10.1126/science.1233137

Brink, H.J.: Periodic signals of the milky way concealed in terrestrial sedimentary basin fills and in planetary magmatism? Int. J. Geosci. **6**(08), 831 (2015)

Calvin, W.H.: A feedback loop able to enlarge the brain for 2.4 Ma without Darwin's selective survival. bioRxiv, p. 053827 (2016)

Campisano, C.J., Cohen, A.S., Arrowsmith, J.R., Asrat, A., Behrensmeyer, A.K., Brown, E.T., Deino, A.D., Deocampo, D.M., Feibel, C.S., Kingston, J.D., Lamb, H.F., Lowenstein, T.K., Noren, A., Olago, D.O., Owen, R.B., Pelletier, J.D., Potts, R., Reed, R.K., Renaut, R.W., Russell, J.M., Russell, J.L., Schäbitz, F., Stone, J.R., Trauth, M.H., Wynn, J.G.: The hominin sites and paleolakes drilling project: high-resolution paleoclimate records from the east African rift system and their implications for understanding the environmental context of hominin evolution. PaleoAnthropology, 1–43 (2017)

Cande, S.C., Kent, D.V.: Revised calibration of the geomagnetic polarity timescale for the Late Cretaceous and Cenozoic. J. Geophys. Res. Solid Earth **100**(B4), 6093–6095 (1995)

Cane, H.V.: Coronal mass ejections and Forbush decreases. Space Sci. Rev. **93**(1–2), 55–77 (2000)

Cane, M.A., Molnar, P.: Closing of the Indonesian seaway as a precursor to east African aridification around 3–4 million years ago. Nature **411**(6834), 157 (2001)

Cane, H.V., Wibberenz, G., Richardson, I.G., Von Rosenvinge, T.T.: Cosmic ray modulation and the solar magnetic field. Geophys. Res. Lett. **26**(5), 565–568 (1999)

Cardillo, M., Tavani, M., Giuliani, A.: The origin of cosmic-rays from SNRs: confirmations and challenges after the first direct proof. Nucl. Phys. B-Proc. Suppl. **256**, 65–73 (2014)

Carslaw, K.S., Harrison, R.G., Kirkby, J.: Cosmic rays, clouds, and climate. Science **298**(5599), 1732–1737 (2002)

Chang, H.Y., Moon, H.K.: Time-series analysis of terrestrial impact crater records. Publ. Astron. Soc. Jpn. **57**(3), 487–495 (2005)

Churchill, S.E., Vansickle, C.: Pelvic morphology in Homo erectus and early Homo. Anat. Rec. **300**(5), 964–977 (2017)

Cita, M.B.: Summary of italian marine stages of the quaternary. Episodes **31**, 251–254 (2008)

Clube, S.V.M., Napier, W.M.: Terrestrial catastrophism—Nemesis or galaxy? Nature **311**(5987), 635–636 (1984)

Cohen, K.M., Gibbard, P.L.: Global chronostratigraphical correlation table for the last 2.7 million years v.2010. In: Subcommission on Quaternary Stratigraphy, International Commission on Stratigraphy. Cambridge (2010). http://www.quaternary.stratigraphy.org.uk/charts/

Curtis, H.J., Smith, H.H.: Corn seeds affected by heavy cosmic ray particles. Science **141**(3576), 158–160 (1963)
Dart, R.A.: Australopithecus africanus: The Man-Ape of South Africa. Nature **115**, 195–199 (1925)
De Schepper, S., Gibbard, P.L., Salzmann, U., Ehlers, J.: A global synthesis of the marine and terrestrial evidence for glaciation during the Pliocene Epoch. Earth Sci. Rev. **135**, 83–102 (2014)
Deino, A.L., Kingston, J.D., Glen, J.M., Edgar, R.K., Hill, A.: Precessional forcing of lacustrine sedimentation in the late Cenozoic Chemeron Basin, Central Kenya Rift, and calibration of the Gauss/Matuyama boundary. Earth Planet. Sci. Lett. **247**(1–2), 41–60 (2006)
Demske, D., Mohr, B., Oberhänsli, H.: Late Pliocene vegetation and climate of the Lake Baikal region, southern East Siberia, reconstructed from palynological data. Palaeogeogr. Palaeoclimatol. Palaeoecol. **184**, 1 (2002). https://doi.org/10.1016/s0031-0182(02)00251-1
Dennell, R.W.: Human migration and occupation of Eurasia. Episodes **31**(2), 207 (2008)
DeSilva, J.M.: A shift toward birthing relatively large infants early in human evolution. Proc. Natl. Acad. Sci. U.S.A. **108**, 1022–1027 (2011)
Ding, Z.L., Derbyshire, E., Yang, S.L., Sun, J.M., Liu, T.S.: Stepwise expansion of desert environment across northern China in the past 3.5 Ma and implications for monsoon evolution. Earth Planet. Sci. Lett. **237**, 45–55 (2005). https://doi.org/10.1016/j.epsl.2005.06.036
Drury, L.O.C.: Origin of cosmic rays. Astropart. Phys. **39**, 52–60 (2012)
Du, A., Zipkin, A.M., Hatala, K.G., Renner, E., Baker, J.L., Bianchi, S., Bernal, K.H., Wood, B. A.: Pattern and process in hominin brain size evolution are scale-dependent. Proc. R. Soc. B. **285**(1873), 20172738 (2018)
Duk-Rodkin, A., Barendregt, R.W.: Stratigraphical record of glacials/interglacials in Northwest Canada. In: Ehlers, J., Gibbard, P.L., Hughes, P.D. (eds.) Quaternary Glaciations: Extent and Chronology: A closer look, pp. 661–698. Elsevier, Amsterdam (2011)
Duplissy, J., Enghoff, M.B., Aplin, K.L., Arnold, F., Aufmhoff, H., Avngaard, M., Baltensperger, U., Bondo, T., Bingham, R., Carslaw, K., Curtius, J., David, A., Fastrup, B., Gagné, S., Hahn, F., Harrison, R.G., Kellett, B., Kirkby, J., Kulmala, M., Laakso, L., Laaksonen, A., Lillestol, E., Lockwood, M., Mäkelä, J., Makhmutov, V., Marsh, N.D., Nieminen, T., Onnela, A., Pedersen, E., Pedersen, J.O.P., Polny, J., Reichl, U., Seinfeld, J.H., Sipilä, M., Stozhkov, Y., Stratmann, F., Svensmark, H., Svensmark, J., Veenhof, R., Verheggen, B., Viisanen, Y., Wagner, P.E., Wehrle, G., Weingartner, E., Wex, H., Wilhelmsson, M., Winkler, P.M.: Results from the CERN pilot CLOUD experiment. Atmos. Chem. Phys. **10**(4), 1635–1647 (2010)
Eichmann, B., Rachen, J.P., Merten, L., Van Vliet, A., Tjus, J.B.: Ultra-high-energy cosmic rays from radio galaxies. J. Cosmol. Astropart. Phys. **02**, 036 (2018). https://doi.org/10.1088/1475-7516/2018/02/036
Englmaier, P., Pohl, M., Bissantz, N.: The Milky Way spiral arm pattern. Memorie della Società Astronomica Italiana **18**, 199–204 (2009)
Erkens, R.H.J.: The less-splendid isolation of the South American continent. Front. Biogeogr. **7**, 89–90 (2015)
Fedorov, A.V., Dekens, P.S., McCarthy, M., Ravelo, A.C., Barreiro, M., Pacanowski, R.C., Philander, S.G.: The Pliocene paradox (mechanisms for a permanent El Niño). Science **312** (5779), 1485–1489 (2006)
Fimiani, L., Cook, D.L., Faestermann, T., Gómez-Guzmán, J.M., Hain, K., Herzog, G., Knie, K., Korschinek, G., Ludwig, P., Park, J., Reedy, R.C., Rugel, G.: Interstellar $^{60}$Fe on the surface of the moon. Phys. Rev. Lett. **116**, 151104 (2016)

Foyle, K., Rix, H.W., Walter, F., Leroy, A.K.: Arm and interarm star formation in spiral galaxies. Astrophys. J. **725**(1), 534–541 (2010)

Frank, M.: Comparison of cosmogenic radionuclide production and geomagnetic field intensity over the last 200,000 years. Philos. Trans. R. Soc. Lond. A **358**, 1089–1107 (2000)

Friedrich, O., Wilson, P.A., Bolton, C.T., Beer, C.J., Schiebel, R.: Late Pliocene to early Pleistocene changes in the North Atlantic current and suborbital-scale sea-surface temperature variability. Paleoceanography **28**, 274–282 (2013)

Froese, D.G., Barendregt, R.W., Enkin, R.J., Baker, J.: Paleomagnetic evidence for multiple Late Pliocene-Early Pleistocene glaciations in the Klondike area, Yukon Territory. Can. J. Earth Sci. **37**, 863–877 (2000). https://doi.org/10.1139/e00-014

Fuss, J., Spassov, N., Begun, D.R., Böhme, M.: Potential hominin affinities of Graecopithecus from the Late Miocene of Europe. PLoS ONE **12**(5), e0177127, 23 (2017). https://doi.org/10.1371/journal.pone.0177127

Galeotti, S., DeConto, R., Naish, T., Stocchi, P., Florindo, F., Pagani, M., Barrett, P., Bohaty, S.M., Lanci, L., Pollard, D., Sandroni, S., Talarico, F.M., Zachos, J.C.: Antarctic ice sheet variability across the Eocene-Oligocene boundary climate transition. Science **352**, 76–80 (2016)

Gallet, Y., Pavlov, V.E.: Three distinct reversing modes in the geodynamo. Izvestiya, Phys. of the Solid Earth **52**(2), 291–296 (2016)

Gehrels, N., Laird, C.M., Jackman, C.H., Chen, W.: Ozone depletion from nearby supernovae. Astrophys. J. **585**, 1169–1176 (2003)

Gibbard, P.L., Head, M.J.: IUGS ratification of the quaternary system/period and the Pleistocene Series/ epoch with a base at 2.58 Ma. Quaternaire **20**, 411–412 (2009)

Gibbard, P.L., Head, M.J., Walker, M., Alloway, B., Beu, A.G., Coltorti, M., Hall, V.M., Liu, J., Knudsen, K.-L., Van Kolfschoten, T., Litt, T., Marks, L., McManus, J., Partridge, T.C., Piotrowski, J.A., Pillans, B., Rousseau, D.-D., Suc, J.-P., Tesakov, A.S., Turner, C., Zazo, C.: Formal ratification of the quaternary system/ period and the Pleistocene Series/Epoch with a base at 2.588 Ma. J. Quat. Sci. **25**, 96–102 (2010)

Gies, D.R., Helsel, J.W.: Ice age epochs and the sun's path through the Galaxy. Astrophys. J. **626**, 844–848 (2005). https://doi.org/10.1086/430250

Gould, S.J.: The Structure of Evolutionary Theory. Belknap Press of Harvard University Press, Cambridge (2002)

Hadly, E.A., Ramakrishnan, U.M.A., Chan, Y.L., Van Tuinen, M., O'Keefe, K., Spaeth, P.A., Conroy, C.J.: Genetic response to climatic change: insights from ancient DNA and phylochronology. PLoS Biol. **2**(10), e290 (2004)

Haug, G.H., Ganopolski, A., Sigman, D., Rosell-Mele, A., Swann, G., Tiedemann, R., Jaccard, S., Bollmann, J., Maslin, M., Leng, M., Eglinton, G.: North Pacific seasonality and the glaciation of North America 2.7 million years ago. Nature **433**, 821–825 (2005). https://doi.org/10.1038/nature03332

Haywood, A.M., Valdés, P.J.: Modelling Pliocene warmth: contribution of atmosphere, oceans, and cryosphere, Earth Planet. Sci. Lett. **218**, 363–377 (2004). https://doi.org/10.1016/S0012-821X(03)00685-X

Helder, E.A., Vink, J., Bykov, A.M., Ohira, Y., Raymond, J.C., Terrier, R.: Observational signatures of particle acceleration in supernova remnants. Space Sci. Rev. **173**(1–4), 369–431 (2012)

Hennissen, J.A.I., Head, M.J., De Schepper, S., Groeneveld, J.: Palynological evidence for a southward shift of the North Atlantic current at ~2.6 Ma during the intensification of late Cenozoic Northern Hemisphere glaciation. Paleoceanography **29**, 564–580 (2014). https://doi.org/10.1002/2013pa002543

Hennissen, J.A., Head, M.J., De Schepper, S., Groeneveld, J.: Increased seasonality during the intensification of Northern Hemisphere glaciation at the Pliocene-Pleistocene boundary 2.6 Ma. Quat. Sci. Rev. **129**, 321–332 (2015)

Herbert, T.D., Ng, G., Peterson, L.C.: Evolution of mediterranean sea surface temperatures 3.5–1.5 Ma: regional and hemispheric influences. Earth Planet. Sci. Lett. **409**, 307–318 (2015). https://doi.org/10.1016/j.epsl.2014.10.006

Herbert, T.D., Lawrence, K.T., Tzanova, A., Peterson, L.C., Caballero-Gill, R., Kelly, C.S.: Late Miocene global cooling and the rise of modern ecosystems. Nat. Geosci. **9**(11), 843 (2016)

Hoorn, C., Flantua, S.: An early start for the Panama land bridge. Science **348**(6231), 186–187 (2015)

Humphreys, R.M., Larsen, J.A.: The Sun's distance above the galactic plane. Astron. J. **110**, 2183 (1995)

Collaboration, I.: Evidence for high-energy extraterrestrial neutrinos at the IceCube detector. Science **342**(6161), 1242856 (2013)

Iqbal, N., Vahia, M.N., Masood, T., Ahmad, A.: Some early astronomical sites in the Kashmir region. J. Astron. Hist. Herit. **12**(1), 61–65 (2009)

Jakob, K.A.: Late Pliocene to early Pleistocene millennial-scale climate fluctuations and sea-level variability: a view from the tropical Pacific and the North Atlantic. Unpublished PhD Dissertation, Naturwissenschaftlich-Mathematische Gesamtfakultät, Universität Heidelberg, p. 212 (2017)

Jansen, E., Sjøholm, J.: Reconstruction of glaciation over the past 6.0 Ma from ice-borne deposits in the Norwegian Sea. Nature **349**, 600–603 (1991)

Jaramillo, C., Montes, C., Cardona, A., Silvestro, D., Antonelli, A., Bacon, C.D.: Comment (1) on "Formation of the Isthmus of Panama" by O'Dea et al. Sci. Adv. **3**(6), e1602321 (2017)

Jiang, D., Wang, H., Ding, Z., Lang, X., Drange, H.: Modeling the middle Pliocene climate with a global atmospheric general circulation model. J. Geophys. Res.: Atmos. **110**(D14) (2005)

Karas, C., Nürnberg, D., Gupta, A.K., Tiedemann, R., Mohan, K., Bickert, T.: Mid-Pliocene climate change amplified by a switch in Indonesian subsurface throughflow. Nat. Geosci. **2** (6), 434 (2009)

Karim, M.T., Mamajek, E.E.: Revised Geometric Estimates of the North Galactic Pole and the Sun's Height Above the Galactic Midplane. Monthly Notices of the Royal Astronomical Society, stw2772 (2016)

Kernthaler, S.C., Toumi, R., Haigh, J.D.: Some doubts concerning a link between cosmic ray fluxes and global cloudiness. Geophys. Res. Lett. **26**(7), 863–865 (1999)

Kirkby, J., Curtius, J., Almeida, J., Dunne, E., Duplissy, J., Ehrhart, S., Franchin, A., Gagné, S., Ickes, L., Kürten, A., Kupc, A., Metzger, A., Riccobono, F., Rondo, L., Schobesberger, S., Tsagkogeorgas, G., Wimmer, D., Amorim, A., Bianchi, F., Breitenlechner, M., David, A., Dommen, J., Downard, A., Ehn, M., Flagan, R.C., Haider, S., Hansel, A., Hauser, D., Jud, W., Junninen, H., Kreissl, F., Kvashin, A., Laaksonen, A., Lehtipalo, K., Lima, J., Lovejoy, E.R., Makhmutov, V., Mathot, S., Mikkilä, J., Minginette, P., Mogo, S., Nieminen, T., Onnela, A., Pereira, P., Petäjä, T., Schnitzhofer, R., Seinfeld, J.H., Sipilä, M., Stozhkov, Y., Stratmann, F., Tomé, A., Vanhanen, J., Viisanen, Y., Vrtala, A., Wagner, P.E., Walther, H., Weingartner, E., Wex, H., Winkler, P.M., Carslaw, K.S., Worsnop, D.R., Baltensperger, U., Kulmala, M.: Role of sulphuric acid, ammonia and galactic cosmic rays in atmospheric aerosol nucleation. Nature **476**(7361), 429 (2011)

Kirkby, J., Duplissy, J., Sengupta, K., Frege, C., Gordon, H., Williamson, C., Heinritzi, M., Simon, M., Yan, C., Almeida, J., Tröstl, J., Nieminen, T., Ortega, I.K., Wagner, R., Adamov, A., Amorim, A., Bernhammer, A.K., Bianchi, F., Breitenlechner, M., Brilke, S., Chen, X., Craven, J., Dias, A., Ehrhart, S., Flagan, R.C., Franchin, A., Fuchs, C., Guida, R., Hakala, J., Hoyle, C.R., Jokinen, T., Junninen, H., Kangasluoma, J., Kim, J., Krapf, M., Kürten, A., Laaksonen, A., Lehtipalo, K., Makhmutov, V., Mathot, S., Molteni, U., Onnela, A., Peräkylä, O., Piel, F., Petäjä, T., Praplan, A.P., Pringle, K., Rap, A., Richards, N.A.D., Riipinen, I., Rissanen, M.P., Rondo, L., Sarnela, N., Schobesberger, S., Scott, C.E., Seinfeld, J.H., Sipilä, M., Steiner, G., Stozhkov, Y., Stratmann, F., Tomé, A., Virtanen, A., Vogel, A.L., Wagner, A.C., Wagner, P.E., Weingartner, E., Wimmer, D., Winkler, P.M., Ye, P., Zhang, X., Hansel, A., Dommen, J., Donahue, N.M., Worsnop, D.R., Baltensperger, U., Kulmala, M., Carslaw, K.S., Curtius, J.: Ion-induced nucleation of pure biogenic particles. Nature **533**(7604), 521 (2016)

Kitaba, I., Hyodo, M., Katoh, S., Dettman, D.L., Sato, H.: Midlatitude cooling caused by geomagnetic field minimum during polarity reversal. Proc. Natl. Acad. Sci. **110**(4), 1215–1220 (2013)

Kitaba, I., Hyodo, M., Nakagawa, T., Katoh, S., Dettman, D.L., Sato, H.: Geological support for the umbrella effect as a link between geomagnetic field and climate. Sci. Rep. **7**, 40682 (2017)

Klaus, J.S., Budd, A.F.: Comparison of Caribbean coral reef communities before and after Plio-Pleistocene faunal turnover: analyses of two Dominican Republic reef sequences. Palaios **18** (1), 3–21 (2003)

Kleiven, H.F., Jansen, E., Fronval, T., Smith, T.: Intensification of Northern Hemisphere glaciations in the circum Atlantic region (3.5–2.4 Ma) - ice-rafted detritus evidence. Palaeogeog Palaeoclimatol. Palaeoecol. **184**, 213–223 (2002)

Klotz, S., Fauquette, S., Combourieu-Nebout, N., Uhl, D., Suc, J.-P., Mosbrugger, V.: Seasonality intensification and long-term winter cooling as a part of the Late Pliocene climate development. Earth Planet. Sci. Lett. **241**, 174–187 (2006)

Knie, K., Korschinek, G., Faestermann, T., Dorfi, E.A., Rugel, G., Wallner, A.: F 60 anomaly in a deep-sea manganese crust and implications for a nearby supernova source. Phys. Rev. Lett. **93**(17), 171103 (2004)

Korte, M., Brown, M., Frank, U., Senftleben, R., Nowaczyk, N.: Global geomagnetic field reconstructions from centuries to excursions. In: Magnetic Fields in the Solar System, pp. 83–110. Springer, Cham (2018)

Kozma, R., Melsted, P., Magnússon, K.P., Höglund, J.: Looking into the past–the reaction of three grouse species to climate change over the last million years using whole genome sequences. Mol. Ecol. **25**(2), 570–580 (2016)

Krassovskij, V.I., Šklovskij, I.S.: Variation of the intensity of cosmic radiation during earth's geological history and their possible influence on life's evolution. Il Nuovo Cimento **8**(2), 440–443 (1958)

Kristjánsson, J.E., Kristiansen, J.: Is there a cosmic ray signal in recent variations in global cloudiness and cloud radiative forcing? J. Geophys. Res. Atmos. **105**(D9), 11851–11863 (2000)

Kuang, Z., Jiang, Y., Yung, Y.L.: Cloud optical thickness variations during 1983–1991: solar cycle or ENSO? Geophys. Res. Lett. **25**(9), 1415–1417 (1998)

Langereis, C.G., Krijgsman, W., Muttoni, G., Menning, M.: Magnetostratigraphy–concepts, definitions, and applications. Newsl. Stratigr. **43**(3), 207–233 (2010)

Langergraber, K.E., Prüfer, K., Rowney, C., Boesch, C., Crockford, C., Fawcett, K., Inoue, E., Inoue-Muruyama, M., Mitani, J.C., Muller, M.N., Robbins, M.M., Schubert, G., Stoinski, T. S., Viola, B., Watts, D., Wittig, R.M., Wrangham, R.W., Zuberbühler, K., Pääbo, S., Vigilant, L.: Generation times in wild chimpanzees and gorillas suggest earlier divergence times in great ape and human evolution. Proc. Natl. Acad. Sci. **109**(39), 15716–15721 (2012)

Laskar, J., Robutel, P., Joutel, F., Gastineau, M., Correia, A.C.M., Levrard, B.: Along-term numerical solution for the insolation quantities of the Earth. Astron. Astrophys. **428**, 261–285 (2004)

Leakey, L.S.B., Tobias, P.V., Napier, J.R.: A new species of the genus Homo from Olduvai Gorge. Nature **202**, 7–9 (1964)

Lebatard, A.E., Bourlès, D.L., Duringer, P., Jolivet, M., Braucher, R., Carcaillet, J., Schuster, M., Arnaud, N., Monié, P., Lihoreau, F., Likius, A., Mackaye, H.T., Vignaud, P., Brunet, M.: Cosmogenic nuclide dating of Sahelanthropus tchadensis and Australopithecus bahrelghazali: Mio-Pliocene hominids from Chad. Proc. Nat. A. Sci. **105**, 3226–3231 (2008)

Lieberman, B.S., Melott, A.L.: Considering the case for biodiversity cycles: re-examining the evidence for periodicity in the fossil record. PLoS ONE **2**(8), e759 (2007)

Lieberman, B.S., Melott, A.L.: Whilst this planet has gone cycling on: What role for periodic astronomical phenomena in large-scale patterns in the history of life? In: Earth and Life, pp. 37–50. Springer, Dordrecht (2012)

Lisiecki, L.E., Raymo, M.E.: A Pliocene-Pleistocene stack of 57 globally distributed benthic $\delta^{18}O$ records. Paleoceanography **20**(1) (2005). https://doi.org/10.1029/2004pa001071

Lisiecki, L.E., Raymo, M.E.: Plio-Pleistocene climate evolution: trends and transitions in glacial cycle dynamics. Quatern. Sci. Rev. **26**(1–2), 56–69 (2007)

Lunt, D.J., Foster, G.L., Haywood, A.M., Stone, E.J.: Late Pliocene Greenland glaciation controlled by a decline in atmospheric $CO_2$ levels. Nature **454**(7208), 1102–1105 (2008)

Mangeard, P.S., Clem, J., Evenson, P., Pyle, R., Mitthumsiri, W., Ruffolo, D., Sáiz, A., Nutaro, T.: Distinct pattern of solar modulation of galactic cosmic rays above a High Geomagnetic Cutoff Rigidity. Astrophys. J. **858**(1), 43 (2018)

Martínez-García, A., Rosell-Mele, A., McClymont, E.L., Gersonde, R., Haug, G.H.: Subpolar link to the emergence of the modern equatorial Pacific cold tongue. Science **328**, 1550–1553 (2010). https://doi.org/10.1126/science.1184480

Maslin, M.A., Li, X.S., Loutre, M.F., Berger, A.: The contribution of orbital forcing to the progressive intensification of Northern Hemisphere glaciation. Quatern. Sci. Rev. **17**(4–5), 411–426 (1998)

McGehee, R., Lehman, C.: A paleoclimate model of ice-albedo feedback forced by variations in Earth's orbit. SIAM J. Appl. Dyn. Syst. **11**(2), 684–707 (2012)

McNulty, P.J., Sparrow, A.H., Schairer, L.A.: Somatic mutations induced in stamen hairs of Tradescantia clone 02 by relativistic muons. Int. J. Radiat. Biol. Relat. Stud. Phys. Chem. Med. **25**(3), 315–320 (1974)

McPherron, S.P., Alemseged, Z., Marean, C.W., Wynn, J.G., Reed, D., Geraads, D., Bobé, R., Béarat, H.A.: Evidence for stone-tool-assisted consumption of animal tissues before 3.39 million years ago at Dikika, Ethiopia. Nature **466**(7308), 857–860 (2010)

Medvedev, M.V., Melott, A.L.: Do extragalactic cosmic rays induce cycles in fossil diversity? Astrophy. J. **664**(2), 879–889 (2007)

Melles, M., Brigham-Grette, J., Minyuk, P.S., Nowaczyk, N.R., Wennrich, V., DeConto, R.M., Anderson, P.M., Andreev, A.A., Coletti, A., Cook, T.L., Haltia-Hovi, E., Kukkonen, M., Lozhkin, A.V., Rosen, P., Tarasov, P., Vogel, H., Wagner, B.: 2.8 million years of Arctic climate change from Lake El'gygytgyn. NE Russ. Sci. **337**, 315–320 (2012). https://doi.org/10.1126/science.1222135

Melott, A.L., Bambach, R.K.: Analysis of periodicity of extinction using the 2012 geological timescale. Paleobiology **40**(2), 177–196 (2014)

Melott, A.L., Thomas, B.C.: Late Ordovician geographic patterns of extinction compared with simulations of astrophysical ionizing radiation damage. Paleobiology **35**(3), 311–320 (2009)

Melott, A.L., Thomas, B.C.: Terrestrial effects of moderately nearby supernovae. Lethaia **51**(3), 325–329 (2018)

Melott, A.L., Thomas, B.C., Kachelrieß, M., Semikoz, D.V., Overholt, A.C.: A Supernova at 50 pc: effects on the Earth's atmosphere and Biota. Astrophys. J. **840**(2), 105 (2017)

Melott, A.L., Marinho, F., Paulucci, L.: Muon radiation dose and marine megafaunal extinction at the end-Pliocene Supernova. arXiv preprint arXiv:1712.09367 (2017b)

Mewaldt, R.A., Cummings, A.C., Stone, E.C.: Anomalous cosmic rays: interstellar interlopers in the heliosphere and magnetosphere. EOS **75**(16), 185–193 (1994)

Micke, A., Smith, H.H., Woodley, R.G., Mashke, A.: Relative cytogenetic efficiency of muons and π-mesons in Zea mays (L.). Proc. Natl. Acad. Sci. **52**(2), 219–221 (1964)

Milankovitch, M.K.: Kanon der Erdbestrahlung und seine Anwendung auf das Eiszeitenproblem. R. Serb. Acad. Spec. Publ. **133**, 1–633 (1941)

Miroshnichenko, L.I.: Solar Cosmic Rays, p. 492. Kluwer Academic Publishers, Dordrecht (2001)

Montes, C., Cardona, A., Jaramillo, C., Pardo, A., Silva, J.C., Valencia, V., Ayala, C., Pérez-Ángel, L.C., Rodríguez-Parra, L.A., Ramírez, V., Niño, H.: Middle Miocene closure of the Central American Seaway. Sci. (New York, NY) **348**(6231), 226 (2015)

Mudelsee, M., Raymo, M.E.: Slow dynamics of the Northern Hemisphere glaciation. Paleoceanography 20 (2005) http://dx.doi.org/10.1029/2005PA001153

Muller, H.J.: X-ray induced mutation of Drosophila virilis. Science **66**, 84–87 (1927)

Naafs, B.D.A., Stein, R., Hefter, J., Khelifi, N., De Schepper, S., Haug, G.H.: Late Pliocene changes in the North Atlantic current. Earth Planet. Sci. Lett. **298**, 434–442 (2010). https://doi.org/10.1016/j.epsl.2010.08.023

Naafs, B.D.A., Hefter, J., Stein, R.: Millennial-scale ice rafting events and Hudson Strait Heinrich (-like) events during the Late Pliocene and Pleistocene: a review. Quat. Sci. Rev. **80**, 1–28 (2013). https://doi.org/10.1016/j.quascirev.2013.08.014

Naish, T., Powell, R., Levy, R., Wilson, G., Scherer, R., Talarico, F., Krissek, L., Niessen, F., Pompilio, M., Wilson, T., Carter, L., DeConto, R., Huybers, P., McKay, R., Pollard, D., Ross, J., Winter, D., Barrett, P., Browne, G., Cody, R., Cowan, E., Crampton, J., Dunbar, G., Dunbar, N., Florindo, F., Gebhardt, C., Graham, I., Hannah, M., Hansaraj, D., Harwood, D., Helling, D., Henrys, S., Hinnov, L., Kuhn, G., Kyle, P., Läufer, A., Maffioli, P., Magens, D., Mandernack, K., McIntosh, W., Millan, C., Morin, R., Ohneiser, C., Paulsen, T., Persico, D., Raine, I., Reed, J., Riesselman, C., Sagnotti, L., Schmitt, D., Sjunneskog, C., Strong, P., Taviani, M., Vogel, S., Wilch, T., Williams, T.: Obliquity-paced Pliocene West Antarctic ice sheet oscillations. Nature **458**(7236), 322–328 (2009). https://doi.org/10.1038/nature07867

Napier, W.M., Clube, S.V.M.: A theory of terrestrial catastrophism. Nature **282**(5738), 455 (1979)

Nguyen, T., Brunson, D., Crespi, C.L., Penman, B.W., Wishnok, J.S., Tannenbaum, S.R.: DNA damage and mutation in human cells exposed to nitric oxide in vitro. Proc. Natl. Acad. Sci. **89** (7), 3030–3034 (1992)

O'Dea, A., Lessios, H.A., Coates, A.G., Eytan, R.I., Restrepo-Moreno, S.A., Cione, A.L., Collins, L.S., de Queiroz, A., Farris, D.W., Norris, R.D., Stallard, R.F., Woodburne, M.O., Aguilera, O., Aubry, M.P., Berggren, W.A., Budd, A.F., Cozzuol, M.A., Coppard, S.E., Duque-Caro, H., Finnegan, S., Gasparini, G.M., Grossman, E.L., Johnson, K.G., Keigwin, L. D., Knowlton, N., Leigh, E.G., Leonard-Pingel, J.S., Marko, P.B., Pyenson, N.D., Rachello-Dolmen, P.G., Soibelzon, E., Soibelzon, L., Todd, J.A., Vermeij, G.J., Jackson, J.B.C.: Formation of the Isthmus of Panama. Sci. Adv. **2**, 1–11 (2016)

Odes, E.J., Randolph-Quinney, P.S., Steyn, M., Throckmorton, Z., Smilg, J.S., Zipfel, B., Tanya, N.A., de Beer, F., Hoffman, J.W., Franklin, R.D., Berger, L.R.: Earliest hominin cancer: million-year-old osteosarcoma from Swartkrans Cave, South Africa. S. Afr. J. Sci. **112**(7-8), 1–5 (2016)

Ogg, J.G., Ogg, G., Gradstein, F.M.: A Concise Geologic Time Scale: 2016, p. 234. Elsevier, Amsterdam (2016)

Orgeira, M.J., Sinito, A.M., Compagnucci, R.H.: The influence of the geomagnetic field in climate changes. In:Gasparini, G. et al. (ed.) Marine Isotope Stage 3 in Southern South America, 60 ka BP–30 ka BP, p. 49–80. Springer, Cham, (2016)

Overholt, A.C., Melott, A.L., Pohl, M.: Testing the link between terrestrial climate change and galactic spiral arm transit. Astrophy. J. Lett. **705**(2), L101 (2009)

Patterson, N., Richter, D.J., Gnerre, S., Lander, E.S., Reich, D.: Genetic evidence for complex speciation of humans and chimpanzees. Nature **441**(7097), 1103 (2006)

Patterson, D.B., Braun, D.R., Behrensmeyer, A.K., Merritt, S., Zliobaite, I., Reeves, J.S., Wood, B.A., Fortelius, M., Bobe, R.: Ecosystem evolution and hominin paleobiology at East Turkana, northern Kenya between 2.0 and 1.4 Ma. Palaeogeogr. Palaeoclimatol. Palaeoecol. **481**, 1–13 (2017)

Philander, S.G., Fedorov, A.V.: Role of tropics in changing the response to Milankovich forcing some three million years ago. Paleoceanography **18**(2), 1045 (2003). https://doi.org/1010.1029/2002PA00083

Pierre Auger Collaboration: Observation of a large-scale anisotropy in the arrival directions of cosmic rays above 8 × 1018 eV. Science **357**(6357), 1266–1270 (2017)

Pimiento, C., Griffin, J.N., Clements, C.F., Silvestro, D., Varela, S., Uhen, M.D., Jaramillo, C.: The Pliocene marine megafauna extinction and its impact on functional diversity. Nat. Ecol. Evol. **1**(8), 1100 (2017)

Prokoph, A., Veizer, J.: Trends, cycles and nonstationarities in isotope signals of Phanerozoic seawater. Chem. Geol. **161**(1–3), 225–240 (1999)

Pross, J., Klotz, S.: Palaeotemperature calculations from the Praetiglian/Tiglian (Plio-Pleistocene) pollen record of Lieth, northern Germany: implications for the climatic evolution of NW Europe. Global Planet. Change **34**, 253–267 (2002). https://doi.org/10.1016/s0921-8181(02)00119-4

Prueher, L.M., Rea, D.K.: Rapid onset of glacial conditions in the subarctic North Pacific region at 2.67 Ma: Clues to causality. Geology **26**(11), 1027–1030 (1998)

Pudovkin, M.I., Veretenenko, S.V.: Cloudiness decreases associated with Forbush-decreases of galactic cosmic rays. J. Atmos. Terr. Phys. **57**(11), 1349–1355 (1995)

Rampino, M.R.: Disc dark matter in the galaxy and potential cycles of extraterrestrial impacts, mass extinctions and geological events. Mon. Not. R. Astron. Soc. **448**(2), 1816–1820 (2015)

Rampino, M.R., Haggerty, B.M.: The "Shiva Hypothesis": Impacts, mass extinctions, and the galaxy. In: Worlds in Interaction: Small Bodies and Planets of the Solar System, pp. 441–460. Springer, Dordrecht (1996)

Rampino, M.R., Stothers, R.B.: Terrestrial mass extinctions, cometary impacts and the Sun's motion perpendicular to the galactic plane. Nature **308**(5961), 709 (1984)

Randolph-Quinney, P.S., Williams, S.A., Steyn, M., Meyer, M.R., Smilg, J.S., Churchill, S.E., Odes, E.J., Augustine, T., Tafforeau, P., Berger, L.R.: Osteogenic tumor in Australopithecus sediba: Earliest hominin evidence for neoplastic disease. S. Afr. J. Sci. **112**(7–8), 1–7 (2016)

Raup, D.M., Sepkoski, J.J.: Periodicity of extinctions in the geologic past. Proc. Natl. Acad. Sci. **81**(3), 801–805 (1984)

Raymo, M.E., Lisiecki, L.E., Nisancioglu, K.H.: Plio-Pleistocene ice volume, Antarctic climate, and the global $\delta^{18}O$ record. Science **313**(5786), 492–495 (2006)

Raymo, M.E., Mitrovica, J.X., O'Leary, M.J., DeConto, R.M., Hearty, P.J.: Departures from eustasy in Pliocene sea-level records. Nat. Geosci. **4**(5), 328 (2011)

Reames, D.V.: The Two Sources of Solar Energetic Particles. Space Sci. Rev. **175**(1–4), 53–92 (2013)

Reed, B.C.: The Sun's displacement from the galactic plane from spectroscopic parallaxes of 2400 OB stars. arXiv preprint astro-ph/0507655 (2005)

Rio, D., Sprovieri, R., Castradori, D., Di Stefano, E.: The Gelasian Stage (Upper Pliocene): a new unit of the global standard chronostratigraphic scale. Episodes **21**, 82–87 (1998)

Robinson, M.M., Dowsett, H.J., Chandler, M.A.: Pliocene role in assessing future climate impacts. Eos, Transact. Am. Geophys. Union **89**(49), 501–502 (2008)

Robson, S.L., Wood, B.: Hominin life history: reconstruction and evolution. J. Anat. **212**(4), 394–425 (2008)

Sabbah, I.: The role of interplanetary magnetic field and solar wind in modulating both galactic cosmic rays and geomagnetic activity. Geophys. Res. Lett. **27**(13), 1823–1826 (2000)

Sagan, C.: The Cosmic Connection an Extraterrestrial Perspective. Anchor Press/ Doubleday, Garden City (1973). ISBN 10: 0385004575, ISBN 13: 9780385004572

Schindewolf, O.H.: Über die möglichen Ursachen der grossen erdgeschichtlichen Faunenschnitte. Neues Jahrbuch für Geologie und Paläontologie, Monatshefre **1954**, 457–465 (1954)

Schönrich, R., Binney, J.: Chemical evolution with radial mixing. Mon. Not. R. Astron. Soc. **396** (1), 203–222 (2009)

Schroeder, L., Ackermann, R.R.: Evolutionary processes shaping diversity across the Homo lineage. J. Hum. Evol. **111**, 1–17 (2017)

Schulreich, M.M., Breitschwerdt, D., Feige, J., Dettbarn, C.: A way out of the bubble trouble?—Upon reconstructing the origin of the local bubble and Loop I via Radioisotopic signatures on earth. Galaxies **6**(1), 26 (2018)

Schwartz, R.D., James, P.B.: Periodic mass extinctions and the Sun's oscillation about the galactic plane. Nature **308**(5961), 712 (1984)

Seki, O., Foster, G.L., Schmidt, D.N., Mackensen, A., Kawamura, K., Pancost, R.D.: Alkenone and boron-based Pliocene pCO2 records. Earth Planet. Sci. Lett. **292**, 201–211 (2010)

Setlow, R.B.: The wavelengths in sunlight effective in producing skin cancer: a theoretical analysis (DNA damage/action spectra/ozone). Proc. Natl. Acad. Sci. U.S.A. **71**, 3363–3366 (1974)

Shaviv, N.J.: Cosmic ray diffusion from the galactic spiral arms, iron meteorites, and a possible climatic connection. Phys. Rev. Lett. **89**(5), 051102 (2002)

Shaviv, N.J.: The spiral structure of the milky way, cosmic rays, and ice age epochs on Earth. New Astron. **8**(1), 39–77 (2003)

Shaviv, N.J., Veizer, J.: Celestial driver of Phanerozoic climate? GSA Today **13**(7), 4–10 (2003)

Shaviv, N.J., Prokoph, A., Veizer, J.: Is the solar system's galactic motion imprinted in the Phanerozoic climate? Sci. Rep. **4**, 6150 (2014)

Simpson, C.: The case for species selection. bioRxiv. (https://doi.org/10.1101/084046) (2016)

Simpson, S.W., Kleinsasser, L., Quade, J., Levin, N.E., McIntosh, W.C., Dunbar, N., Semaw, S., Rogers, M.J.: Late Miocene hominin teeth from the Gona Paleoanthropological Research Project Area, Afar, Ethiopia. J. Hum. Evol. (2015). https://doi.org/10.1016/j.jhevol.2014.07.004

Singer, B.S.: A quaternary geomagnetic instability time scale. Quater. Geochronol. **21**, 29–52 (2014)

Slater, G.J., Goldbogen, J.A., Pyenson, N.D.: Independent evolution of baleen whale gigantism linked to Plio-Pleistocene ocean dynamics. Proc. R. Soc. B **284**(1855), 20170546 (2017)

Sloan, T., Wolfendale, A.W.: Testing the proposed causal link between cosmic rays and cloud cover. Environ. Res. Lett. **3**(2), 024001 (2008)

Spoor, F., Leakey, M.G., Gathogo, P.N., Brown, F.H., Antón, S.C., McDougall, C., Kiarie, F., Manthi, K., Leakey, L.N.: Implications of new early Homo fossils from Ileret, east of Lake Turkana. Kenya. Nature **448**(7154), 688 (2007)

Steph, S., Tiedemann, R., Prange, M., Groeneveld, J., Nürnberg, D., Reuning, L., Schulz, M., Haug, G.H.: Changes in Caribbean surface hydrography during the Pliocene shoaling of the Central American Seaway. Paleoceanography **21**(4) (2006)

Stephenson, F.R.: Historical records of supernovae. Handbook of Supernovae, pp. 1–14. Springer, Berlin (2016)

Sun, B., Bradley, R.S.: Solar influences on cosmic rays and cloud formation: A reassessment. J. Geophys. Res. Atmos. **107**(D14), 4211 (2002). https://doi.org/10.1029/2001JD000560

Sun, Y., Clemens, S.C., An, Z., Yu, Z.: Astronomical timescale and palaeoclimatic implication of stacked 3.6-Ma monsoon records from the Chinese Loess Plateau. Quat. Sci. Rev. **25**, 33–48 (2006). https://doi.org/10.1016/j.quascirev.2005.07.005

Sun, Y., An, Z., Clemens, S.C.: Non-stationary response of Plio-Pleistocene East Asian winter monsoon variation to ice volume forcing. Geol. Soc. Lond. Spec. Publ. **342**, 79–86 (2010a). https://doi.org/10.1144/SP342.7

Sun, Y., An, Z., Clemens, S.C., Bloemendal, J., Vandenberghe, J.: Seven million years of wind and precipitation variability on the Chinese Loess Plateau. Earth Planet. Sci. Lett. **297**, 525–535 (2010b). https://doi.org/10.1016/j.epsl.2010.07.004

Svensmark, H.: Imprint of Galactic dynamics on Earth's climate. Astron. Nachr. Astron. Notes **327**(9), 866–870 (2006)

Svensmark, H.: Evidence of nearby supernovae affecting life on Earth. Mon. Not. R. Astron. Soc. **423**(2), 1234–1253 (2012)

Svensmark, H.: Cosmic rays, clouds and climate. Europhys. News **46**(2), 26–29 (2015)

Svensmark, H., Friis-Christensen, E.: Variation of cosmic ray flux and global cloud coverage—a missing link in solar-climate relationships. J. Atmos. Solar Terr. Phys. **59**(11), 1225–1232 (1997)

Svensmark, H., Enghoff, M.B., Shaviv, N.J., Svensmark, J.: Increased ionization supports growth of aerosols into cloud condensation nuclei. Nat. Commun. **8**(1), 2199 (2017)

Tardieu, C., Hasegawa, K., Haeusler, M.: How did the pelvis and vertebral column become a functional unit during the transition from occasional to permanent Bipedalism? Anat. Rec. **300**(5), 912–931 (2017)

Thierens, M., Browning, E., Pirlet, H., Loutre, M.F., Dorschel, B., Huvenne, V.A.I., Titschack, J., Colin, C., Foubert, A., Wheeler, A.J.: Cold-water coral carbonate mounds as unique palaeo-archives: the Plio-Pleistocene Challenger Mound record (NE Atlantic). Quat. Sci. Rev. **73**, 14–30 (2013)

Thomas, B.C.: Photobiological effects at Earth's surface following a 50 pc Supernova. Astrobiology **18**, 5 (2018). https://doi.org/10.1089/ast.2017.1730

Thomas, B.C., Engler, E.E., Kachelrieß, M., Melott, A.L., Overholt, A.C., Semikoz, D.V.: Terrestrial effects of nearby Supernovae in the Early Pleistocene. Astrophys. J. Lett. **826**, L3 (2016). https://doi.org/10.3847/2041-8205/826/1/L3

Veizer, J., Ala, D., Azmy, K., Bruckschen, P., Buhl, D., Bruhn, F., Carden, G.A.F., Diener, A., Ebneth, S., Godderis, Y., Jasper, T., Korte, C., Pawellek, F., Podlaha, O.G., Strauss, H.: $^{87}Sr/^{86}Sr$, $\delta^{13}C$ and $\delta^{18}O$ evolution of Phanerozoic seawater. Chem. Geol. **161**(1–3), 59–88 (1999)

Venti, N.L., Billups, K.: Stable-isotope stratigraphy of the Pliocene-Pleistocene climate transition in the northwestern subtropical Pacific. Palaeogeogr. Palaeoclimatol. Palaeoecol. **326**, 54–65 (2012)

Versteegh, G.J.M., Brinkhuis, H., Visscher, H., Zonneveld, K.A.F.: The relation between productivity and temperature in the Pliocene North Atlantic at the onset of northern hemisphere glaciation: a palynological study. Glob. Planet. Change **11**, 155–165 (1996). https://doi.org/10.1016/0921-8181(95)00054-2

Vieira, C.L.Z., Janot-Pacheco, E., Lage, C., Pacini, A., Koutrakis, P., Cury, P.R., Shaodan, H., Pereira, L.A., Saldiva, P.H.N.: Long-term association between the intensity of cosmic rays and mortality rates in the city of Sao Paulo. Environ. Res. Lett. **13**(2), 024009 (2018)

Wagner, G., Livingstone, D.M., Masarik, J., Muscheler, R., Beer, J.: Some results relevant to the discussion of a possible link between cosmic rays and the Earth's climate. J. Geophys. Res. Atmos. **106**(D4), 3381–3387 (2001)

Wallner, A., Feige, J., Kinoshita, N., Paul, M., Fifield, L.K., Golser, R., Honda, M., Linnemann, U., Matsuzaki, H., Rugel, G., Tims, S.G., Steier, P., Yamagata, T., Winkler, S.R., Merchel, S.: Recent near-Earth supernovae probed by global deposition of interstellar radioactive $^{60}Fe$. Nature **532**, 69 (2016)

Wara, M.W., Ravelo, A.C., Delaney, M.L.: Permanent El Niño-like conditions during the Pliocene warm period. Science **309**(5735), 758–761 (2005)

Ward, C.V., Feibel, C.S., Hammond, A.S., Leakey, L.N., Moffett, E.A., Plavcan, J.M., Skinner, M.M., Spoor, F., Leakey, M.G.: Associated ilium and femur from Koobi Fora, Kenya, and postcranial diversity in early Homo. J. Hum. Evol. **81**, 48 (2015)

Wei, K.: Stratophenetic tracing of phylogeny using SIMCA pattern recognition technique: A case study of the late Neogene planktic foraminifera Globoconella clade. Paleobiology **20**(1), 52–65 (1994). https://doi.org/10.1017/S0094837300011131

Weissman, P.R.: The Solar system and its place in the galaxy. In: Encyclopedia of the Solar System, 3 edn., pp. 3–28 (2014)

Wickramasinghe, J.T., Napier, W.M.: Impact cratering and the Oort cloud. Mon. Not. R. Astron. Soc. **387**(1), 153–157 (2008)

Willeit, M., Ganopolski, A., Calov, R., Robinson, A., Maslin, M.: The role of $CO_2$ decline for the onset of Northern Hemisphere glaciation. Quater. Sci. Rev. **119**, 22–34 (2015)

Williams, S.A., Ostrofsky, K.R., Frater, N., Churchill, S.E., Schmid, P., Berger, L.R.: The vertebral column of Australopithecus sediba. Science **340**(6129), 1232996 (2013)

Young, J.R., Bown, P.R., Lees, J.A.: Nannotax3 website. International Nannoplankton Association (2017). http://www.mikrotax.org/Nannotax3. Accessed 21 Apr 2017

Zachos, J.C., Dickens, G.R., Zeebe, R.E.: An Early Cenozoic perspective on greenhouse warming and carbon-cycle dynamics. Nature **451**, 279–283 (2008)

# Neogene Continental Deposits of Bahia Blanca, Argentina. Palaeoenvironmental and Palaeoclimatic Inferences

Silvia C. Grill[1,2](✉), Mauro L. Gómez Samus[3], and Ana L. Fernández[1]

[1] Department of Geology, Universidad Nacional del Sur (UNS), Av. Alem 1253, Bahía Blanca 8000, Argentina
sgrill@criba.edu.ar, ana.fernandez@uns.edu.ar
[2] INGEOSUR (CONICET), Bahía Blanca, Argentina
[3] Instituto de Ciencias Polares, Ambiente y Recursos Naturales, Universidad Nacional de Tierra del Fuego (ICPA – UNTDF), Fuegia Basket 251, Ushuaia, Argentina
mlgomezsamus@untdf.edu.ar

**Abstract.** The present contribution involves deposits of continental Neogene age that outcrop in the Bahía Blanca region. The studies that were carried out in order to contribute to the paleoenvironmental and chronostratigraphical knowledge involve sedimentological, paleomagnetic and magnetic susceptibility analyses. One of the outcrops studied (Cueva de Los Leones) is mainly composed of facies of fine to thick sandstones, largely pedogenized and with important amounts of calcrete, whereas the other (Loma Sarmiento) has a predominance of loess-like silt facies, reddish brown in colour, an abundance of rhizoconcretions and evidence of pedogenic processes. Both outcrops culminate in an important layer of calcrete of regional scope, with particular morphologies at each of the sites. The paleomagnetic analysis leads to the conclusion that the sedimentation of the deposits underlying the calcareous crust at the two sites was not synchronous. Reverse polarity was recorded in the Cueva de Los Leones in contrast to Loma Sarmiento, where normal polarity is predominant. The paleomagnetic samples of calcareous crust on the top presented anomalous polarity directions in both sections, with very high values of magnetic intensity. The reverse directions in the Cueva de Los Leones can be assigned to the Gilbert Cron (6.0–3.6 Ma), without discarding younger ages in the Pliocene. In contrast, Loma Sarmiento seems to be older (Late Miocene). The susceptibility records also indicated differences between the two sites that would be linked to pedogenic development; they were higher in Loma Sarmiento and judging by the F% factor, in Cueva de los Leones the superparamagnetic particles contribution is higher. Petrographic, X-Ray difractometry, paleontological and soil micromorphology analyses, in some places at the studied sites, also contributed to the paleoenvironmental/paleoclimatic reconstruction carried out.

**Keywords:** Neogene · Bahía blanca · Sedimentology · Magnetostratigraphy


M. M. Collantes et al. (Eds.): *Advances in Geomorphology and Quaternary Studies in Argentina*, SPRINGEREARTH, pp. 40–57, 2020.
https://doi.org/10.1007/978-3-030-22621-3_2

# 1 Introduction

The continental Neogene presents a wide distribution in the extra-Andean region of Argentina, which includes the SW and SE of Buenos Aires province (Folguera and Zárate 2009). In the Bahía Blanca region, these deposits are found in the urban (Loma Sarmiento) and peri-urban sectors of the city (quarries and the natural outcrop of "Cueva de Los Leones") (Fig. 1a). Many stratigraphic and biostratigraphic studies have been carried out in Loma Sarmiento and in several quarries (Deschamps 2005, Fidalgo 1983, Verzi et al. 2008, Folguera and Zárate 2009, among others). In Cueva de Los Leones, the only preliminary geomorphological and magnetostratigraphic studies were performed by Grill et al. (2014a, 2014b, 2014c) and Gómez Samus et al. (2013). In particular, the work of Moretti et al. (2017), about the soil micromorphology, is partially used in this publication to complement the studies of the southern sector of the Cueva de los Leones outcrop. There is no available dating yet and no fossils have been exhumed from this outcrop. However, by regional stratigraphic correlation and visual analysis of satellite imagery, it is assumed that the calcareous crust that crowns the outcrop belongs to the cuspidal deposit of the "general planation surface" assigned to the Pliocene-Pleistocene (González Uriarte 1984) and that the underlying sediments would form part of the Miocene-Pliocene record of the region. This paper aims to contribute to the knowledge of these deposits from sedimentological and magnetostratigraphic analyses in the outcrops of Loma Sarmiento and Cueva de Los Leones (Fig. 1a).

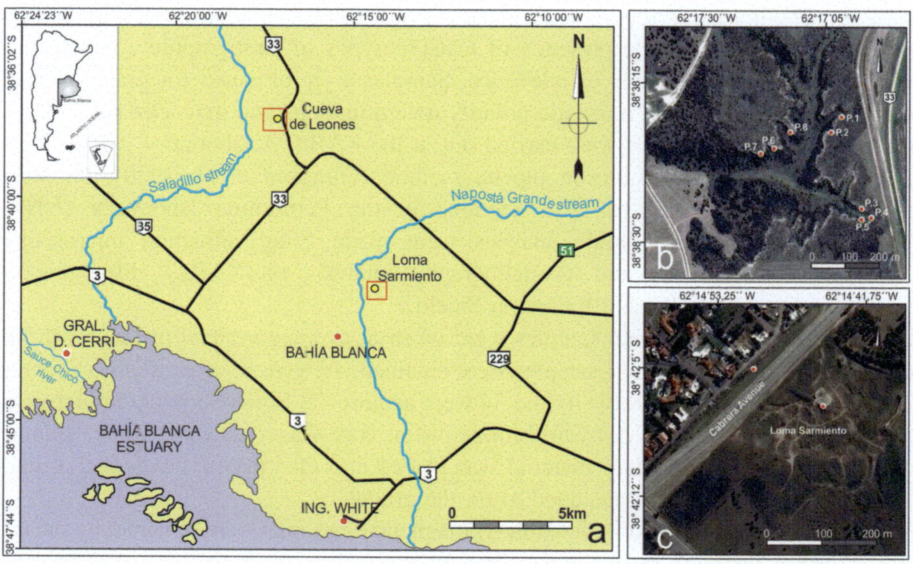


**Fig. 1.** (a) Study area. (b) Cueva de los Leones. (c) Loma Sarmiento.

# 2 Geological Framework

The study area involved in the present work is associated with stratigraphic sequences represented by the continental deposits that occurred after the withdrawal of the "Paranaense Sea" (Malumián 1999). These deposits, made up of four sedimentation cycles that culminate with important banks of calcareous crust (Zárate 2005), are involved in the period extending from the Late Miocene to the Late Pleistocene and Holocene. Generically, these units have been called “Pampean sediments”, later receiving different names according to their lithological and fossiliferous composition (Zárate 2005). In the outcrops of Cueva de Los Leones and Loma Sarmiento, part of the mentioned continental filling is exposed. The first of them, of around 0.30 $km^2$ in extension, is composed of relicts of plains, dissected by flat bottom valleys and steep escarpments that reach an average elevation of 86 m a.s.l.; the maximum thickness of the sedimentary material is 2 m. The morphology of the calcareous crust that crowns this slightly undulating outcrop exceeds 1.70 m in some sectors and it presents a scanty sedimentary cover. The outcrops of Loma Sarmiento are of around 0.10 $km^2$ in extension. The northwest sector reaches 64 m in altitude, and a thickness of the sedimentary material of 6.50 m, whereas the outcrop of the southeast sector has an average altitude of 60 m a.s.l. and it is 4.90 m thick. Both sequences finish off with a calcareous crust layer around 1.10 m thick.

## 3 Methodology

Ten stratigraphic profiles were analysed, eight in Cueva de Los Leones, the profiles 1, 2, 6, 7 and 8, in the northern sector, and the profiles 3, 4 and 5, in the southern sector (Fig. 1b), and two in Loma Sarmiento, in the northwestern and southeastern sectors (Fig. 1c). The field descriptions were made providing lithological aspects, sedimentary and pedogenetic structures, presence of $CaCO_3$, types of stratigraphic contacts and other morphological features of relevance. Samples were obtained for granulometric and petrographic analysis, magnetic polarity determinations and magnetic parameters. The granulometric analyses were carried out in the LABSPA laboratory (CERZOS – CONICET), using the Robinson pipette method (Olmsterd et al. 1930). Previous grinding and preparation of thin sections in the Petrotomía laboratory (UNS-INGEOSUR), the petrographic analysis were made using refraction microscopy (Zeiss, Axiolab model) and crystallography analysis of X Ray difractometry (XRD) were made by Rigaku Radiation Shield.

The magnetostratigraphic samples were taken at average vertical distances of 5.0 and 20.0 cm. Seventy-four 74 samples were obtained; 40 were so at the southern sector of Cueva de los Leones (Profile 4) and 34 were acquired at Loma Sarmiento outcrop in both sides of the avenue. A hydro-bronze piston core was used (see Gómez Samus et al. 2017). The sedimentary material was placed in 8 $cm^3$ plastic boxes and a non-magnetic adhesive liquid was added to fix the grains.

The declination (D) and inclination (I) magnetic directions were obtained by means of a Minispin Spinning magnetometer and an alternating magnetic field (AF) demagnetizer, both of the Molspin Ltd. Company. The analyses started by applying low fields (2.5 mT) to a maximum peak field of 80 mT, in at least 10 steps of demagnetization. The data were processed in the software SUPERIAPD2000 and the values of D and I characteristic of the samples (ChRM) were obtained and the maximum angular deviation (MAD) was calculated. Besides, magnetic susceptibility records were obtained at 470 ($\chi_{lf}$) and 4700 Hz ($\chi_{hf}$) by Bartington equipment (MS2 and MS3) with a MS2B sensor and the frequency-dependent susceptibility factor was determined [F % = 100 x ($\chi_{lf}$- $\chi_{hf}$)/($\chi_{lf}$)].

# 4 Results and Discussion

## 4.1 Characteristics of the Outcrops

**Cueva de Los Leones** (Fig. 2a)

**Fig. 2.** (a) Cueva de los Leones outcrop, general view. (b) Structure in lattice in facies of coarse reddish sandstones (Upper section, Northern outcrops). (c) *Scoyenia* ichnofacies (Profile 7). (d) "Tafone" in calcareous culminate crust (Profile 4).

In the northern sector (Fig. 3; profiles 1, 2, 6, 7 and 8) the outcrops corresponding to Profiles 6, 7 and 8 comprise the complete sedimentary sequences and the largest variety of sedimentary facies, with the greatest thickness (3.0 m) being recorded in Profile 8. In general terms, the succession in both sectors was divided into three stratigraphic sections: lower, middle and upper sections.

**Northern Sector**

Lower Section: it starts without an exposed base facies developed of fine to medium sandstones with silt and little clay, followed by facies of sandy siltstones and claystones that reach 30 cm in thickness. As a whole, they form an association of facies of decreasing grain size, with a uniform brown colour and a massive structure, from which

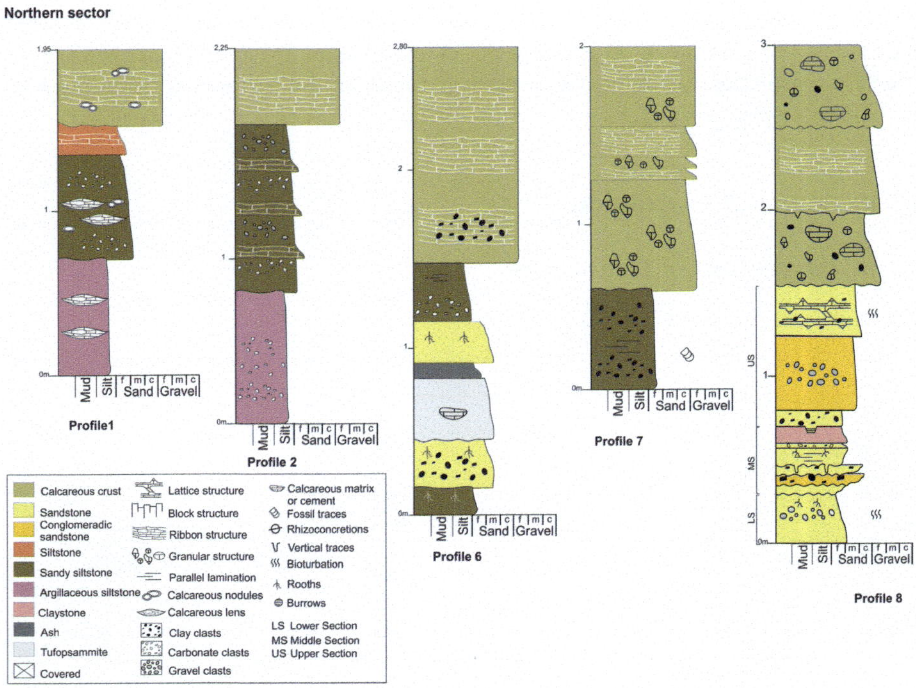


**Fig. 3.** Cueva de Los Leones northern profiles.

we can infer continental environment deposits. Towards the upper part of this association, fine clay sheets are arranged, clay clasts of varying size and sub-rounded gravel clasts homogeneously dispersed, that allowed us inferring continental environment deposits by dense flow action. The presence of fossil roots and bioturbation shows that the accumulation is in a floodplain paleoenvironment next to a fluvial channel (ephemeral stream).

Lower Middle section: it begins with 18 cm of fine sandstones facies with silt, conglomerate sandstones with normal gradation and coarse sandstones. All these facies have an erosion base, massive structure and brown colour. The characteristics described suggest an association of facies of fluvial environments of moderate energy. The presence of clay clasts in the roof of the succession of facies, whose major axes are oriented, would sustain deposits generated by unidirectional flows.

Upper Middle section: on an erosion base, it begins with a fine to medium sandstone facies (5–10 cm) of laminated structure, followed by facies of fine to medium sandstones with some clasts of very thick sand and massive structure (3 cm). Over the sandstone, massive claystone facies with ochre colours and limited lateral continuity are arranged (thickness: 2 cm). The association of sandstone facies presents a greyish colour that clearly differentiates it from the remaining facies of the profile; it could indicate deposits of poorly drained floodplains. The ochre colours of the clay facies would assume an increase in the aerobic influence in the sedimentation zone. This, linked to grain size and structural characteristics of the facies and the presence of paleoedaphic features (fossil roots) is interpreted as deposits of levees and overflows accumulated in a floodplain environment following the criteria exposed by Bridge (2003).

In relation to the origin of the materials, the petrographic analysis carried out in this section of the sedimentary sequence indicates clasts (sub-angular to sub-rounded), with moderate to scarce sorting, in a matrix that occupies more than 15% (Fig. 4a) which

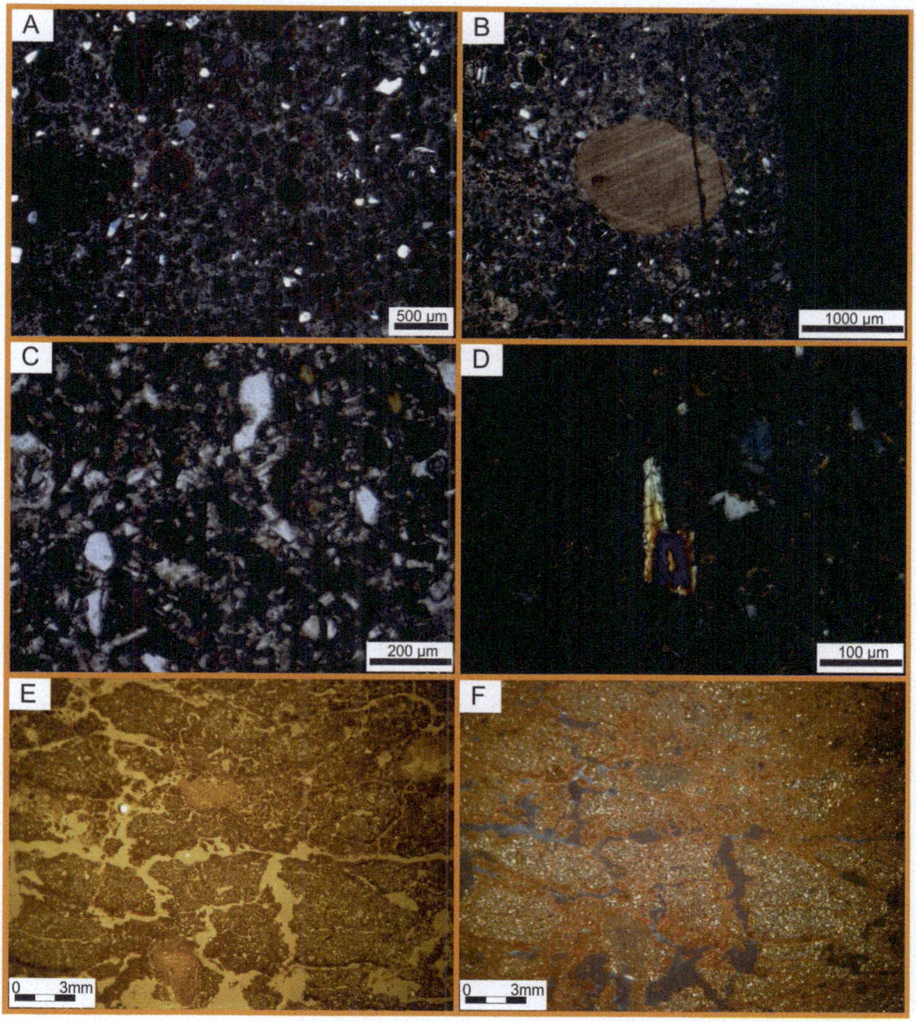


**Fig. 4.** Cueva de Los Leones microphotographs. (a) General texture (Middle section, Profile 8). (b) Sedimentary lithic (Middle section, Profile 8). (c) General texture (Middle section, Profile 4). (d) Clinopyroxene crystal (Middle section of Profile 4). (e) Microphotography of "fragmented laminar structure" under natural light and (f) in polarized light (Southern Sector, Upper section, Profile 4) modified from Moretti et al. (2017).

allow inferring that these are inmature rocks whose constituent materials travelled short distances before their deposition. The clasts of lithic fragments correspond to sedimentary lithic fragments (20%) and volcanic fragments (10%) and the monomineral clasts to plagioclase (andesine) (34%), and monocrystalline and polycrystalline quartz (26%). The composition of the sedimentary lithic clasts (Fig. 4b) (size ~2 mm) corresponds to a rock composed by quartz, plagioclase, pyroxene and biotite, while the volcanic fragments (probably andesite) (size around 0.1 mm) correspond to rocks mainly formed by plagioclase phenocrysts (andesine). In the case of quartz, the indented contacts between the sub-grains of the polycrystalline quartz suggest that their origin is from metamorphic rocks, probably from the nearby Sierras Australes (quartzites). The predominant grain size (fine to medium sand) and the petrographic characteristics (presence of subangular to sub-rounded clasts, sparse matrix and presence of volcanic fragments) allowed us inferring a transport by eolian action (Johnsson et al. 1991). Sediments of these characteristics are often retained in the floodplains of the fluvial systems; thus, a fluvio-eolian genesis is assumed for these plains.

Lower Upper section: it starts on a clear-cut base, with a succession of 85 cm of a thick facies, coarsening upwards. The sequence starts with fine to medium sandstones facies (10 cm) of greyish colour, with clay clasts 1–15 mm in length. It continues with fine sandy conglomerate facies (35 cm) with gravel clasts (1–25 mm in length) and coarse reddish sandstones facies with a few clay clasts (up to 1 mm in length), evidence of incipient bioturbation, and calcium carbonate with a lattice structure (40 cm) (Fig. 2b). The difference in colour would indicate a poorly drained floodplain, which evolved into a more aerobic one. This inference agrees with bioturbation recorded in the upper facies. All these facies present massive structure. Deposits by unidirectional flows are inferred for this facies association, in a continental environment of low to moderate energy, probably linked to an association of channel bars.

Laterally, the middle and upper sections of this profile (Profile 8) are graded in Profile 6 (total thickness: 270 cm; Fig. 3) as sandy siltstone facies (thickness: 17 cm) which grade upwards to fine sandstones facies (thickness: 30 cm). Over this facies, a lens of cineritic material occurs (2 m of length and a maximum thickness of 47 cm). The lower 37 cm correspond to a tufopsammite with a massive structure, whitish in colour with dusty calcium carbonate. The top 10 cm are composed of coarse ash (Grill et al. 2014b). Continuing the sequence, fine to medium sandstones facies (25 cm thick) were recorded.

The three facies have massive structure, brown colour and numerous fossil roots; this allowed us inferring that the deposits are in a continental environment of low to moderate energy with biological activity. Its tabular geometry, grain size and structure suggest the occurrence of channeled flows and a floodplain paleoenvironment.

The facies of tufopsammite and coarse ash make up a lenticular deposit in clear erosion unconformity in the sedimentary sequence. Its geometry, massive structure and grain size allowed us linking them with channeled flows of moderate energy. The mineralogy (relatively high contents of volcanic glass) and the morphology of the deposit were interpreted as the product of fluvial reworking of pre-existing pyroclastic deposits in the relatively higher areas of the terrain (Grill et al. 2014b).

Laterally, in Profile 7 (Fig. 3) (total thickness: 200 cm), in consolidated sandy siltstone facies (60 cm thick), with incipient lamination and carrying clay clasts (diameter: 1–2 mm), some fossil traces were recorded that are associated with *Beaconites* and *Taenidium* ichnogenera, which show the *Scoyenia* ichnofacies (Fig. 2c). The fossil traces observed are distributed along the entire section that underlies the calcrete (Grill et al. 2014a).

The *Scoyenia* ichnofacies generally occurs in sandy to clayey, moist sediments in low energy environments, characteristic of subaquatic sediments that periodically suffer subaerial exposure or vice versa (Frey and Pemberton 1987, Pemberton et al. 1992c). Other authors (Buatois and Mangano 1995, among others) associated it with continental floodplain, lake margins and humid interdune environments. These ichnogenera and the characteristics of the hosted sedimentary rocks allowed us inferring floodplain paleoenvironments (Grill et al. 2014a).

Specimens of *Taenidium* ichnogenus have also been recorded in fine silty sandstones facies in the Seminario and Sanitary Landfill quarries, located in the vicinity of the city of Bahía Blanca (Deschamps et al. 1998) and in the Cerro Azul Formation (La Pampa province) (Sostillo et al. 2013). Ichnofacies were also recorded in facies of sandy siltstones towards the sector of the present valley of the Saladillo creek (the only perennial runoff in the area). These strata were followed upwards by fine sandstones with planar stratification and cross-bedded stratification in two directions.

The southern Sector is represented by profiles 3, 4 and 5 (Fig. 5). The best outcrops are located in Profile 4 (total thickness: 3.40 m).

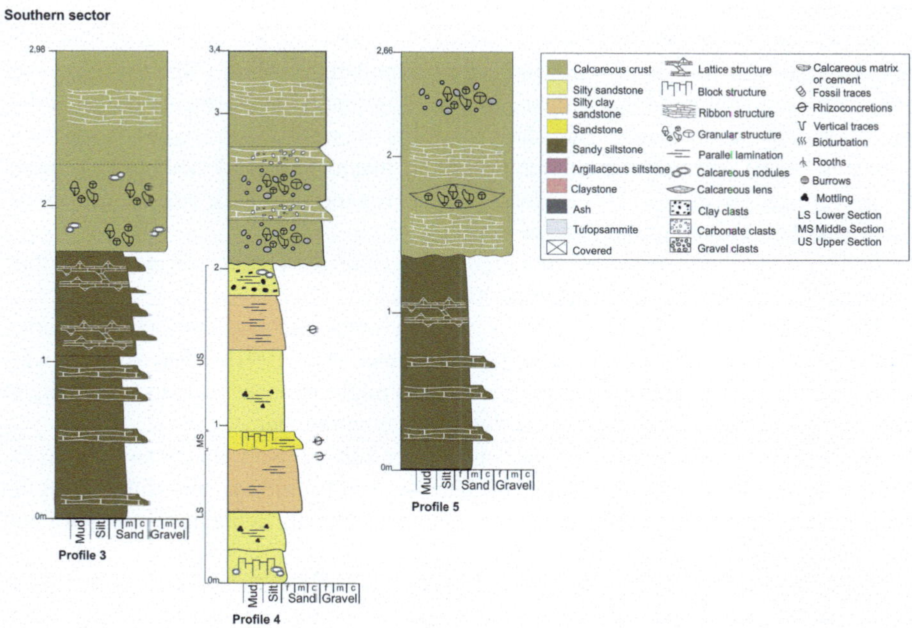


**Fig. 5.** Cueva de Los Leones southern profiles.

**Southern Sector**

Lower section: the sequence starts, without exposed base, with facies of silty sandstones (thickness: 20 cm), with blocky to granulated soil structure, with carbonate irregular nodules (diameter: 1 cm) and accumulation of organic matter in small fissures. Two layers of silty sandstones (thickness: 25 cm) and silty-clay sandstones (thickness: 40 cm) continue upwards, the first with a thick primary laminated and secondary blocky structure, with evidence of hydromorphism (mottled grey-green). The second one corresponds to a pedogenic level with primary laminated structure, subangular blocky structure and large calcareous rhizoconcretions, concentrated in the last 15 cm (Moretti et al. 2017). The three facies described are brown to light brown in colour.

The mineralogical composition determined from XRD shows equivalent proportions of quartz and feldspar (potassium feldspar and plagioclases), as well as micas and absence of calcite. The Ca carbonate is found in rhizoconcretions and nodules, probably with opal (Moretti et al. 2017).

The association of facies and the paleoedaphic features described suggest paleoenvironments linked to floodplains. The deposits would have developed during episodes of an overflowing of the channels (described for the northern sector) which would have flowed towards lowlands and remote areas and the pedological characteristics recorded are a consequence of the subaerial exposure.

Middle section: it is represented by facies of thin sandstones that form a bank, 10 cm thick, with a slightly erosion contact at the base, greyish colour, and thick laminar structure that breaks into blocks. The micromorphological analysis indicates an important porosity, clay coatings and biological activity, the edaphological features being similar to those of the upper section (Moretti et al. 2017).

In relation to the origin of the materials, the petrographic analysis shows subangular clasts with low sphericity, indicating a moderate sorting (Fig. 4c). Forty per cent of the total volume of the sample corresponds to matrix and cement, the remaining percentage represents clasts. The clasts of lithic fragments correspond to volcanic fragments (21%) and the monomineral clasts to plagioclase (andesine) (50%), mono and polycrystalline quartz (29%), biotite, clinopyroxene (Fig. 4d), amphiboles (hornblende and lamprobolite) and opaque minerals.

As in the northern sector of the outcrop, the indented contacts in polycrystalline quartz suggest an origin from metamorphic rocks.

The predominant grain size (fine to medium sand) and the petrographic characteristics allowed us inferring wind transport (Johnsson et al. 1991), whereas the greyish colour and the paleoedaphic features (fossil roots) might indicate deposits belonging to levees or overflows, accumulated in a floodplain environment.

This part of the sedimentary sequence is correlated with that recorded in the northern sector of the outcrop (Upper Middle Section, Profile 8), only differing in the composition of the lithic, sedimentary and volcanic materials in the northern sector and volcanic material in the southern sector.

Upper section: on a clear-cut base, a facies succession records finning in an upper succession of around 100 cm thickness. The section begins with silty sandstones facies (thickness: 50 cm) and it is graded to silty clay sandstones facies (thickness: 35 cm).

Both are dark brown in colour, with laminar structure, thinner than in the lower section, with higher proportion of calcareous rhizoconcretions than the rest of the sequence and motted Mn (in the lower layers of this section). The section culminates with silty clay sandstones facies (thickness: 20 cm) with a fragmented laminar structure (Fig. 4e and f), and paleoedaphic features as clay aggregates, carbonate veins and abundant Mn nodules (diameter: 2–3 mm) (Moretti et al. 2017).

Regarding to the XRD analysis, it suggests a greater proportion of quartz than of feldspar, in addition to mica and the presence of calcite, which decreases towards the base (Moretti et al. 2017).

From the characteristics described above, paleoenvironments of the fluvio-eolian plain are again associated with greater pedogenesis than that in the lower section, which would imply longer periods of exposure.

All the outcrop culminates, prior to the erosion contact, with an important encrustation conformed by granular calcrete (or dusty one) which underlies calcrete in the form of a ribbon (or lamina) that crowns the outcrop. The greatest thickness is recorded in the northern sector of the outcrop, being ~ 1.50 m on average. The facies of granular calcrete are associated with a conglomerate, with coarse clasts (1 mm to 5 cm in length of the major axis) and an alternating composition between reddish clay and light brown sand. In general, these clasts are sub-rounded to angular and they are arranged without any gradation or orientation. The matrix is composed of a coarse to fine sand of different composition and rounded grains. In some sectors tabular facies of dusty calcrete of a laminated structure, calcium carbonate intraclasts of various sizes and sometimes gypsum layers of a few mm in thickness have developed, corresponding with this type of crust. The facies of laminar calcrete arranged above the granular (or pulverulent) calcrete and upholstering the outcrop, presents a large amount of gravel-sized clasts cemented with calcium carbonate, which decrease in quantity and size towards the top of the sequences. Because it is strongly consolidated and it conserves scarce features of the primary rock, the original lithological and structural characteristics cannot be recognised from the macroscopic analysis. It usually shows overlapping of at least two clearly differentiated layers. The presence of interspersed silcretes (~ 5 cm thick) give it greater hardness in relation to the granular calcrete; honeycombs (Fig. 2d) are an observable feature in these types of crusts, as well as "gnammas" of different sizes on the surface of the outcrop, some of them filled with very fine sediments.

**Loma Sarmiento** (Fig. 6a)

Lower section: in the northwest outcrops of the Cabrera Avenue (Fig. 7), the sequence starts with red sandy siltstone facies (30 cm thick) with massive structure and reduced lateral extension.

There is an association of facies in clear-cut contact, 0.90 m thick, formed by diamictite sediments and facies of sandy siltstones, light brown in colour (in its basal part), with a variable content of reddish clay clasts of different sizes (diameter: < 3 cm), rounded and arranged randomly. Fossils assigned to the Huayqueriense stage (Late Miocene) have been recorded in these facies (Verzi et al. 2008). Towards the upper layers, facies of argillaceous siltstones are recorded. They hosted small sized

**Fig. 6.** (a) Loma Sarmiento outcrop general view. (b) Block structure in facies of reddish sandy claystone (Upper section). (c) Conglomerate sedimentite facies (Upper section).

clasts with a high proportion of carbonates, this part of the sequence culminates with layers of similar grain size covered by a cutt of Mn oxides.

Middle section: in clear-cut contact and overlapping the previous sequence, there is a new association of facies, 1.30 m thick, composed of clay siltstone facies, light brown in colour, which in some sectors have thin sheets of calcium carbonate arranged vertically and horizontally acquiring the typical "in lattice" morphology. Towards the top, the sequence becomes a highly bioturbated brown siltstone, with smaller proportions of sand and clay (presence of cavities of different shapes and sizes), and the carbonate content increases in this sector as well.

Upper section: the outcrop in this sector of the avenue culminates with facies of sandy claystone (~2.70 m thick), reddish, coarsely stratified, partially broken into blocks due to the formation of many rhizoconcretions by carbonate precipitation (Fig. 6b). Laterally, in erosion contact, sandstones facies with the same thickness are recorded. These levels are clast supporting and formed by intraclasts of different sizes, with intense bioturbation and reddish colour.

The XRD analysis corresponding to this part of the sequence allowed the identification of quartz, feldspar (probably from the albite group) and a large amount of volcanic glass; other minerals of doubtful presence are gypsum and kaolinite.

The outcrop located in the southeast sector of the avenue begins with a sequence equivalent to that mentioned for the last part of the northwest sector (Upper section). It is associated with facies of sandy siltstones (~2.70 m thick), with only 29% of clay.

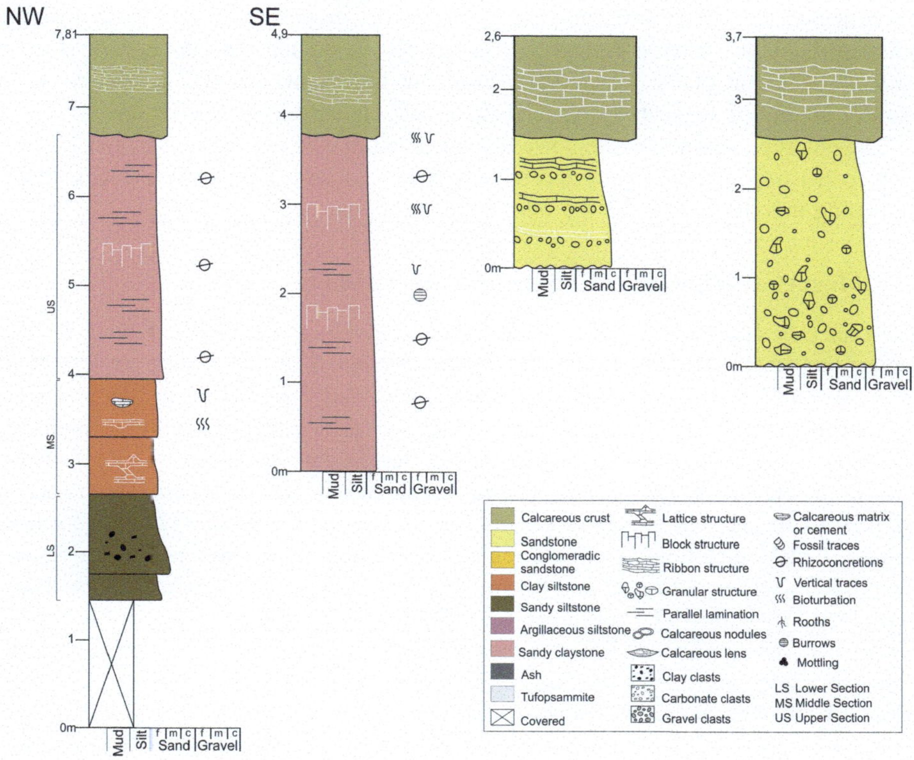


**Fig. 7.** Loma Sarmiento profiles.

It continues keeping the brownish/reddish tone and presenting an important quantity of calcareous rhizoconcretions of different shapes and sizes at the base. Towards the top, these bioturbations increase markedly. In this sector, the sedimentary rocks also present a coarse stratification and secondary blocky structure due to the precipitation of carbonates.

On the contrary, a paleochannel ($\sim$1.50 m depth) filled with a sandy sediment with incipient stratification was identified in the outcrops on both sides of the avenue, formed by an alternation of reddish layers with clasts (diameter $\leq$ 2.50 cm) and fine layers of carbonates. Burrows (Fig. 6c) also filled with volcanic ash and conglomerate sedimentary facies (thickness $\sim$2.60 m) are recorded, mainly bearing clasts of carbonate composition (coarse, size < 1 cm to 20/30 cm), with a clast supporting matrix (at the base) and sandy (towards the top).

Judging by the grain size recorded (predominantly silty) and the brownish/reddish tones in a large part of the sediments that make up both outcrops, they could be linked to continental paleoenvironments associated with fluvio-eolian plains. On the other hand, the recurrence of paleo-edaphological features (rhizoconcretions and blocky structures) allowed us inferring that mature pedogenetic processes developed during exposure episodes of these plains. Towards the end of some deposition cycles, the events would have been drier, judging by the higher proportion of carbonate nodules that they contain. Other morphologies of calcretes (laminar, in lattice) would also involve oscillation of the water table. Brief but important episodes associated with sporadic streams evidenced by paleochannels that have dissected the plains.

The outcrop culminates with a heavily cemented mantle of calcareous crust 1.10 m thick, consisting of thin superimposed sheets of carbonated material and a slightly undulating morphology, as on the other side of the avenue.

## 4.2 Magnetostratigraphy and Magnetic Susceptibility

The values of Magnetic Intensity mass-normalized (J°) at both sites (Cueva de los Leones and Loma Sarmiento) show a great dispersion. The lower ones are less than $25 \times 10^{-6}$ m$^2$/kg, while the higher ones are greater than $300 \times 10^{-6}$ m$^2$/kg, reaching the maximum values in the upper part of both sections, coinciding with the regional calcareous crust (Figs. 8 and 9). Characteristic Remanent Magnetism (ChRM) directions were determined after erasing viscous components from peak field which varied from 2.5 to 10 mT. The values of the Maximum Angular Deviation (MAD) were less than 12. ChRM directions under the regional calcareous crust have opposite orientations in both sections (Figs. 8 and 9). In Cueva de los Leones the reverse ones dominate (D = 179.5°; I = 57.3°; α95 = 5,6°; MDA = < 5.6°) and in Loma Sarmiento the normal ones do so (D = 353.5°; I = −43.3°; α95 = 13.7°; MAD = < 12°), although with some intervals of thin thickness (< 20 cm) of anomalous polarity (i.e., those values of magnetic declination that do not correspond with those of the inclination expected for the Southern Hemisphere). In both sections, anomalous polarity values are recorded at the top, coinciding with the regional calcareous crust and very high J° values (Figs. 8 and 9). This allowed us assuming that the deposition of the sediments underlying the regional calcareous crust in Cueva de los Leones and Loma Sarmiento

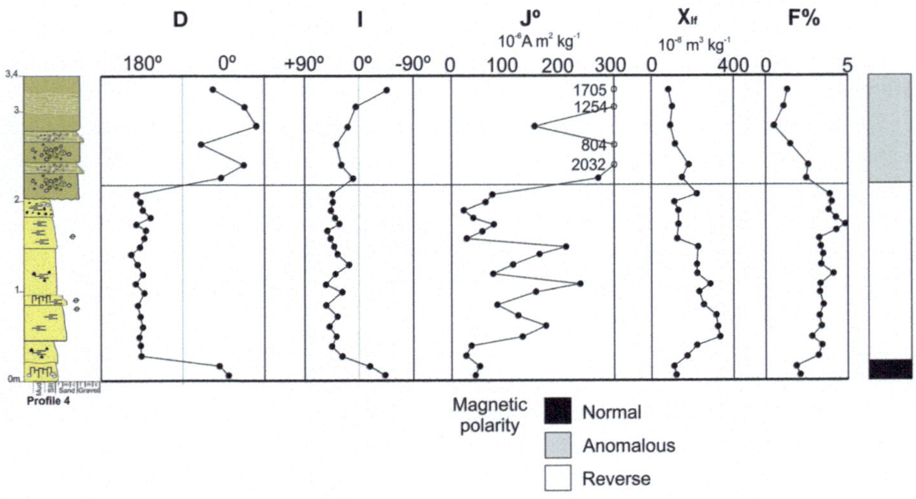


**Fig. 8.** Magnetic polarity of Cueva de los Leones.

was not synchronous. Thus, it is possible to sketch a first hypothesis: the carbonation events in both sections would respond to the same event, although affecting sediments of different ages.

Although geochronological data are not enough to serve for the moment as an accurate reference to establish the magnetostratigraphic column of the analyzed sites, taking into account the regional correlations as well as the reported fossil remains, it is possible to approximate the age of these deposits. Tentatively, the reverse directions of Cueva de los Leones can be assigned to the Gilbert Chron (6.0–3.6 Ma), Early Pliocene, although it could be of a younger age within the Pliocene (e.g., the Mammoth subchron, 3.21–3.33 Ma). The Loma Sarmiento succession would be even older, probably deposited during the C4 to C3A Chrons, in the Late Miocene.

Magnetic susceptibility values ($\chi_{lf}$) (Figs. 8 and 9), which depend upon the concentration, type and grain size of the magnetic minerals, are relatively high, indicating a signal dominated by minerals of the titanomagnetite series. The values of this parameter are higher in Loma Sarmiento, reaching values close to $400 \times 10^{-8}$ m$^3$/kg (average $310 \times 10^{-8}$ m$^3$/kg), while in Cueva de los Leones they were below than $350 \times 10^{-8}$ m$^3$/kg (average $200 \times 10^{-8}$ m$^3$/kg) (Fig. 10a). The values of F% (Figs. 8

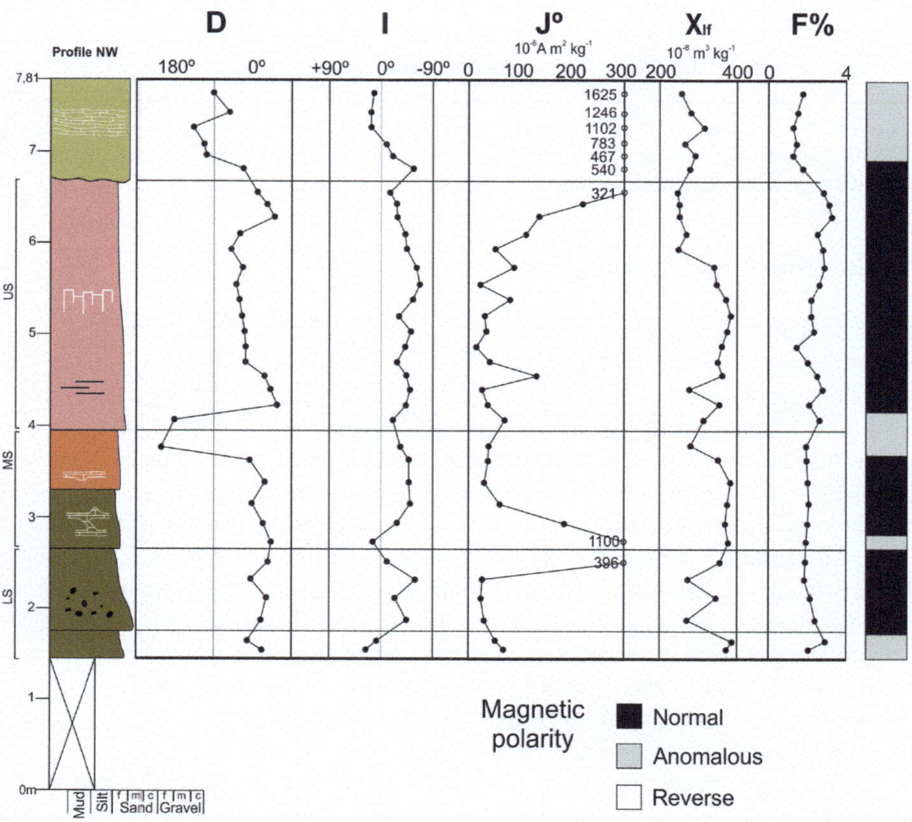


**Fig. 9.** Magnetic polarity of Loma Sarmiento.

and 9) were higher in Cueva de los Leones than in Loma Sarmiento (Fig. 3b). In Cueva de los Leones they reach values of 6% (average 3.2%), while in Loma Sarmiento they are lower than 3.4% (average 2.3%). Thus, in Cueva de los Leones there is a greater relative contribution of superparamagnetic particles than in Loma Sarmiento (Fig. 10 b). As a first approximation, the differences, both in the values of $\chi_{lf}$ and F%, would be linked to the pedogenetic degree, which reach a higher maturity in the Cueva de los Leones.

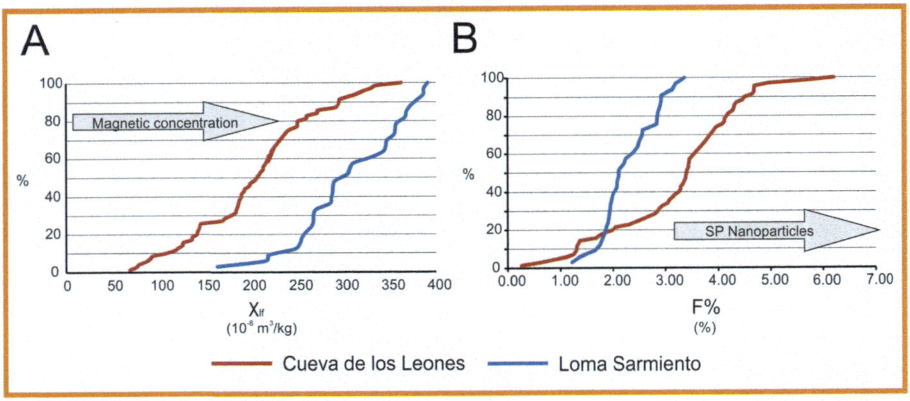


**Fig. 10.** Cumulative curves of magnetic susceptibility at 470 Hz (χ) and the frequency-dependent susceptibility factor (F%).

## 5 Summary and Conclusions

The sedimentological and magnetostratigraphic analyses carried out in the urban and peri-urban sectors of Bahía Blanca (Loma Sarmiento and Cueva de Los Leones) made it possible to contribute to the knowledge of the paleoenvironmental and paleoclimatic conditions in southwest Buenos Aires province, Argentina, during the Late Cenozoic.

In Cueva de Los Leones the analysis of sedimentary facies, complemented with the information provided by other "proxy data" (volcanic ash, fossil ichnites and micromorphology of soils), for the northern sector of the outcrop, allowed us inferring that a stream similar to the current Saladillo creek would have drained in the area, generating deposits such as channels, bars, levees and alluvial plains; the latter would have extended towards the southern sector. In these paleoenvironments, periods of topographic surface stability, associated with low deposition rates and a relatively low phreatic level, would have favoured pedogenesis and bioturbation. In particular, the carbonate concretions would be linked to drier episodes, where the well-aerated paleosols would have supported a meagre plant cover.

The magnetostratigraphic analysis allowed us assuming that these deposits correspond to the Pliocene. Taking into account also the palaeoclimatic and paleoenvironmental records, it is possible to link these deposits with the climatic amelioration that occurred globally during the Middle Pliocene (between 3.3 and 3.0 Ma). It should be noted that Deschamps and Tomassini (2016) proposed this same interval for the Monte Hermoso Formation, considering the fossiliferous content. Based on this, a possible correlation between the deposits of Cueva de los Leones and the Monte Hermoso Formation may be established.

In Loma Sarmiento, the sedimentological analyses carried out, complemented with chemical, grain size, XRD and fossil analyses, allowed us linking the stratigraphic sequences analysed with fluvio-eolian plains in which the pedogenetic processes would have been recurrent, while restricted deposits (paleochannels) would reflect wetter, brief and sporadic episodes. These characteristics and the variations in the facies (sandy and clay) would reflect paleoenvironmental conditions characteristic of semiarid climates. The recorded fossil remains support the paleoecological inferences formulated from the geological data. According to Verzi (2001; in Quattrocchio et al. 2008), the trend towards an increase of hypsodontics, observed in the lineage of the genus *Xenodontomys* and in other rodents, would be related to adaptation to open environments within the Late Miocene climate deterioration. The sedimentary sequence belonging to this outcrop was assigned by Verzi et al. (2008) to the Saldungaray Formation, but subsequently Zárate and Folguera (2009) assigned it to the Cerro Azul Formation.

The magnetic parameters recorded in Loma Sarmiento coincide with the data mentioned above, corresponding to semi-arid paleoenvironmental conditions, in agreement with paleoclimates that occurred previously to the Gilbert Chron, during the Late Miocene.

In relation to the origin of the materials that have filled this part of SW Buenos Aires province, the macroscopic, microscopic and XRD analyses in both outcrops suggest an important contribution of volcanic clastic material coming from the Andean ranges (ash and volcanic glass, plagioclase (andesine), among others) as well as the Sierras Australes (metamorphic rocks with abundant quartz). Consequently, and considering the predominant grain size in both outcrops (fine sandstones in Cueva de Los Leones and loess-like sediments in Loma Sarmiento), it is assumed that the latter (Loma Sarmiento) would represent distal facies in relation to their origin (the Andes mountains and/or the Sierras Australes).

It can be also concluded that although the sedimentary sequences analysed represent fragmentary records of the geological past, they would reflect part of the two most important sedimentary cycles that characterize the continental Neogene of the region. In the city of Bahía Blanca and its surroundings, such cycles partially overlap (Folguera and Zárate 2009). Therefore, the depocenter of the Late Miocene, that would have been more extensive, would be reflected in the outcrop of Loma Sarmiento and the Pliocene cycle in the Cueva de Los Leones.

In conclusion, and in relation to the calcareous crust that covers both outcrops, its thickness and magnetic signal are similar, although their morphology is slightly dissimilar. This might indicate that this important regional incrusting was synchronous in both outcrops.

**Acknowledgements.** We thank Heraclio Ortiz, Carlos Bade, Federico Mateos, Andrea Montoya, Lic. Guillermo Pera Vallejos, Dr. Verónica Gil and Dr. Jorge Gentile, for their collaboration in the field tasks. This work was carried out through the PGI 24/ZH21 subsidised by SECYT (Universidad Nacional del Sur). The paleomagnetic analyses were carried out at LEMIT (City of La Plata) and the magnetic susceptibility was determined in the IFAS (City of Tandil).

## References

Bridge, J.S.: Rivers and foodplains. Form, processes and Sedimentary Records. Blackwell, Oxford (2003). 491 páginas

Buatois, L.A., Mángano, M.G.: The paleoenvironmental and paleoecological significance of the lacustrine Mermia ichnofacies: an archetypical subaqueous nonmarine tracess fossil assemblage. Ichnos **4**, 151–161 (1995)

Deschamps, C.M.: Late Cenozoic mammal stratigraphy in southwestern Buenos Aires province, Argentina. Ameghiniana **42**, 733–750. Buenos Aires (2005)

Deschamps, C.M., Tomassini, R.L.: Late Cenozoic vertebrates from the southern Pampean region: systematic and bio-chronostratigraphic update. In: Martínez, M., Olivera, D. (eds.) Palinología del Meso-Cenozoico de Argentina-Volumen en homenaje a Mirta Elena Quattrocchio. Publicación Electrónica de la Asociación Paleontológica Argentina vol. 16 (2), pp. 202–225 (2016)

Deschamps, C.M., Tonni, E., Verzi, D., Scillato Yané, G., Zavala, C., Carlini, A., Di Martini, V.: Bioestratigrafía del Cenozoico Superior continental en el área de Bahía Blanca, provincia de Buenos Aires. V Jornadas Geológicas y Geofisícas Bonaerenses La Plata, 49–58. La Plata (1998)

Fidalgo, F.: Geología y Geomorfología del área White – Cerri y los alrededores de Bahía Blanca. Comisión de Estudios de Suelos White – Cerri. Ministry of Public Works, Buenos Aires Province, unpublished report, La Plata 32 (1983)

Folguera, A., Zárate, M.: La sedimentación neógena continental en el sector extrandino de la Argentina central. Revista de la Asociación Geológica Argentina **64**, 692–712 (2009)

Frey, R.W., Pemberton, S.G.: The Psilonichnus ichnocoenose, and its relationship to adjacent marine and nonmarine ichnocoenoses along the Georgia coast. Bull. Can. Pet. Geol. **35**, 333–357 (1987)

Gómez Samus, M.L., Grill, S., Bidegain, J.C., Gil, V., Gentili, J.O.: Paleomagnetismo en sedimentos del Cenozoico tardío de Cueva de los Leones, Bahía Blanca, provincia de Buenos Aires (Argentina). Segundo Simposio del Mioceno-Pleistoceno del Centro y Norte de Argentina. Resúmenes, pp. 14–15 (2013)

Gómez Samus, M.L., Chimento, N.R., Löffler, J., Rico, Y., Sierra, L., Bidegain, J.C.: Neogene-Quaternary in Tandilia, South America: litho- bio- magnetostratigraphy. J. Iberian Geol. **43**, 559–581 (2017). ISSN 1698-6180, https://doi.org/10.1007/s41513-017-0037-z

González Uriarte, M.: Características geomorfológicas de la porción continental que rodea la Bahía Blanca, provincia de Buenos Aires. In: 9° Congreso Geológico Argentino Actas vol. 3, pp. 556–576 (1984)

Grill, S., Garrone, M., Manera, T.: Hallazgo de trazas fósiles en la Cueva de los Leones. In: 24° Congreso Argentino de la Ciencia del Suelo and II Reunión Nacional "Materia Orgánica y Sustancias Húmicas". Published in CDRom (2014a)

Grill, S., Reinhardt, C.J., Garrone, M., Ortiz, H.: Bahía Blanca. Análisis de cenizas volcánicas en el área de Bahía Blanca. In: 24° Congreso Argentino de la Ciencia del Suelo and II Reunión Nacional "Materia Orgánica y Sustancias Húmicas". Published in CDRom (2014b)

Grill, S., Cesaretti, N., Gil, V., Gentili, J.: Morfología y Petrología de la tosca de la región de Bahía Blanca. In: 24° Congreso Argentino de la Ciencia del Suelo and II Reunión Nacional "Materia Orgánica y Sustancias Húmica". Published in CDRom, Bahía Blanca (2014c)

Johnsson, M.J., Stallard, R.F., Lundberg, N.: Controls on the composition of fluvial sands from a tropical weathering environment: sands of the Orinoco river drainage basin, Venezuela and Colombia. Geol. Soc. Am. Bull. **103**, 1622–1647 (1991)

Malumián, N.: La sedimentación y el volcanismo terciario en la Patagonia extrandina. In: Camino, R. (ed) Geología Argentina, Instituto de Geología y Recursos Minerales, Anales vol. 29 (18), pp. 557–578. Buenos Aires (1999)

Moretti, L.M., Morrás, H.J.M., Grill, S.C.: Micromorfología y mineralogía de niveles edafizados subyacentes al calcrete regional de la llanura subventánica, provincia de Buenos Aires. In: XX Congreso Geológico Argentino (Tucumán). Sesión Técnica 3. Geomorfología y Suelos: 32–37 (2017)

Olmstead, L.B., Alexander, L.T., Middleton, H.E.: A pipette method of mechanical analysis of soils based on im-proved dispersion procedure. Technical Bulletin no.170 (1930)

Pemberton, S.G., Frey, R.W., Ranger, M.J., Maceachern, J.A.: The conceptual framework of ichnology. In: Applications of ichnology to petroleum exploration—A core workshop, SG Pemberton, editor, Society of Economic Paleontologists and Mineralogists, Core Workshop vol. 17, pp. 1– 32 (1992)

Quattrocchio, M., Borromei, A.M., Deschamps, C., Grill, S., Zavala, C.: Landscape evolution and climate changes in the Late Pleistocene – Holocene of southern pampas (Argentina). Evidence from palynology, mammals and sedimentology. Quatern. Int. **181**, 123–138 (2008)

Sostillo, R., Cardonatto, R., Montalvo, C., Visconti, G., Melchor, R.: New findings of invertebrate trace fossils in a Late Miocene loess paleosol succession, Cerro Azul Formation, La Pampa. Argentina. Segundo Simposio Latinoamericano de Icnología, Santa Rosa, province of La Pampa (2013)

Verzi, D.H.: Phylogenetic position of Abalosia and the evolution of the extant Octodontinae (Rodentia, Caviomorpha, Octodontidae). Acta Theriol. **46**(3), 243–268 (2001)

Verzi, D.H., Montalvo, C., Deschamps, C.M.: Biostratigraphy and biochronology of the late Miocene of central Argentina: evidence from rodents and taphonomy. Geobios **41**, 145–155 (2008)

Zárate, M.: El Cenozoico tardío continental de la provincia de Buenos Aires. In: de Barrio, R.E., Etcheverry, R.O., Caballé, M.F., Llambías, E. (eds.) 16° Congreso Geológico Argentino Relatorio de la Geología y Recursos Minerales de la provincia de Buenos Aires, Relatorio, 4, 139–158. La Plata (2005)

# Perched Dunes in the Fuegian Steppe, Southern Argentina: Archeological Reservoirs of Holocene Information

Andrea Coronato[1,2(✉)], Mónica Salemme[1,3], Jimena Oría[1], Florencia Mari[4], and Ramiro López[1]

[1] Centro Austral de Investigaciones Científicas (CADIC-CONICET), Bernardo Houssay 200 (9410), Ushuaia, Tierra del Fuego, Argentina
acoronato@cadic-conicet.gob.ar

[2] ICPA-Universidad Nacional de Tierra del Fuego, Walanika 250, (9410), Ushuaia, Tierra del Fuego, Argentina

[3] ICSE-Universidad Nacional de Tierra del Fuego, Walanika 250, (9410), Ushuaia, Tierra del Fuego, Argentina

[4] Laboratorio de Radiocarbono, Centro de Investigaciones Geológicas (CONICET-UNLP), Diagonal 113 N° 275 (B1904DPK), La Plata, Argentina

**Abstract.** The description of the genesis and the geomorphological characteristics of three perched dunes, named as O'Connor, Amalia and Arturo is presented herein. They are located in the periphery of the shallow lakes of such name, which have a semi-perennial nature in the Fuegian steppe (lat. 53° S). These landforms are characterized according to their location, orientation and quantity, as well by color, thickness and textural composition of the eolian origin units and the units with soil development. These units, considered as paleosoils, are indicators of stable environments that would have diminished the eolian accumulation which is responsible for the formation of eolian units of variable thickness. A chronological sequence, based upon radiocarbon dating on organic matter belonging to each one of the units transformed into soils, fits quite well with already published absolute dates and it allows the interpretation of the evolution of these dunes formed between the Late Glacial and the Late Holocene. An eolian accumulation rate for the Amalia and Arturo dunes is suggested, based upon the thicknesses of each unit and the obtained radiocarbon dates. The tephras contained in the Amalia and Arturo dunes are described and compared with the Middle Holocene eruption of Mt. Burney (Chile). The characteristics of archeological record and its state of conservation are analyzed, whether they are lithic materials or bone remains. The chronology obtained for the paleosoil sequence is discussed in relation to paleoenvironmental events. These events are determined on the base of certain proxies, both in the studied soils as in lacustrine cores obtained in shallow lakes from the surroundings. The development of some of the paleosoils is in contrast with certain regional examples that appear in the literature. From the archeological point of view, the development of dune geomorphological processes such as auto-erosion or "cannibalism" and re-deposition exposed the possible existence of taphonomic traps and palimpsests is highly significant; these are generally visible in surface contexts, which are interpreted as archeological reservoirs. In this kind of sites, the resolution is low and usually, the degree of mixture could be high.


M. M. Collantes et al. (Eds.): *Advances in Geomorphology and Quaternary Studies in Argentina*, SPRINGEREARTH, pp. 58–91, 2020.
https://doi.org/10.1007/978-3-030-22621-3_3

**Keywords:** Perched dunes · Palaeosoils · Archaeological reservoirs · Holocene

## 1 Introduction

The Fuegian steppe is dominated by an extensive, grassy-shrubby plant cover, over hills, plains, valley bottoms and glaciofluvial fans. The landforms of eolian origin developed mainly in the periphery of ephemeral lakes, which are transformed into pans when, during the drier summer season, the water body disappeared, and the silty-clayey bottom is exposed with salty surfaces resulting from evaporation (Villarreal and Coronato 2017).

Among the various types of eolian landforms, the perched dunes become relevant due to their important thickness and the inter-stratification between them and edaphic layers and their constant erosion-deposition activity. The levels with edaphic processes found in the perched dunes of the Arturo shallow lake were described as paleosoils (Coronato et al. 2011a, b) due to the recognition of soil features. Studies of the environmental magnetism allowed the interpretation of their genesis as the sedimentary result of climatic periods wetter than today (Orgeira et al. 2012), that is, precipitation values larger than 300 mm per year. These edaphic layers are very important to increase the knowledge of the human occupation in the area, since they represent unique conditions for the concentration and preservation of archeological materials in this landscape. The identification of paleosoils of similar age in dune systems of different sites allows also its use as stratigraphic markers of environmental conditions appropriate for human settlement in zones previously considered as of a marginal character. According to Favier Dubois (2001), the paleosoils show a high potential to concentrate the archeological signal, either for showing periods of low eolian erosion or due to relative landscape stability (Holliday 1992) and, therefore, representing moments favoring human settlement from an environmental point of view.

From the archeological point of view, the shallow lakes of the steppe of northern Tierra del Fuego are spaces in which the archeological signal is much more intense than in other sectors or landforms of the same steppe environments (Oría 2014). In the hills, fluvial terraces and valleys, the archeological visibility is almost null due to a thicker plant cover. Contrarily, the erosion features in the periphery of the shallow lakes, such as gullies or depressions generated on eolian deposits, expose artifacts and faunal remains, a product of human settlements.

Infrequent eolian landforms, of restricted location in a sector of the Fuegian steppe, are described in this paper (Fig. 1). It is herein proposed to highlight their role as reservoirs of archeological contexts and taphonomic traps in the steppe region.

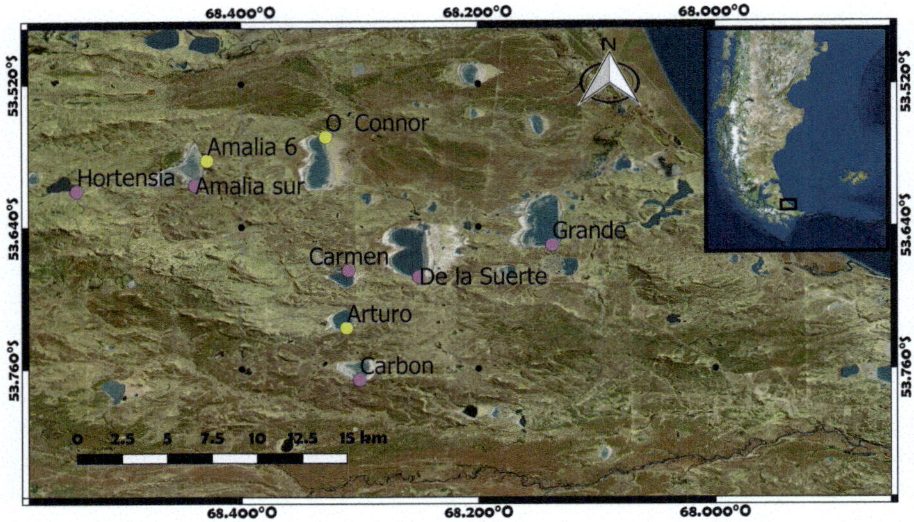


**Fig. 1.** Location of the studied area in northern Tierra del Fuego, Argentina. Yellow dots point out the location of the three studied perched dunes. Violet dots point out the location of perched dunes described in Table 1.

## 2 Methodology

The presence of perched dunes was recognized in various sites of the Fuegian steppe by means of field surveys that satisfied different research objectives. For this paper, dunes located in the periphery of the O'Connor, Amalia and Arturo shallow lakes were selected due to visibility, accessibility and relative easiness to perform the sampling processes.

Stratigraphy cross sections were prepared, calculating the thickness of the sedimentary beds using a Pentax R-326 Total Station. In Amalia dune the survey was completed by means of an aluminum ladder and metric tape, because the intensity of the wind was so strong that it was impossible to keep the needed stability for the measurement with the cited equipment. Samples of each sedimentary layer were obtained to be analyzed with the techniques mentioned below.

The sedimentology analysis was visually performed in a descriptive form and by textural analysis with a Malvern 2000 grain-size particle laser analyzer at CADIC laboratories. The definition of color in dry conditions was based upon a Munsell (1973) color table.

The estimation of organic matter was performed by the loss-on-ignition method following Davies (1974) and Page (1982), over 5 g of soil dried to 105 °C during 24 h for the calculation of hygroscopic moisture, then weighted again and heated up to 430 °C during 24 h. These analyses were performed at CADIC laboratories.

The sediment accumulation rates for Amalia and Arturo dunes sequences were estimated using the radiocarbon age obtained on organic matter. The thickness (cm) of each pair of eolian unit and its corresponding edaphic soil was divided by the age difference in radiocarbon years (yr) of consequent edaphic units to derive the accumulation rate in cm per radiocarbon year (cm $yr^{-1}$).

The location of the archeological materials scattered in the O'Connor dune was surveyed with GPS Garmin Etrex-Vista whereas the Arturo 1 and Amalia 6 sites were so with a Pentax R-326 Total Station. The three-dimensional position (x, y, z coordinates) of all the archeological materials (bones and lithic artifacts), were recorded. The Yowen Ko site was excavated following the usual methods, since it has a stratified context.

The lithic materials were analyzed taking into consideration the variables related to the degree of preservation and integrity of the archeological sets. In addition to the artifact category, the size in mm was considered, according to length, width and thickness, weight in grams and degree of weathering of the artifacts. In this case, the polishing and abrasion of the surfaces was taken into account, using the qualitative scale for abrasion degree: smooth (w1), moderate (w2) and intense (w3), and for polishing: null, low, moderate and intense (Borrazzo 2010). Both variables are useful to understand the action of water and wind on the artifacts. For flakes and artifacts formatized on flakes, the degree of fragmentation was also checked. The size of the artifacts was classified for each 5 mm of increment according to the methodology proposed by Franco (2002). It was considered also the size both in complete and fractured tools, to evaluate the dimensions of the materials that are preserved, independently of considering knapping techniques or destruction processes.

The bone remains were analyzed from the anatomical and taxonomic perspective. Complete and fragmented bones were found in varying quantities except in O'Connor dune; some specimens present striations, desiccation cracks and exfoliation of the crust surface. The minimum number of specimens (NISP) was calculated following Grayson (1984) and Lyman (1984, 2008). Likewise, a taphonomic analysis was developed taking into account the degree of weathering according to Behrensmeyer (1978), the root marks (Binford 1981; Schiffer 1987), rodent traces (Binford 1981; Mengoni Goñalons 2010), carnivore signals (Binford 1981), and trampling and fracture patterns (Binford 1981; De Nigris 1994).

The radiocarbon dating was performed on bulk sediments, at the laboratories of the AMS Facility of the University of Arizona (USA) and at LATYR (Laboratorio de Radiocarbono del Centro de Investigaciones Geológicas, CONICET-Universidad Nacional de La Plata, La Plata, Argentina). The calibration of the dates were done using the CALIB Rev 7.0.4 program (http://calib.qub.ac.uk/calib/) and the calibration curve for the Southern Hemisphere named as Shcal13.$^{14}C$ (Hogg et al. 2013). Other dates were obtained from faunal bones buried in the sediments. Some of the radiocarbon dates obtained were checked against those presented by Riches (2018), obtained in the AMS Laboratory Facility at the University of California at Irvine, UCIAMS (USA).

The chronostratigraphic interpretation of the sedimentary sequences was based upon the time boundaries proposed for the Holocene in the International Chronostratigraphic Chart (Cohen et al. 2018).

## 3 Perched Dunes

The perched dunes are sediment accumulations transported at a short distance and deposited over cliffs. They form elongated and asymmetrical hills located at the top of active cliffs eroded by lake waves. The main source of the materials comes from the cliff itself on which the dunes are places, generally formed by till, agglomerates and/or weathered sedimentary rocks, as it is the case of the sites herein presented. To a lesser degree, the deflation material coming from nearby deflation hollows provides the finer debris (Fig. 2). The type localities for these dunes have been described in the cliffy coasts of the Great Lakes of North America (Marsh and Marsh 1987), and they have been regionally recognized at the cliffs of Lake Fagnano, located in the forested area of Tierra del Fuego (Coronato et al. 2009). In the Fuegian steppe, the perched dunes are located along the cliffy coasts of the seasonal shallow lakes or pans (Fig. 1). These landforms are exposed towards the N, W and S quadrants, from where the dominant winds come, and they have a thickness between 2.3 and 23.0 m (Table 1) and an alternating sequence of one to nine edaphic layers. They are developed over cliffs of sandy conglomerates, sandstones and friable siltstones. The edaphic layers are interpreted as paleosoils (herein named as Ps from now onwards) due to certain physico-chemical characteristics that differentiate them from the underlying and superposed eolian deposits (Coronato et al. 2011a, b; Orgeira et al. 2012). The faces of the dunes exposed by erosion in which these sedimentary characteristics were surveyed correspond to the up-wind or down-wind sectors or some of their sides (Table 1).

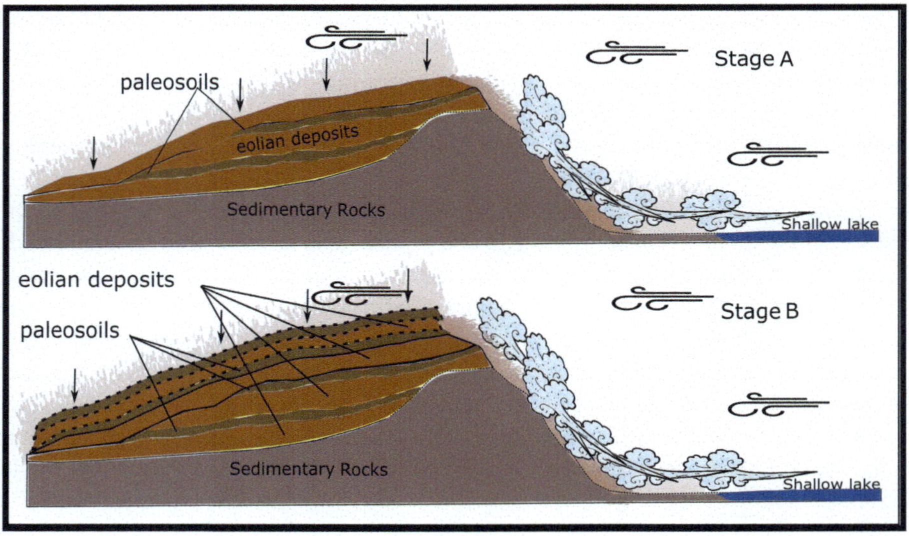


**Fig. 2.** Perched dunes formation including paleosoils or edaphized levels on the top of each eolian unit. A: initial stage. B: advanced stage with paleosoil development.

**Table 1.** Perched dunes distribution in northern Tierra del Fuego and their characteristics.

| Dunes | Geographical position | | Exposure | Number of edaphized levels | Thickness (m) | Source material | Studied dune side |
|---|---|---|---|---|---|---|---|
| Hortensia Chica | 53°36′ 10′S | 68°34′ 36″W | NE | 4 | 12 (estimated) | Not visible | Windward |
| Amalia-Vista | 53°36′ 18.07″ S | 68°26′ 14,01″ W | N | 2 | 23 (estimated) | Conglomerates and sandstones | Lateral |
| Amalia | 53°34′ 39′S | 68° 25′59″ W | W | 8 | 7,4 | Conglomerates, siltstones and sandstones | Lateral |
| O'Connor | 53°33′ 41′S | 68°20′ 13″W | SW | 2 | 13,7 | Conglomerates and siltstones | Windward |
| Carmen | 53°40′ 27′S | 68°18′ 48″W | S | 1 | 2,3 | Conglomerates and sandstones | Leeward |
| Arturo | 53°43′ 28′S | 68°18′ 51″ | NW | 8 | 20 | Conglomerates and sandstones | Windward and lateral |
| Carbón | 53°46′ 09.2″S | 68°18′ 02″W | NNW | 2 | 3 (estimated) | Siltstones and claystones | Windward |
| Tres Marías Occidental | 53°48′ 04.5″S | 68°12′ 12.07″ W | NNW | 2 | 11 (estimated) | Conglomerates and sandstones | Windward |

The up-wind surface is very steep, almost vertical in some cases, and it is exposed to eolian erosion and mass movements, whereas the down-wind slope does not exceed 15°. The lateral sides of the dunes may be eroded by gullies formed from the edge of the cliffs, a product of receding water erosion. Once formed, the gullies work as wind tunnels in which deflation forms wind hollows or blowouts, and they partially discover the sediment that forms the dunes. The up-wind slopes receive the material from the cliff by means of twisters or whirlwinds, creep, and/or saltation, in all cases in ascending trajectories. The down-wind slopes receive material by deflation from the pans, as in other sectors of the same dune, by means of deflation and saltation processes, respectively. In the crests of certain dunes blowouts are formed, from where material may be mobilized to form secondary active dunes, semi-circular in shape.

The deposits that form these dunes are composed of alternating massive sedimentary strata and edaphic layers and, in some cases, with inter-bedded tephra beds. The edaphic layers are interpreted as paleosoils of incipient development, generated during stable and humid periods and buried by massive eolian deposits resulting from the reactivation of the dunes during dryness periods.

## 4 The Studied Landforms and the Archeological Record

The perched dunes associated to the semi-permanent shallow lakes of this steppe environment generated taphonomic trap conditions. These are spaces where remains of natural or cultural origin may be retained (Oría et al. 2016a). These traps also form potential spaces for the preservation of archeological contexts with relative good integrity. The erosion processes that take place in this type of dunes expose archeological materials at the surface, which generates palimpsests.

In this work, sets of artifacts and bones from the three sites and isolated findings in the studied dunes are analyzed (Table 2). The archeological record of the O'Connor shallow lake is limited to isolated and dispersed findings in an area of 11,600 m$^2$ in the SE sector of the dune (Oría et al. 2014). At the Amalia shallow lake, the Amalia 1, 2, 4, 5, 6, Yar Hayen, Yowen Ko and Cantera sites have been found (Oría 2016); two of them, Amalia 6 and Yowen Ko, are located in the studied dune. In the Arturo shallow-lake the Arturo 1 to 4 sites are located (Oría and Salemme 2016; Oría et al. 2017), but only the Arturo 1 site is directly related to the studied dune.

**Table 2.** Number of archeological items considered the studied locations.

| Assemblages | Lithic artifacts | Faunal remains |
|---|---|---|
| O'Connor isolated finds | 40 | 0 |
| Amalia 6 site | 144 | 2018 |
| Yowen Ko (context 3) site | 1 | 572 |
| Arturo 1 site | 148 | 152 |

## 5 The O'Connor Dune

The O'Connor dune is developed in the northern coast of the shallow lake of such name (Figs. 1 and 3), over a siltstone cliff of 17 m in thickness, underlying a thin, sandy gravel bed of less than 0.25 m in thickness. Above it, a massive eolian deposit (0.56 m) covers it, on which a darker layer occurs, of similar clayey-silty texture, 0.75 m in thickness, named as the Paleosoil 1 (Ps1), Fig. 4. On top of it, an eolian deposit of more than 9.0 m thick is found, on which a second edaphic layer, named as Paleosoil 2 (Ps2), 0.42 m in thickness, is placed. The profile is completed by 2.42 m of massive eolian deposits. The textural composition of the eolian layers and the edaphic beds is quite similar: 85–90% silt, 10–15% clays and 10% sand (Fig. 4). However, the color varies from olive green (5Y 5/6) to dark brown (10YR 3/3), from the base to the top of the sequence of edaphic layers. The content of organic matter varies from 4.25% at Ps1 to 7.1 at Ps2 (Table 3). They present a columnar, prismatic structure, mottled areas and root marks.

The archeological record of this dune is composed of isolated findings at the surface. This is a set of 40 lithic pieces (Table 4), coming from a sector of the NE coast of the shallow lake of this name (Fig. 1). In all cases, the basic form of the cores corresponds to cobbles, which are available at the lake (Oría et al. 2014); the predominant artifacts are flakes (55%), which were the pieces more disturbed by fragmentation.

The range in size of the recovered materials goes from 20 to 130 mm, but with scarce representation above 65 mm, with some of the sizes absent from the record (Fig. 5). Between 20 and 55 mm the largest concentration of the record took place

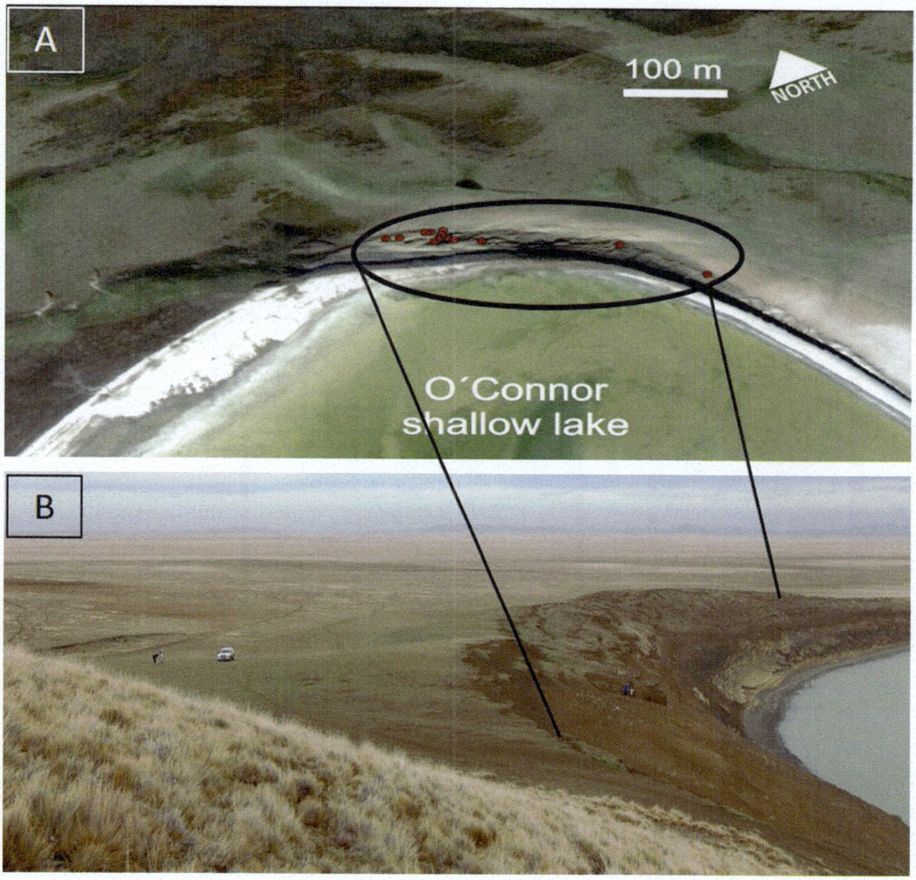


**Fig. 3.** O'Connor dune. A: The northern coast of O'Connor lake and the perched dune formed over the cliff (DTM from Google Earth™); green dots point out the location of archeological findings; B: View to the N showing the up-wind side of the dune towards the shallow lake and the down-wind side with low slope value and covered by grasses. They are permanently buried by eolian sediments eroded from the up-wind side of the dune.

(67.5%) (Fig. 5). All the artifact sizes show certain degrees of abrasion and/polished surfaces, and it has been interpreted that the weathering of the artifact surfaces is not related to its size (Oría et al. 2014).

The largest part of the set (67.5%) is found within a soft degree of abrasion (w1). Also, 45% of the pieces (n = 18) shows soft abrasion and slight polishing, and 20% (n = 8) exposes soft abrasion and moderate polishing. Likewise, 10% (n = 4) of the artifacts present a w2 grade of abrasion whereas 15% (n = 6) has the same degree of abrasion, but with moderate polishing and 7% (n = 3) present intense polishing (Table 5).

The whole group of recovered lithic artifacts has a weight equal or smaller than 200 gr, although 50% of the sample is found within the range of 10 to 100 gr (Fig. 6).

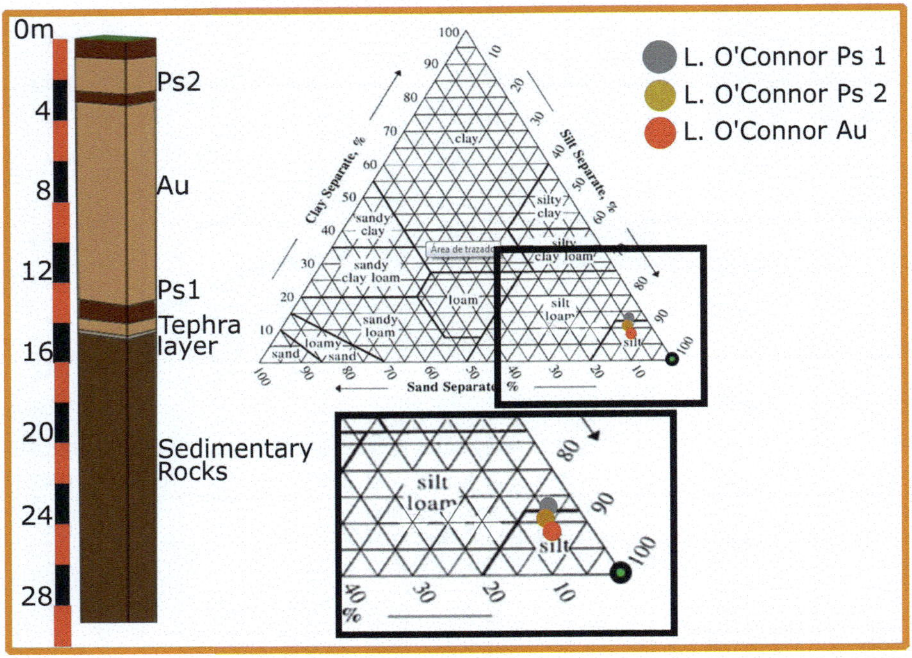


**Fig. 4.** Eolian sedimentary profile of the O'Connor dune and grain-size tri-axial diagram.

**Table 3.** Color (after Munsell 1973) and organic matter content (%) of edaphized levels in dunes.

| Dunes stratigraphy | Colour | | OM (%) |
|---|---|---|---|
| O'Connor | | | |
| PS2 | 10YR 3/3 | Dark brown | 7,1 |
| PS1 | 5Y 5/6 | Pale olive | 4,25 |
| Amalia 6 | | | |
| PsS9 | 2.5Y 5/6 | Light olive brown | 5.41 |
| PS8 | 2.5Y 5/6 | Light olive brown | 5.86 |
| PS7 | 2.5Y 5/6 | Light olive brown | 4.1 |
| PS6 | 2.5Y 5/6 | Light olive brown | 5.05 |
| PS5 | 2.5Y 4/4 | Olive brown | 4.15 |
| PS4 | 2.5Y 5/6 | Light olive brown | 4.73 |
| PS3 | 2.5Y 4/4 | Olive brown | 4.55 |
| PS2 | 2.5Y 4/4 | Olive brown | 4.29 |
| PS1 | 2.5Y 4/4 | Olive brown | 4.49 |
| Arturo | | | |
| PS8 | 5Y 5/4 | Olive | 4.87 |
| PS7 | 5Y 5/4 | Olive | 4.86 |

*(continued)*

**Table 3.** (*continued*)

| Dunes stratigraphy | Colour | | OM (%) |
|---|---|---|---|
| PS6 | 5Y 5/4 | Olive | 3.2 |
| PS5 | 2.5Y 4/4 | Olive brown | 3.59 |
| PS4 | 10YR 3/2 | Very dark graiysh brown | 4.18 |
| PS3 | 10YR 3/2 | Very dark graiysh brown | 4.12 |
| PS2 | 10YR 3/2 | Very dark graiysh brown | 4.38 |
| PS1 | 10YR 2/2 | Very dark brown | 5.68 |

**Table 4.** Artifactual categories studied in each lithic assemblage from the three dunes.

| Artifactual category | | O'Connor | | Arturo 1 | | Amalia 6 | | Yowen Ko context 3 | |
|---|---|---|---|---|---|---|---|---|---|
| | | N | % | N | % | N | % | N | % |
| Artifact with secondary formatization | Retouched flake | 2 | 5 | 2 | 1.35 | 4 | 2.78 | - | - |
| | Side scraper | - | - | - | - | 7 | 4.86 | - | - |
| | End scraper | - | - | - | - | - | - | - | - |
| | Projectile point | - | - | - | - | 1 | 0.69 | - | - |
| Artifact with initial formatization | Flake | 22 | 55 | 100 | 67.57 | 89 | 61.81 | 1 | 100 |
| | Undifferentiated debris | 7 | 17.5 | 10 | 6.76 | 17 | 11.81 | - | - |
| | Nodule | 2 | 5 | 7 | 4.73 | | 0.00 | - | - |
| | Core | 7 | 17.5 | 29 | 19.59 | 25 | 17.36 | - | - |
| Artifact modified by use | Percussor/hammerstone | - | - | - | - | 1 | 0.69 | - | - |
| Total | | **40** | **100** | **148** | **100** | **144** | **100** | **1** | **100** |

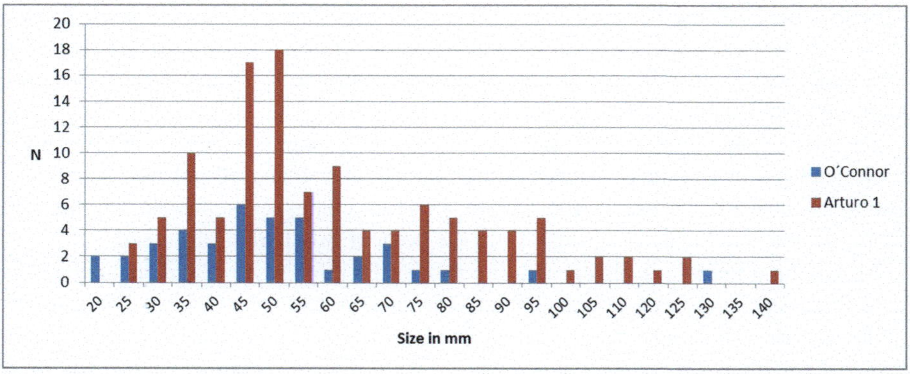


**Fig. 5.** Size distribution of lithic artifacts from the O'Connor dune and the Arturo 1 archeological site.

**Table 5.** Frequency of artifacts with altered surfaces in the O'Connor dune.

| Alteration on the lithic surfaces | | Polishment | | | | Total |
|---|---|---|---|---|---|---|
| | | Null | Light | Moderate | Intense | |
| Abrasion | w1 | 1 | 18 | 8 | - | 27 |
| | w2 | - | 4 | 6 | 3 | 13 |
| | w3 | - | - | | - | - |
| Total | | **1** | **22** | **14** | **3** | **40** |

## 6 The Amalia Dune

The Amalia Dune is developed along the eastern coast of the shallow lake of that name, exposed to the westerly winds. This dune seats on an active cliff of 19 m in thickness, composed of weathered sandstones, silty-claystones, a layer of sandy conglomerate of 3 m in thickness (Figs. 1 and 7). The top of the dune is placed 50 m above the level of the lake coast. The down-wind slope extends for 500 m; at its northern face, this dune is dissected by receding erosion gully, which acts as a wind tunnel, in which wind gasps of up to 100 km $h^{-1}$ have been recorded. The intense winds eroded the northern face of the dune and formed a blowout at the foot of the slope. The uncovered face exhibits the alternation of nine beds of massive eolian deposits, interbedded with other nine edaphic levels, in a total of 12 m of sedimentary deposits (Fig. 8). The mean thickness of the edaphic layers is 0.45 m, with a maximum value for Ps4 of 0.74 m and a minimum value of 0.16 m for Ps5, whereas the thickness of the eolian levels has an average value of 0.74 m with extreme values of 0.33 and 1.19 m. The sequence starts with Ps1 which is the thickest but a without a visible base. The overlying eolian deposit contains a layer of whitish tephra of 0.05 m in thickness.

The edaphic levels have a silt texture (Fig. 8), with a silt content of 90% to 95% and with a sand content between 5% to 10%. The color varies from olive brown (2.5Y 4/4) to light olive brown (2.5Y 5/6) and the contents of organic matter ranges between 4.15% and 5.86% (Table 3). They have prismatic structure, evidence of root penetration and iron oxide mottled.

The preliminary results of the sites Amalia 6 and Yowen Ko (Fig. 7) are presented in this paper. The former is a dense scattering of lithic material and bones remains; Yowen Ko, instead, is a site in stratigraphy with four contexts associated with different edaphic layers previously described; only the context 3 is herein discussed.

At Amalia 6 faunal remains predominate over the lithic assemblage; the material is found dispersed in an area of 5,610.6 $m^2$. The lithic set is composed of 144 artifacts (Tables 2 and 4). It should be also noted, that 94.44% of the recovered artifacts have a weight of less than 200 gr, within a range of 10 to 100 gr, being this range the most frequent (56.95%) (Fig. 6).

The archeofaunal assemblage (n = 2018) presents a larger diversity than other sets (Tables 2 and 6); however, this group is mainly composed of individuals of *Lama guanicoe* (guanaco, a South American native camelid). In some of them, marks of anthropic action such as perimeter grooves and cut marks have been recorded; in

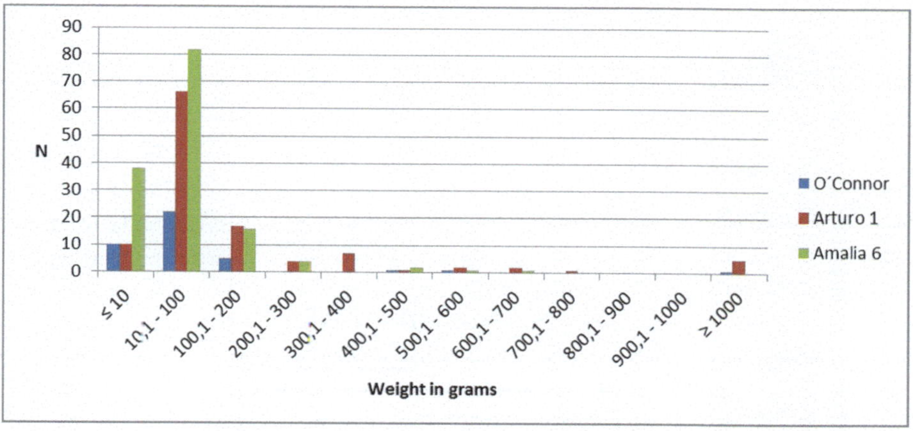


**Fig. 6.** Weight distribution of lithic artifacts from the O'Connor dune and the Arturo 1 and Amalia 6 archeological sites.

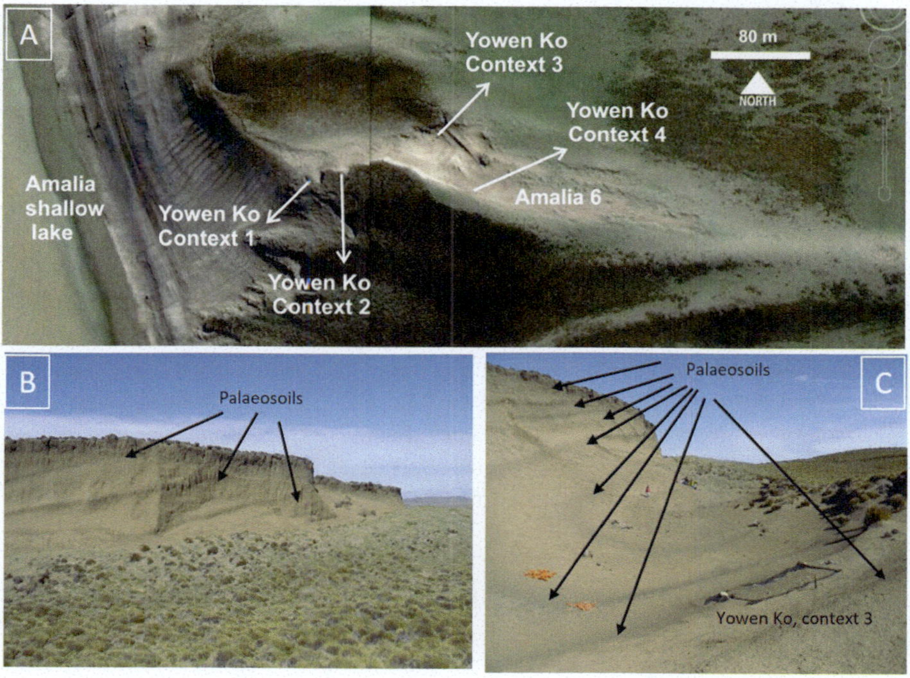


**Fig. 7.** Amalia dune view from Google Earth™ terrain model. A: location of archeological contexts of the Amalia 6 site. Yowen Ko, context 3, is the one analysed in this chapter. B: Amalia dune lateral side exposing eolian units (light brown) and the upper paleosoils (dark lines). C: Amalia dune seen towards west, paleosoils could be seen both in the dune and in the blowout, where Yowen Ko context 3 is placed. The black rectangle shows the *Lama guanicoe* bone concentration.

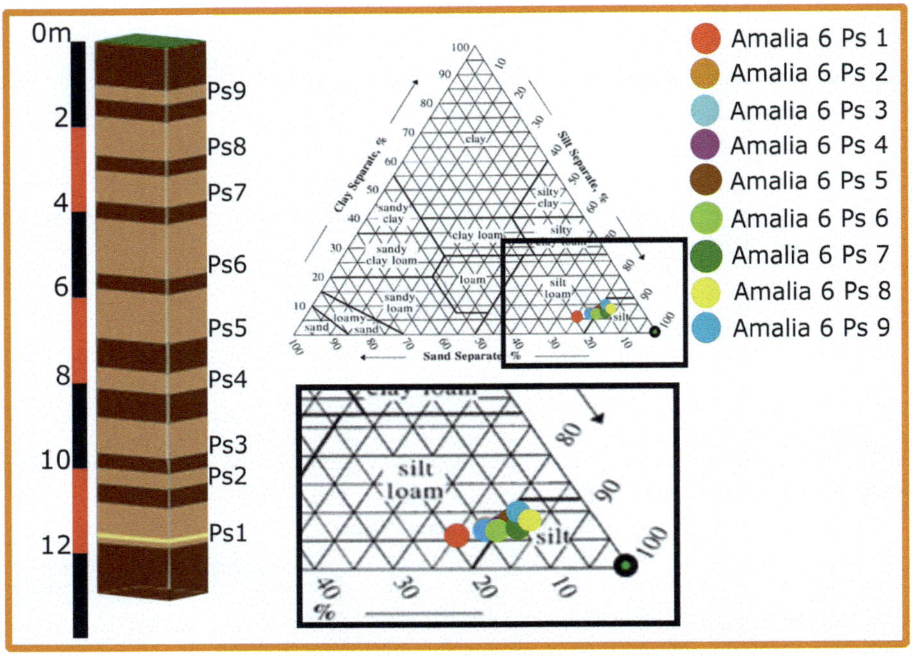


**Fig. 8.** Eolian sedimentary profile of the Amalia dune and grain-size tri-axial diagram. The yellow line indicates the tephra layer position.

**Table 6.** Faunal diversity in the studied sites.

| Taxon | O'Connor | Amalia 6 | Yowen Ko context 3 | Arturo 1 |
|---|---|---|---|---|
| *Lama guanicoe* | - | 1066 | 446 | 134 |
| Artiodactyla | - | 164 | - | - |
| *Ovis aries* | - | 50 | - | 5 |
| Aves undetermined | - | 14 | - | - |
| Canidae | - | 7 | - | - |
| Rodentia | - | 625 | 30 | - |
| *Ctenomys sp.* | - | - | - | 7 |
| Undetermined | - | 92 | 96 | 6 |
| Total (N) | - | **2018** | **572** | **152** |

addition, their association with lithic artifacts allows the defense of the anthropic origin at least for part of them. In a representative sample of the guanaco from in Amalia 6, a 35% of its components expose an intermediate weathering stage (stage 3), 12.45% is included in the highest weathering stage (stage 5), whereas only 3.8% appears unweathered (stage 0) (Fig. 9, Oría et al. 2016a).

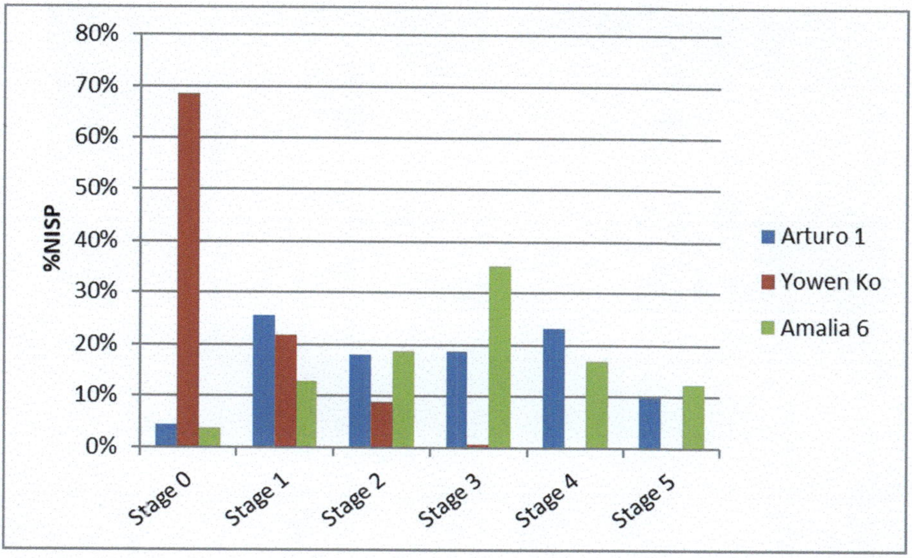


**Fig. 9.** Weathering stages in faunal remains of the Amalia 6, Yowen Ko and Arturo 1 archeological sites.

Yowen Ko context 3 is contained within the Ps6 of the Amalia Dune, visible in the deflation hollow. In an excavated surface of 6 m$^2$ bone remains and one single lithic artifact were recovered (Tables 2 and 4). The species better represented is guanaco (Table 6) with 78% of the entire faunal assemblage. The more abundant skeletal parts are head bones and the appendicular skeleton, whereas the elements of the column and ribs show a lower frequency (Vázquez et al. 2016; Oría and Mari in press). The weathering condition of the group is low, since 90.5% of the components are found within the stages 0 and 1; no cases belonging to the highest stages (4 and 5) are recorded, an indication of good preservation (Fig. 9). In various proportions, there is evidence of action of other natural agents, such as roots (70.8%), rodent (16.7%) and carnivore marks (6.2%). In addition, 54% of the group shows evidence of anthropic action, such as cutting marks, impact blows or perimeter grooves (Vázquez et al. 2016).

In this context, it was also observed the presence of abundant charcoal fragments, rodent bones and some lithic small pieces dispersed in the sediments. The remnant density was estimated in 95.3 findings per square meter and a radiocarbon date on guanaco bone yielded an age of 1020 ± 50 yr. B.P. (LP-3337) (Vázquez et al. 2016; Oría and Mari in press).

The low vertical dispersion and the low weathering stage of the remnants of this context of the Yowen Ko site suggest a high preservation and resolution. Nevertheless, the presence of carnivores and rodents marks advocates about a more complex taphonomic history, which may have disturbed the integrity of the record. Although no evidence of caves was detected, abundant articulated rodent remains suggest that these animals died in situ (Vázquez et al. 2016).

**Fig. 10.** Arturo dune. A: 3-D view from Google Earth™ terrain mode showing the location of the Arturo 1 archeological site. B: Arturo dune panoramic view towards the south, eolian units, paleosoils and rocks are pointed out. C: Arturo 1 site, bones and lithic materials are spread out forming a palimpsest; D: articulated *Lama guanicoe* vertebrae deposited in Ps7.

# 7 The Arturo Dune

The Arturo Dune is located over the top of a cliff exposed towards the NW, in the southern coast of the shallow lake of that name (Figs. 1 and 10). The cliff is composed of a silty-clayey sedimentary sequence underlying a sandy conglomerate of approximately 15 m of total thickness. A sequence of nine eolian deposits is superimposed to these rocks, with eight edaphic levels, adding 19 m more to the thickness of the whole sequence (Fig. 11). In the thicker eolian layer, a 4 cm thick tephra bed is found, pertaining to a Holocene volcanic eruption of Mt. Burney (Coronato et al. 2011a), one of the Chilean volcanoes that compose the Austral Volcanic Zone (Stern 2008). The average thickness of the eolian layers is 1.6 m, varying between 0.43 and 9 m in the case of the deposit that originated the Ps5. The edaphic layers have a mean thickness of

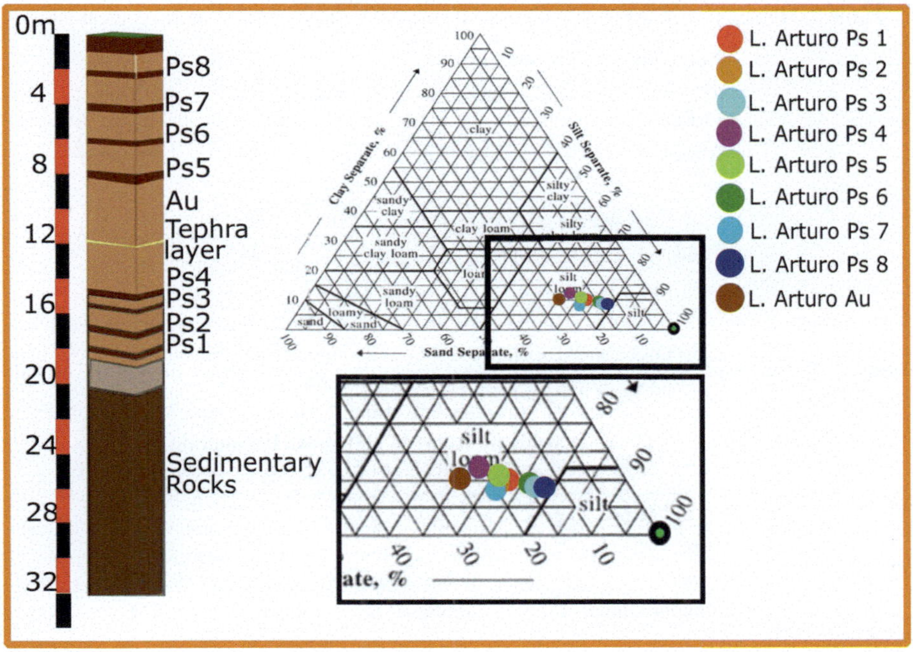


**Fig. 11.** Eolian sedimentary profile of the Arturo dune and grain-size tri-axial diagram. The yellow line indicates the tephra layer position.

0.33 m, ranging from 0.15 (Ps3) and 0.51 (Ps5). The textural composition of the eolian deposits is a silty loam (Fig. 11). The content of organic matter varies between 5.68% in Ps1 to 3.2% in Ps6, being in most of the cases higher than the present soil (3.8%) (Table 3). The color takes lighter tones going from the older edaphic layers to the younger ones, varying from very dark brown (10YR 2/2), to olive (5Y 5/4, Table 3). The edaphic structures identified in the different paleosoils are granular and block structures, vertical cracking, pellets and macroscopic organic matter. In this dune, both the eolian deposits and most of the edaphic layers contain fragments of *Ctenomys sp.*, *Lama guanicoe* and a canid bone, the latter found uniquely in Ps3 (Coronato et al. 2011a).

The Arturo 1 archeological site is located in a blowout eroded in the eastern sector of the dune (Fig. 10). The surface deposit covers approximately 1004 $m^2$ and contains lithic artifacts (flakes, debris, and cores) and faunal remains (Table 2 and 4).

The lithic assemblage contains 148 pieces (Tables 2 and 4), where the flakes represent the highest frequency (n = 100). In addition, cores (n = 29) and nodules (n = 7) complete this assemblage. Only two side-scrapers are formatized artifacts (Oría 2012).

**Table 7.** Frequency of artifacts with altered surfaces in the Arturo 1 archeological site.

| Alteration on the lithic surfaces | | Polishment | | | | Total |
|---|---|---|---|---|---|---|
| | | Null | Light | Moderate | Intense | |
| Abrasion | w1 | 0 | 57 | 35 | 1 | 93 |
| | w2 | 0 | 11 | 6 | 3 | 20 |
| | w3 | 0 | 1 | 1 | 0 | 2 |
| Total | | **0** | **69** | **42** | **4** | **115** |

**Table 8.** Taphonomic agents affecting the faunal assemblage from Arturo 1 site (after Coronato et al. 2011a).

| Taxa | Rodents | Roots | Carnivores | Weathering |
|---|---|---|---|---|
| | % | % | % | % |
| *Lama guanicoe* | 4.57 | 37.25 | 3.26 | 81.04 |
| *Ctenomys* sp. | 0 | 0 | 0 | 1.96 |
| *Ovis aries* | 0.65 | 0.65 | 0.65 | 1.3 |
| Mammalia Undet. | 0 | 0.65 | 0 | 3.92 |

Among the artifacts recovered in this site, a range of sizes from 25 to 140 mm was observed (Fig. 5). The more frequent sizes vary between 35 and 60 mm (57% of the set). In general, all sizes of artifacts present some degree of abrasion and polishing, indicating that the artifact surface alteration does not correlate with their size (Oría 2012).

In general terms, the assemblage has a soft degree of abrasion (w1). 49.5% of the pieces (n = 57) has soft abrasion and slight polishing, whereas 30% (n = 20) of the set has soft abrasion and moderate polishing; moderate abrasion (w2) and light polishing are represented in 9.5% of the artifacts (Table 7). In less than 1.0% a w3 abrasion grade is recorded with light and moderate polishing. Most of the recovered lithic artifacts have a weight smaller or equal to 200 gr, although 43.9% of the sample is grouped in the range of 11 to 100 gr, 11.48% of them weighs less than 101 and 200 gr and 6.75% weighs less than 10 gr and the rest is distributed in weights less than 1,000 gr (Fig. 6).

The archeofaunal assemblage is composed of 152 bone remains, assigned to three taxa, of which guanaco is the best represented (NISP = 134). The remnants that could not be taxonomically identified are indicated as "undetermined" (Table 6). Most of the specimens of guanaco are long bones, generally complete or almost complete (Coronato et al. 2011a); less than 40% correspond to other parts of the skeleton.

Starting from the taphonomic analysis, root marks were identified in 37% of the specimens and, in most of them, they cover practically the entire surface of the bones, suggesting a prolonged burial. Rodents and carnivores have left their marks also but in a smaller proportion (Table 8). In the first case, marks are distributed in all sectors of the bones, suggesting that those with a larger number of marks are those which have been available and exposed for a longer time to the action of these agents (Oría et al. 2016b).

The specimens of *Lama guanicoe* and those undetermined show evidence of weathering, mainly in the low to medium stages. Among guanaco, 80% of them present stages 1 to 4 in the Behrensmeyer (1978) scale, whereas there are few of them within stages 0 to 5 (Oría et al. 2016b). In some of these specimens, the crust surfaces are exfoliated, particularly in those sectors previously disturbed by roots. The remains of sheep (*Ovis aries*, brought from Europe in the XIXth. century) show low degrees of weathering between 0 to 1, and those of *Ctenomys* sp. do not present weathering signals or any other kind of marks (Table 8).

The anthropic marks are very scarce. A cutting mark in a phalanx and some negative flaking in other two long bones were found. The few cases of recorded refitting are more related to taphonomic fractures than to anthropic action (Oría et al. 2016b).

The low representation of the axial skeleton and of smaller elements of the appendicular skeleton, added to the carnivore marks and the recorded weathering profiles may explain the destruction and/or loss of elements of smaller resistance (Oría et al. 2016b).

The characteristics of this group show the co-existence of faunal materials of natural origin with others resulting from human activities, which is sustained on the variability of the weathering degrees as well as on the low frequency of anthropic marks.

## 8 Chronological Frame

$^{14}C$ ages were obtained from bulk organic matter contained in all the edaphic units or paleosoils for chronological purposes. According to Stein (1992), organic matter is a tool to be used in archeological research in the context of dating, landscape reconstruction and site definition. Soil organic matter is considered as the result of accumulation of plants and animals remains with signs of weathering and pedogenetic processes along time. It is assumed that the obtained radiocarbon ages are time markers for soil formation and landscape stability. Although, the geomorphological processes which form the perched dunes stratigraphy involve erosion and re-deposition of sediments which could also include re-deposition of organic matter coming from ancient edaphic levels. In this case, organic matter must be considered as transported organic matter and could trend to erroneous chronological interpretations.

The dates obtained in the different edaphic levels of the three dunes indicate that these landforms were developed from the Pleistocene-Holocene boundary until recent times in northern Tierra del Fuego (Fig. 12 and Table 9). There is neither similarity in the number of edaphic levels among the three dunes nor exact synchrony between their ages. However, the obtained dates allow interpreting that regional processes permitted the soil formation and their posterior burial due to eolian action.

Two edaphic levels assigned to the Late Pleistocene are recorded, the Ps1 of the O'Connor dune and the Ps1 of the Arturo dune (Table 9, Fig. 12); these paleosols developed between 15,300 and 12,800 cal. yr. B.P. However, a second date obtained at the Ps1 of the O'Connor dune yielded an age that placed it at the Pleistocene-Holocene boundary (Table 9, Fig. 12).

During the Early Holocene the Ps2 developed in the Arturo dune (Table 9, Fig. 12). In the Middle Holocene, four edaphic levels are recorded in the Amalia dune (Ps1 to 4) and other four ones in the Arturo dune (Ps 3 to 6). For the Late Holocene,

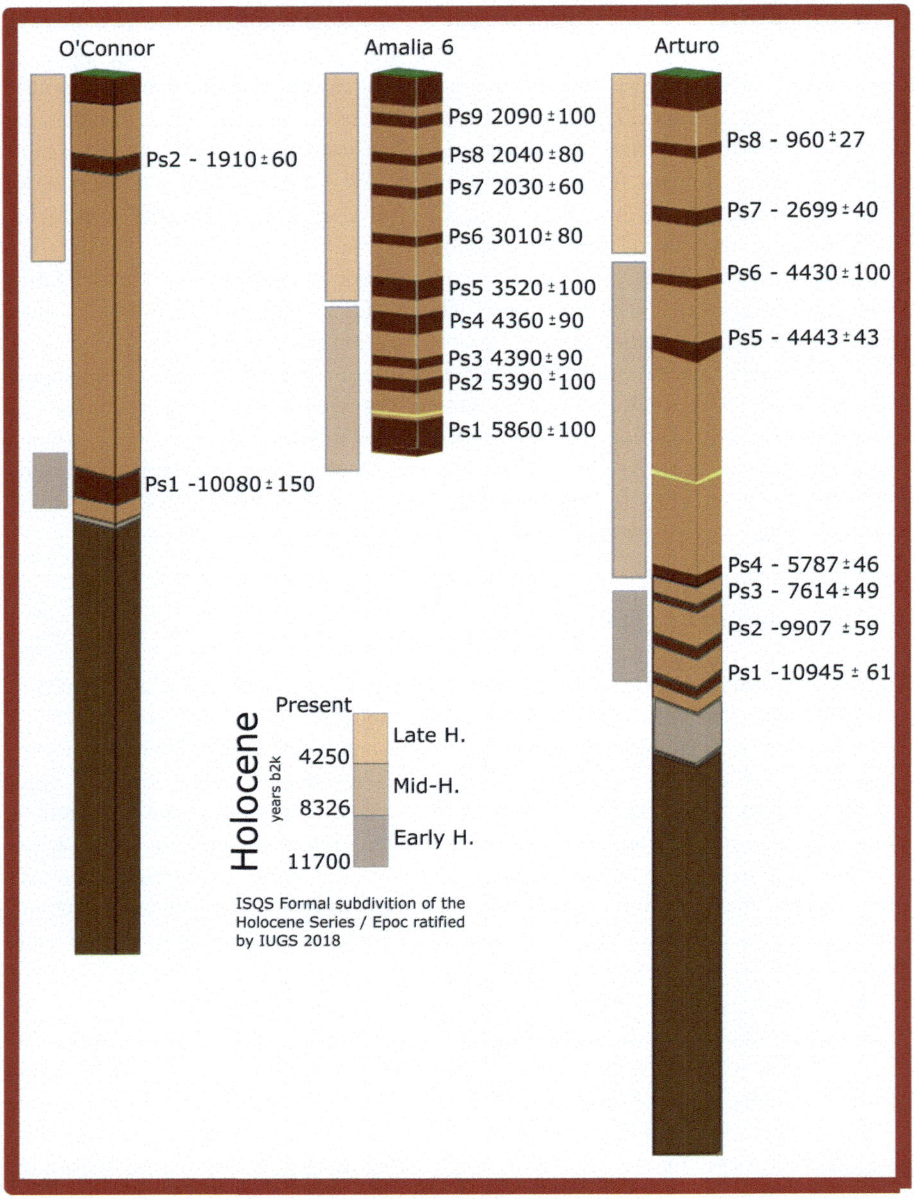


**Fig. 12.** Sedimentary profiles and radiocarbon dates ($^{14}$C yr B.P.) of paleosoils in O'Connor, Amalia and Arturo dunes.

one edaphic level has been recorded at the O'Connor dune (Ps2), five in the Amalia dune (Ps5 to 9) and two edaphic levels in the Arturo dune (Ps7 and 8) (Table 9, Fig. 12).

**Table 9.** Radiocarbon dates for paleosoils (Ps) levels and eolian units (AU) in O'Connor, Amalia and Arturo dunes.

| Lab. code | Level | Material | 14C age BP (1 σ) | Date cal BP (68.2% prob.) | Date cal BP (95.4% prob.) |
|---|---|---|---|---|---|
| O'Connor | | | | | |
| LP-3228 | Ps2 | Organic matter | 1910 ± 60 | 1734–1875 | 1698–1933 |
| LP-3301 | Ps1 | Organic matter | 10080 ± 150 | 11280–11773 | 11183–12073 |
| Beta-346886 | Ps 1 | Organic matter | 12,960 ± 50[a] | 15304–15546 | 15221–15672 |
| Amalia | | | | | |
| LP-3410 | Ps 9 | Organic matter | 2090 ± 100 | 1896–2149 | 1818–2310 |
| LP-3404 | Ps 8 | Organic matter | 2040 ± 80 | 1869–2051 | 1741–2154 |
| LP-3464 | Ps 7 | Organic matter | 2030 ± 60 | 1877–2010 | 1817–2095 |
| LP-3317 | Ps 6 | Lama bone | 1020 ± 50[b] | 804–930 | 773–961 |
| LP-3351 | Ps 6 | Organic matter | 3010 ± 80 | 2999–3240 | 2924–3359 |
| LP-3460 | Ps 5 | Organic matter | 3520 ± 100 | 3630–3874 | 3478–3987 |
| LP-3348 | Ps 4 | Organic matter | 4360 ± 90 | 4816–5046 | 4613–5283 |
| LP-3246 | Ps 3 | Organic matter | 4390 ± 90 | 4828–5053 | 4807–5296 |
| LP-3518 | AU 3 | Lama bone | 3490 ± 90[b] | 3595–3833 | 3477–3962 |
| LP-3264 | Ps 2 | Organic matter | 5390 ± 100 | 6002–6268 | 5914–6309 |
| LP-3369 | Ps1 | Lama bone | 4090 ± 90[b] | 4420–4800 | 4295–4828 |
| LP-3240 | Ps 1 | Organic matter | 5860 ± 100 | 6497–6734 | 6404–6863 |
| Arturo | | | | | |
| AA95362 | Ps8-top | Organic matter | 960 ± 27 | 773–904 | 863–904 |
| AA95363 | Ps8 base | Organic matter | 3556 ± 27 | 3721–3839 | 3694–3883 |
| AA95364 | Ps7-top | Organic matter | 2229 ± 39 | 2150–2305 | 2091–2325 |

*(continued)*

**Table 9.** (*continued*)

| Lab. code | Level | Material | 14C age BP (1 σ) | Date cal BP (68.2% prob.) | Date cal BP (95.4% prob.) |
|---|---|---|---|---|---|
| AA95635 | Ps7-base | Organic matter | 2699 ± 40 | 2744–2838 | 2731–2855 |
| AA 108180 | Ps6-top | Organic matter | 4424 ± 26 | 4872–4972 | 4861–5040 |
| LP-3368 | Ps6-base | Organic matter | 4430 ± 100 | 4849–5268 | 4815–5313 |
| AA 89590 | Ps5-top | Lama bone | 4871 ± 59[c] | 5474–5614 | 5328–5663 |
| AA95367 | Ps5-base | Organic matter | 4443 ± 43 | 4876–5039 | 4854–5276 |
| AA85452 | Ps4 | Lama bone | 5800 ± 64[c] | 6483–6640 | 6406–6679 |
| AA95368 | Ps4-base | Organic matter | 5787 ± 46 | 6486–6631 | 6434–6659 |
| UCIAMS101824 | Ps4-base | Organic matter | 5690 ± 20[d] | 6402–6466 | 6341–6494 |
| UCIAMS101822 | Au4 | Organic matter | 5710 ± 15[d] | 6411–6469 | 6398–6501 |
| UCIAMS101823 | Ps3-middle to top | Organic matter | 6640 ± 20[d] | 7458–7556 | 7433–7564 |
| AA89592 | Ps3 | Canidae bone | 6333 ± 67[d] | 7156–7316 | 7006–7334 |
| UCIAMS101825 | Ps3 middle to base | Organic matter | 6230 ± 20[d] | 7023–7120 | 6987–7173 |
| AA95369 | Ps3-middle to base | Organic matter | 7614 ± 49 | 8346–8416 | 8296–8459 |
| AA95370 | Ps2-base | Organic matter | 9907 ± 59 | 11208–11324 | 11176–11409 |
| AA89589 | Ps1 | charcoal | 9941 ± 59[c] | 11230–11391 | 11202–11499 |
| AA95371 | Ps1-top to middle | Organic matter | 10945 ± 61 | 12712–12810 | 12695–12941 |

LP = Radiocarbon Laboratory (LATYR) of the Centro de Investigaciones Geológicas-CONICET-UNLP, Argentina; Beta = Beta Analytic, Miami, USA; AA = AMS Facility Laboratory of the University of Arizona, USA; UCIAMS = University of California Irvine AMS Facility, U.S.A.

[a]Oría et al. 2014; [b]Oría y Mari, en prensa; [c]Coronato et al. 2011a; [d]Riches 2018.

**Table 10.** Accumulation rate for the Amalia dune, based on radiocarbon ages obtained for paleosoils 1 to 8. Data for Ps8 and Ps4 -in boxes- is doubtful due to the similarity on the radiocarbon dates between consecutives paleosoils. Yellow comprises Middle Holocene ages while light yellow comprises the late Holocene ones.

| Thickness difference (cm) | Sedimentary unit | 14C-date | 14C date difference | Accumulation rate ($cmy^{-1}$) |
|---|---|---|---|---|
| 134 | Present soil | | | |
| | Aeolian 10 | | | |
| 168 | Ps 9 | 2090 | -10 | |
| | Aeolian 9 | | | |
| 99 | Ps 8 | 2040 | 10 | 9,9 |
| | Aeolian 8 | | | |
| 159 | Ps 7 | 2030 | 980 | 0,162244898 |
| | Aeolian 7 | | | |
| 174 | Ps6 | 3010 | 510 | 0,341176471 |
| | Aeolian 6 | | | |
| 57 | Ps 5 | 3520 | 840 | 0,067857143 |
| | Aeolian 5 | | | |
| 154 | Ps 4 | 4360 | 30 | 5,133333333 |
| | Aeolian 4 | | | |
| 69 | PS 3 | 4390 | 1000 | 0,069 |
| | Aeolian 3 | | | |
| 196 | Ps2 | 5390 | 470 | 0,417021277 |
| | Aeolian 2 | | | |
| | Ps 1 | 5860 | | |

The O'Connor dune, exposed to the SW winds is the one that preserved the smaller record of edaphic levels. Taking into consideration the radiocarbon dating, the basal level is one of the more ancient of the region, ca. 13 and 10 $^{14}$C ky. B.P. (Table 9), exposing that the eolian deposition and soil formation processes took place during Late Glacial times.

The O'Connor and Arturo dunes were studied on its up-wind slope, that is, the one in which the accretion/erosion process takes place from the basal cliff and/or from its own deposits. For these reasons, it would reasonable to find that the sedimentary record would be controlled by both (i) erosion of the lower strata, and (ii) re-deposition within the upper levels, or (iii) even cannibalization of their own deposits. This could explain some inconsistencies in the radiocarbon dating obtained in organic matter from some of

**Table 11.** Accumulation rate for the Arturo dune, based on radiocarbon ages obtained for paleosoils 1 to 8. Data for Ps8 -in box- is doubtful due to the similarity of its radiocarbon date with the underlying paleosoil. Brown color comprises Late Pleistocene-Early Holocene ages, yellow comprises Middle Holocene ages and light yellow does so for the Late Holocene ones.

| Thickness difference (cm) | Sedimentary unit | 14C-date | 14C date difference | Accumulation rate (cmy$^{-1}$) |
|---|---|---|---|---|
| 222 | Present soil | | | |
| | Aeolian 9 | | | |
| 202 | Ps 8 | 960 | 1269 | 0,159180457 |
| | Aeolian 8 | | | |
| 232 | Ps 7 | 2229 | 2195 | 0,105694761 |
| | Aeolian 7 | | | |
| 191 | Ps6 | 4424 | 16 | 11,9375 |
| | Aeolian 6 | | | |
| 752 | Ps 5 | 4440 | 1347 | 0,558277654 |
| | Aeolian 5 | | | |
| 84 | Ps 4 | 5787 | 853 | 0,098475967 |
| | Aeolian 4 | | | |
| 131 | PS 3 | 6640 | 3267 | 0,040097949 |
| | Aeolian 3 | | | |
| 163 | Ps2 | 9907 | 1038 | 0,157032755 |
| | Aeolian 2 | | | |
| | Ps 1 | 10945 | | |

the edaphic layers. An example of this is the radiocarbon dating of the Ps3 (middle to base), estimated in 7614 ± 49 $^{14}$C yr. B.P. (AA95369, Table 9), which provides an older age than the ages yielded by organic matter of this paleosoil, between ca. 6,600–6,200 $^{14}$C yr. B.P. (Table 9). Likewise, the date on bone obtained in this same paleosoil (Table 9) is consistent with the cited radiocarbon ages. Another case in this dune is Ps8, which yielded a basal age of 3556 ± 27 $^{14}$C yr. B.P. (AA95363), that would provide an older age than that of the underlying Ps7 (Table 9). In the Ps7 and Ps5 of the Amalia dune were also observed problems of old-age of the sedimentary sequence according to the obtained ages of 5,430 ± 90 $^{14}$C yr. BP (LP3386) for Ps7 and 5930 ± 110 $^{14}$C yr. B.P. (LP 3379) for Ps5, which were discarded when compared to the results obtained and presented in Table 9.

The radiocarbon dating of these faunal remains has provided certain ages unlike those obtained in total (bulk) organic matter. In the case of the Arturo dune, the date of 434 ± 43 $^{14}$C yr. B.P. on a *Lama guanicoe* bone remain (Fig. 10D), coming from Ps7 and the date of 6333 ± 67 $^{14}$C yr. B.P. (Table 9) on a Canidae remain from Ps3 (Coronato et al. 2011a), rejuvenated the chrono-stratigraphic sequence; now this is better adjusted with their respective dating on organic matter as indicated in Table 9. Besides, the date on *Lama guanicoe* in Ps5 (4871 ± 59 $^{14}$C yr. B.P., Table 9) appears too old with respect to the dating sequence performed in bulk organic matter. Contrarily, the date on the Ps4 (5800 ± 64 $^{14}$C yr. B.P., Table 9), also published by Coronato et al. (2011a) on guanaco bone is statistically equal to the other dates from the base of Ps4, obtained on bulk organic matter (Table 9). These differences could be related to the incorporation of humus substances coming from the soil that may have not been appropriately removed during pre-treatment of the dated bone during the collagen extraction.

At Amalia dune, several cases of inconsistency are present between dates on bulk organic matter and those on guanaco bones, as recorded in archeological contexts. In Ps1, the date obtained in bone with respect to the one obtained on organic matter for this unit, could be interpreted as too young in ca. 1,800 radiocarbon years. In the eolian unit 3, underlying Ps3, the date on bone is 900 radiocarbon years younger than that obtained on bulk organic matter, and in Ps6, the date yielded by bone is 2000 radiocarbon years younger than that one on organic matter, and they are even around 1000 radiocarbon years younger than the overlying paleosoils, Ps7, Ps8 and Ps9 (Table 9).

## 9 Accumulation Rate

The achievement of a complete chronological sequence for the Amalia and Arturo dunes allow the estimation of sedimentary accumulation rates in these eolian landforms for the Holocene (Tables 10 and 11). Given the erosion processes that take place on the up-wind face of these dunes, it should be expected that greater erosion of the cliffs would provide a larger material availability for its deposition in the dune; therefore, a larger accumulation rate is interpreted as a greater eolian activity during arid periods. The upper level of each eolian deposition unit would have been under the influence of edaphic processes during periods of environmental improvement and stabilization.

Due to the existence of erosion processes in the formation of these deposits and the problems of record and dating of the organic matter coming from them, the calculation of the accumulation rate is impacted for the reliability of the radiocarbon dates obtained for the chronostratigraphic sequence. Values of accumulation rate on those stratigraphic units in which the radiocarbon dates showed inconsistencies were not calculated. In those cases of dates with small difference between the dates, this calculation is not precise and it is not reliable. For this reason, the proposed accumulation rates herein presented should be a matter of permanent revision.

Table 10 shows the values of accumulation rates obtained in the lateral section of the Amalia dune. The maximum obtained value is 5.13 cm yr$^{-1}$ for the eolian unit 4 and the Ps4 (Table 10), a value that results from the deposition of 159 cm of material towards the end of the Middle Holocene. Notwithstanding, since that Ps3 and Ps4 are statistically identical, that is, they would be synchronous, and thus, this accumulation rate value is not fully reliable. The same problem occurs for the accumulation rate value obtained for the Ps8, of 9.9 cm $yr^{-1}$, considering that the Ps7 and Ps8 are also statistically undifferentiated. The rest of the accumulation rates obtained oscillate between 0.41 (Ps2) and 0.067 (Ps7) cm $yr^{-1}$ during the Middle and Late Holocene, respectively, being much smaller than the maximum value.

Table 11 presents the values obtained for the Arturo dune. The maximum value reached is 11.93 cm $yr^{-1}$, which represents the accumulation rate for the eolian levels and the Ps6, formed during the Middle Holocene. Such very high value corresponds to the very short difference in time existing between the obtained dates for the cited paleosoil and the underlying one, but it also results from the thickness of the deposit (191 cm). The remaining values stand between 0.04 and 0.5 cm $yr^{-1}$. The latter corresponds to the accumulation rate of the eolian unit and the Ps5, the thickest one of the whole sequence (752 cm), deposited during 1347 radiocarbon years, almost at the end of the Middle Holocene.

When comparing the sedimentary accumulation rates between both dunes (Tables 10 and 11), it becomes clear that the larger rates are achieved between 4360 $^{14}C$ yr. B.P. (Amalia dune) and 4424 $^{14}C$ yr. B.P. (Arturo dune), that is, in a period shorter than one century. It should be noted that the scarce time difference between consecutive sedimentary units is recorded in both stratigraphic sequences, which would suggest that the radiocarbon dates are not inconsistent. Contrarily, it may be inferred that in those hundred years the environmental conditions were very favorable for sediment deposition and dune accretion. This is interpreted as the existence of environmental arid conditions with very intense winds, interrupted by short stability cycles that allowed the development of edaphic levels.

## 10 Discussion

The perched dunes do not only expose the wind modelling capacity in a semi-arid region of cold-temperate climate; they also play the role of information reservoir. Unlikely the dunes of this kind described in their type localities in North America (Marsh and Marsh 1987), one of the processes that originate the dunes of the Fuegian steppe is deflation of the pan surfaces. This provides a greater richness to the eolian sedimentary record, since these records may include remains of lacustrine organic matter (Coronato et al. 2011b), thus the sedimentary record of the dunes may be comparable with that of the lacustrine cores. The repeated alternation of eolian levels with a certain degree of soil formation indicates the occurrence of periods with intense eolian activity, and therefore interpreted as arid periods. These arid periods were followed by moments in which soil formation process were generated at the surface of the eolian deposits. These paleosoils are interpreted as humid periods that allowed the stabilization of the sediments and their pedogenesis. The edaphic levels are understood as soils with poor horizon development interpreted, in which only the A and C horizons may be differentiated, the latter generated from the underlying eolian deposit as parental material. They are rich in organic matter, in some cases with values that duplicate the present soil values. Based upon the study of the Arturo dune it is proposed that these cycles would have developed along the Holocene, partly due to the existence of more humid climatic periods (Coronato et al. 2011a), later exposed by the occurrence of paramagnetic mineralization in the Ps2, Ps4, Ps5 and Ps7 (Orgeira et al. 2012). Taking into consideration the radiocarbon dates of those edaphic levels, it was suggested the occurrence of climatic phases with greater precipitation or environmental moisture than today during the Early and Middle Holocene (Orgeira et al. 2012).

From a palynological point of view, Musotto et al. (2018) determined the predominance of shrubby communities (Asteraceae subfam. Asteroideae) with scarce development of pastures (Poaceae) and herbs (Acaena, Caryophyllaceae, Brassicaceae, Rubiaceae) in the 6, 7 and 8 paleosoils. These conditions differ from the present ones, in which *Festuca gracillima* predominates as a species representative of the steppe grasslands that have developed over the dunes.

When the favorable conditions for soil formation recorded in the Arturo dune are considered in regional terms, the genesis of the Ps1 of the O'Connor dune and the Ps1, 3 and 4 of the Amalia dune are justified (Table 9, Fig. 12). The similarity between the radiocarbon dates of the Ps3 and 4 of the Arturo dune and the Ps5 and 6 of the Amalia dune permit to infer the occurrence of wetter than today regional conditions until the end of the Middle Holocene, around 4.3 and 4.4 $^{14}$C kyr. B.P. (Table 9, Fig. 12). The upper edaphic levels of the three dunes were formed during the Late Holocene, although their radiocarbon dates do not show synchrony among them. The greater similitude is recorded between the Ps2 of the O'Connor dune and the Ps7, 8 and 9 of the Amalia dune, being the Ps8 of the Arturo dune the most recent one (Table 9).

Fernández et al. (2018) interpreted the Ps2 of the Arturo dune as the result of a wetter cycle when its development period is considered during the age range between 10,500–8200 cal yr. B.P., obtained from the core taken at the shallow lake of that name. For this period, a high frequency of *Thalassiosira patagonica* was recorded, a diatom species that indicates wetter and/or fresh water environments. This chronology is younger than new dates obtained for this edaphic level (Table 9). Although this range of calibrated ages is not recorded in the chrono-stratigraphic sequence presented in this chapter, it is herein maintained the possibility that such wetter cycle had been responsible of the soil formation on the eolian deposits that covered the Ps2 and formed the Ps3 in this dune.

Besides, the multi-proxy analysis of a lacustrine core in the Carmen shallow lake (Fig. 1), located 5 km to the north of the Arturo dune, indicates the alternation of wetter and drier events in the Holocene climate of the Fuegian steppe (Borromei et al. 2018). Four wetter stages were determined by these authors based upon the expansion of herbs and aquatic plants, and the diminution of halophyte species and of ostracods that indicate salty waters. Some of these wetter periods fit quite well with the range of radiocarbon dates assigned to certain edaphic levels studied. The wetter period between 3600 and 3000 cal. yr. B.P., identified by Borromei et al. (2018), included the time of formation of the Ps5 and 6 in the Amalia dune (Table 9), which corroborates what has been sustained from the environmental magnetism studies by Orgeira et al. (2012). The determination of another wetter period between 2100 and 1900 cal. yr. B.P. (Borromei et al. 2018) correlates with the Ps9 of the Amalia dune and the Ps7 of the Arturo dune (Table 9). Besides, the lacustrine wetter period observed between 100 and 700 cal. yr. B.P. provides a regional framework to the formation of the Ps8 of the Arturo dune.

These dates could confirm the relationship between lacustrine dynamics and edaphization of the perched dunes during humid periods. The erosion eolian activity would have diminished on the cliffs and the deflation of the lacustrine dry beds. Therefore, the dune accretion would have permitted the development of edaphic levels as consequence of environmental stability. These conditions may suggest that the soil formation periods had also been favorable to human occupation.

According to Tuhkanen (1992), the SW winds have a larger frequency and intensity due to the wind provenance from the southern flank of the South Pacific Ocean Anticyclone, attracted by the lower pressure system of the Polar Front. If these conditions occurred in the same manner along the Holocene, it should be expected that the dunes, whose up-wind faces were exposed to the SW quadrant, would have experienced more intense erosion processes than the dunes formed by winds coming from other quadrants. This would explain the feeble paleo-edaphic record of the O'Connor dune, in spite of the thickness of such sedimentary deposit.

The sedimentary accumulation rate recorded in the Amalia and Arturo dunes is highly variable along the entire Holocene. However, its noted increment between ca. 4.3 and 4.4 $^{14}$C kyr. B.P. is recorded in both dunes, forming the thickest sedimentary units. It is not possible to establish a detailed correlation between the values herein obtained and those presented for the Arturo dune by Coronato et al. (2011a), due to the difference in the considered stratigraphic units, as it was mentioned in the chronological frame paragraph of this paper. Coronato et al. (2011a) postulated a noted increment in sedimentation between 6.5 and 5.5 $^{14}$C kyr. B.P., whereas in the chrono-stratigraphic sequence presented here, the largest sedimentary accumulation rate is recorded approximately 1,000 years later. Nevertheless, the Middle Holocene is, in both cases, the period in which the highest sedimentary accumulation rate took place, and therefore, the largest accretion of these eolian landforms.

The presence of a thin, whitish tephra layer may be observed in the Amalia and Arturo dunes. This layer is found within the eolian level 2 in the first case and in the eolian level 5 in the second one (Fig. 12). The geochemical analysis of the Arturo dune tephra, presented by Coronato et al. (2011a), confirms its provenance from Mt. Burney, a volcano located 500 km to the NW. It has been recognized that this volcano has had volcanic activity of the Plinian type during the Early Holocene at 8425 ± 500 $^{14}$C yr. B.P. (Stern et al. 2008) and between 9009 and 9175 cal. yr. B.P., according to Kilian et al. (2003). Another eruption of the same kind has been identified for the Middle to Late Holocene at 3830 ± 390 $^{14}$C yr. B.P. after Stern et al. (2008) or in 4250 ± 120 cal. yr. B.P. after Kilian et al. (2003). Both contributions recognized that the cinder plume would have reached the western coast of the Isla Grande de Tierra del Fuego, establishing an isopach > 1 cm only for the Chilean sector of this island.

In the Amalia dune, the tephra layer overlies the Ps1, dated in 5860 ± 100 $^{14}$C yr. B.P. or in the range of 6497–6734 cal. yr. B.P. (1 $\sigma$) and it underlies the Ps2, the latter dated in 5390 ± 100 $^{14}$C yr. B.P. or in the range of 6002–6268 cal. yr. B.P. (1 $\sigma$) (Table 9, Fig. 12). This discrepancy exposes the possibility that this tephra does not represent the Mt. Burney Middle Holocene eruption or, alternatively, the ages of the paleosoils are too old, due to re-deposition of organic matter along the formation process of the perched dune.

In the Arturo dune, the tephra layer is found in the eolian unit 5, the thickest one in the stratigraphic sequence (Fig. 12). As indicated in Table 9, the formation of the Ps4, dated in 5787 ± 46 $^{14}$C yr. B.P. or in the range of 6486–6631 cal. yr. B.P. (1 σ), predated the deposition of the tephra, whereas the Ps5, dated in 4443 ± 43 $^{14}$C yr. B. P. or in the range of 4876–5039 cal. yr. B.P. (1 σ) which post-dates it. Considering the interval between the tephra deposition and that of the eolian sediments which later were edaphized, the dates herein presented would be in agreement with those of the Mt. Burney eruption during the Middle Holocene, as proposed by Kilian et al. (2003) and Stern et al. (2008). The new radiocarbon date, obtained on organic matter for the Ps4 (Table 9), is better adjusted to the proposed age by the cited authors than the radiocarbon date published by Coronato et al. (2011a), which was obtained in bone, and the reason for which these latter authors suggested the discordant chronology with the dates presented in the literature.

Assuming that the tephra layers identified in the Amalia and Arturo dunes effectively belong to the Middle to Late Holocene eruption of Mt. Burney, the dispersal of the pyroclastic plume in Tierra del Fuego widens approximately 100 km eastwards compared with the estimations of Kilian et al. (2003). Likewise, the thickness of the tephra in both dunes is larger than the values presented by Stern et al. (2008) for the Chilean sector of Tierra de Fuego.

In other landforms of the Fuegian steppe, the development of a soil with regional continuity was observed. In three sedimentary sequences of eolian origin studied in the northeastern coast of Tierra del Fuego (Bahía San Sebastián) a buried soil is noted which overlies eolian layers bearing archeological materials (bones, lithics, shells, charcoal) dated between ca. 1600 and 900 $^{14}$C yr. B.P. (Favier Dubois 2007). The development of this paleosoil of the Mollisol type, dated in ca. 1000 $^{14}$C yr. B.P., suggests the interruption of the arid climate regional conditions and an improvement of the environmental conditions during the end of the Late Holocene (Favier Dubois 2007). This soil could be associated to the Warm Medieval Period (WMP). Besides, in another locality of the same area, two archeological contexts assigned to the Middle and early Late Holocene were recorded at the base of a sandy eolian deposit. This is the case of the Cerro Bandurrias site, a level of shells and cultural materials (ca. 5700 $^{14}$C yr. B.P.; Favier Dubois and Borrero 2005) and guanaco remains at Cabeza de León 4 (ca. 3700 $^{14}$C yr. B.P.; Martín and Borella 1999). In these sites, the above mentioned paleosoil was recorded at 2.5 m and 1.0 m, over each group of materials, respectively (Favier Dubois 2007). The site Cabeza de León 4, located in a talus deposit of the hill of such name, contains bones buried by because of taphonomic rain that present very low weathering indicating a very fast burial.

The dynamics of reactivation of the erosion-deposition processes generates evidence of about the archeological materials that permit to interpret the incidence of agents such as the wind in the processes of formation of the archeological contexts, both in the surface and in stratigraphy. Some taphonomic variables such as position, size and weight, or the completeness of the bones have been useful for establish whether the bones buried in the paleosoils are in situ or are intrusive, resulting in these case, reliable indicators of the ages older and younger than the evolution of these soils (Favier Dubois 2007). Besides, the dynamics of the dunes offers the possibility of a fast burial of the materials, thus allowing a good preservation of them, being either organic or inorganic (Borrazzo and Borrero 2016). Inversely, these dynamics may act over the remains, provoking displacements, loss, transport and alteration in their surfaces (Oría et al. 2014; Montes et al. 2016; Borrazzo and Borrero 2016) or leaving exposed materials of diverse provenance and/or chronologies. In these areas, the palimpsests are frequent, due to the alternation of erosion cycles and sediment deposition that imply burial, exposition and partial re-burial of the remains (Oría et al. 2016b). In terms of opportunities, this same dynamics becomes an advantage for the detection of isolated materials, concentrations or archeological sites (Santiago and Oría 2007; Santiago 2013; Santiago and Salemme 2010, 2016; Cruz and Borrazzo 2016).

During the periods of aridity and dominance of eolian action, alternating processes of deflation and sediment accumulation are produced. In the first case, archeological remains are exposed whereas in the second one, these materials are buried or covered again. This alternation generates alteration on the surfaces of the lithic and bony materials. The analysis of these effects explains the degree of alteration of the artifacts; if the degree of abrasion and/or polishing that they present, and the weathering, root, carnivore or rodent marks on the bones, histories of recurrent deposition and exposition may be recognized. Besides, starting from the presence/absence of materials of different sizes and weights, and of the conditions of their surfaces, the degree of integrity and resolution of the contexts (Oría et al. 2014, 2016b). In the set recovered at the O'Connor dune, the degrees of abrasion and polishing recorded in their surfaces indicate that 65% of them have soft abrasion and polishing from slight to moderate. This suggest a very good preservation of this set of isolated findings. The stratigraphic provenance of these artifacts is unknown, since materials of this kind were not recorded either in the exposed profile of the dune nor buried in their paleosoils. However, along the coast and the bottom of the shallow lake of such name erratic boulders and gravels of varied lithology exist, which may have been the primary matter for the manufacturing of the artifacts. Two nodules and seven cores plus the high frequency of flakes could suggest a local use of these resources.

Among the lithic materials, the absence of artifacts smaller than 25 mm in the Arturo 1 site may be assigned to the wind capacity as transport agent of small or light particles, considering the gasp intensity in the region, as it has been observed in other contexts in the area, such as the Amalia 4, Amalia 5 or Tres Marías sites (Oría et al. 2011, 2014, 2016b). In relation to the weight of the artifacts in the O'Connor and Amalia 6 sites, more than 80% weigh less than 100 gr, whereas those of less than 10 gr are represented between 19 and 26%, respectively. In the case of the Arturo 1 site, the artifacts with weight less than 100 gr are only 66%, whereas those of less than 10 gr represent only 8.7%. In this last context, the loss of very light elements is more significant, particularly if the Arturo 1 site is considered a workshop, in agreement with the present artifact categories and the high index of refitting (sensu Coronato et al. 2011a, b). For this reason, it would be expected a higher frequency of light debitage, which is absent. In the same Arturo locality, in the Arturo 4 site, more than 60% of the artifacts exceeds 100 gr (Oría et al. 2018). Since it is an area of supply materials, where the first stages of knapping, it would not be expected to find a microflake record.

In the case of the faunal sets, the analysis of the taphonomic aspects is a key issue to understand its integrity. The set of the Arturo 1 site, as well as Amalia 6 and Yowen Ko context 3, show similarities and differences. In all of them, *Lama guanicoe* is the better represented species, and, although in a different degree, it presents evidence of anthropic action. The presence of root marks, differential color and diverse degrees of weathering in some specimens indicate burial and posterior exposition in the three sites. For some bones with a lower weathering degree, as it happens in specimens of the Amalia 6 site, a relatively recent exposition is interpreted (Oría et al. 2016a). The presence of sheep remains, the introduced species *Ovis* sp. which is presently risen in the area, and the variability in the weathering of the different specimens, expose not only the present taphonomic rain but also the generation of palimpsests where materials of natural and cultural origin are mixed (Oría et al. 2016a). They are generated in different times and it results impossible to establish such diachronism.

Different is the case of the Yowen Ko context 3 and two other stratigraphic contexts not described in this paper, that are assumed as the result of rapid burial, given the good conservation of the remains. These contexts were considered as different moments of human occupation in the dune (Oría and Mari in press) and it is from this information that the Amalia dune is presented as a well-preserved, context reservoir, although the available results are still of preliminary nature. Materials coming from two contexts of stratigraphy, close to Yowen Ko context 3, have been dated in the Middle Holocene (Oría and Mari in press). The first one was detected below the tephra layer and dated in 4090 ± 90 $^{14}C$ yr. B.P. (LP 3369, Table 9) and the other deposited between the edaphic levels 2 and 3 and dated in 3960 ± 100 $^{14}C$ yr. B.P. (LP 3381, Table 9). However, in the Arturo 1 site, various edaphic levels as well as eolian layers contain bone remains, but lithic materials were not identified yet in any of them. The formation of the blowout in the dune generated a palimpsest, then it is difficult to identify the provenance sector of the recovered materials (Coronato et al. 2011a, Oría et al. 2016b).

## 11 Final Comments

The perched dunes are landforms of specific localization, composed of materials provided from short distance-cliffs and pans bottoms- and possibly, re-deposited from its own lower levels disturbed by deflation, run off, or mass movements. The processes of auto-erosion or cannibalization that contribute to the accretion of this type of dunes should be considered in front of possible inconsistencies in the dating of the chrono-stratigraphic sequences.

The dunes of this type, present in most of the seasonal, shallow lakes of the Fuegian steppe, confirm the occurrence of arid and wet periods since the end of the Late Pleistocene-Early Holocene until the historic times in which the eolian processes and the pedogenesis dominated, respectively.

The calculation of the accumulation rates offered a high variability along the Holocene. However, the Middle Holocene is highlighted as the period in which the largest eolian accumulation occurred, specifically along the century that took place between 4.4 and 4.3 $^{14}C$ kyr. B.P. It is then inferred the existence of favorable conditions for the eolian erosion and deposition induced by the existence of strong winds, but that culminated with periods of stability, favorable to pedogenesis.

Some wetter cycles determined by the development of edaphic levels are in agreement with the results obtained by other multi-proxy studies, developed in paleosoils or lacustrine cores of the region. Thus, the radiocarbon data reveal wetter periods well defined between 3600–3000, 2100–1900 and 700–100 calibrated years, that is, from the Middle Holocene to historical times. This includes the Ps2 of the O'Connor dune and the Ps5 to 9 in the Amalia dune, among which Ps6 is bearing an archeological context whose chronological location is coherent with these events of environmental improvement (Table 9). Besides, although the sedimentary composition of this type of dunes generates conditions of taphonomic traps (Oría et al. 2016a, and b), these landforms, located in peculiar emplacements, behave as reservoirs of archeological contexts with good resolution and possibly with a very low degree of mixing. This is very relevant for a region of thick plant cover and low archeological visibility, and in which the archeological records in stratigraphic position are very rare. In opposite, the geomorphological dynamics of auto-erosion and re-deposition forces a lack of context and generates palimpsests, generally visible in surface contexts.

The geomorphological, sedimentological, paleoenvironmental and archeological study in landforms of this kind is very useful to interpret the environmental conditions under which the human occupation would have developed during the Holocene in northern Tierra del Fuego. The micro-morphological analysis will be performed on the edaphic layers, which will provide more detailed precision on the environmental conditions that would have favored the pedogenetic processes that contributed to the time stabilization of the dunes.

Northern Isla Grande de Tierra del Fuego offers excellent perspectives for research and geoarcheological prospection in steppe environments. There, the Fuegian perched dunes offer excellent opportunities for the preservation of the archeological record since their edaphic levels may be considered as guide beds of environmental conditions less arid and windy that today, that is, environmental stability and better conditions for human occupation.

**Acknowledgements.** M. Vázquez and V. Bártoli (CADIC-CONICET) participated in fieldwork and the archeological analysis in the laboratory. I. Magneres (CADIC-CONICET) did the textural determinations by means of a laser grain-size particle analyzer. C. Di Lello (CIG-CONICET-UNLP) collaborated in the sampling for radiocarbon dating. A. Di Bastiano (CIG-FCNyM-UNLP) worked in the processing of the samples for radiocarbon dating. Many colleagues and advanced students participated in the archaeological field work. The managers and personnel of the Flamencos and San Julio ranches facilitated the access to the study sites located in their private properties. EARG-Río Grande facilitated lodgment and logistics. This work was done in the framework of the following projects: PICT ANPCYT 2012-0628 to A. Coronato, PICT 2012-1944 and PIP 0409-2013 to M. Salemme and PICT 2013–1997 to J. Oría. Dr. John Clague (Simon Fraser University, Canada) gently provided radiocarbon ages for the Arturo dune included in the unpublished PhD thesis by J. Riches (2018). J. Rabassa (CADIC-CONICET) made valuable suggestions and helped with the translation of this manuscript into English. We thank the editors of this volume for their invitation to publish our research in this volume.

## References

Behrensmeyer, A.K.: Taphonomic and ecologic information from bone weathering. Paleobiology **4**(2), 150–162 (1978)

Binford, L.R.: Bones: Ancient Men and Modern Myths. Academic Press, New York (1981)

Borrazzo, K.: Arqueología de los esteparios fueguinos. Tecnología y tafonomía lítica en el norte de Tierra del Fuego. Unpublished doctoral thesis, Facultad de Filosofía y Letras, Universidad de Buenos Aires (2010)

Borrazzo, K., Borrero, L.: Caracterización del modo tafonómico dunas en Fuego-Patagonia: aportes de la localidad Chorrillos (Tierra del Fuego, Argentina). In: Actas del XIX Congreso Nacional de Arqueología Argentina, Serie Monográfica y Didáctica, vol. 54, pp. 1588–1593. Facultad de Ciencias Naturales & Instituto Miguel Lillo, San Miguel de Tucumán (2016)

Borromei, A.M., Candel, M.S., Musotto, L.L., Cusminsky, G., Martínez, M.A., Coviaga, C.A., Ponce, J.F. Coronato, A.: Late Holocene wet/dry intervals from Fuegian steppe at Laguna Carmen, southern Argentina, based on a multiproxy record. Palaeogeogr. Palaeoclimatol. Palaeoecol. **499**, 56–71 (2018). https://doi.org/10.1016/j.palaeo.2018.03.008

Cohen, K.M., Finney, S.C., Gibbard, P.L., Fan, J.X.: The ICS International Chronostratigraphic Chart. Episodes **36**, 199–204 (2013)

Coronato, A., Seppälä, M., Ponce, J.F., Rabassa, J.: Sand dunes on lake Fagnano cliffs, Tierra del Fuego, Argentina. 4to. Congreso Argentino de Cuaternario y Geomorfología. Resúmenes, 342. Buenos Aires (2009)

Coronato, A., Fanning, P., Salemme, M., Oría, J., Pickard, J., Ponce, J.F.: Aeolian sequence and the archaeological record in the Fuegian steppe, Argentina. Quat. Int. **245**, 122–135 (2011a)

Coronato, A., Salemme, M., Moretto, A., Oría, J., Ponce, J.F., Orgeira, M.J., Onorato, M.R., Vásquez, C., Osterrieth, M., Fernández, M., Laprida, C., Benvenuto, M.L.: Paleoduna de Laguna Arturo: registro de variaciones climáticas y contextos arqueológicos durante el Holoceno en la estepa fueguina, Argentina. In: Leanza, H.A., Franchini, M., Impiccini, A., Pettinari, G., Sigismondi, M., Pons, J., Tunik, M. (eds.) Actas del XVIII Congreso Geológico Argentino, Neuquén, Argentina, CR Rom-pp. 1204–1205 (2011b). ISBN 978-987-22403-4-9

Cruz, I., Borrazzo, K.: Tafonomía y procesos de formación del registro arqueológico en ambientes de dunas: hacia la definición de un modo tafonómico. In: Actas del XIX Congreso Nacional de Arqueología Argentina, Serie Monográfica y Didáctica, vol. 54, pp. 1582–1584. Facultad de Ciencias Naturales and Instituto Miguel Lillo, San Miguel de Tucumán (2016)

Davies, B.E.: Loss-on-ignition as an estimate of soil organic matter 1. Soil Sci. Soc. Am. J. **38**(1), 150–151 (1974)

De Nigris, M.: Patrones de fragmentación de huesos largos en el sitio Cerro de los Indios I, Lago Posadas, Provincia de Santa Cruz, Argentina. Unpublished Graduation thesis in Anthropological Sciences, Universidad de Buenos Aires, 122 (1994)

Favier Dubois, C.: Análisis geoarqueológico de los procesos de formación del registro, cronología y paleoambientes, en sitios arqueológicos de Fuego-Patagonia. Unpublished Doctoral thesis, Universidad de Buenos Aires, Buenos Aires (2001)

Favier Dubois, C.: Late Holocene climatic fluctuations and soil genesis in southern Patagonia: effects on the archaeological record. J. Archaeol. Sci. **30**, 1657–1664 (2003)

Favier Dubois, C.: Soil genesis related to medieval climatic fluctuations in southern Patagonia and Tierra del Fuego (Argentina): Chronological and paleoclimatic considerations. Quatern. Int. **162–163**, 158–165 (2007)

Favier Dubois C., Borrero, L.: Playas de acreción: Cronología y procesos de formación del registro arqueológico en la costa central de la bahía San Sebastián, Tierra del Fuego (Argentina). Magallania **33**, 83–98 (2005)

Fernández, M., Maidana, N.I., Ponce, J.F., Oría, J., Salemme, M., Coronato, A.: Palaeoenvironmental conditions for human settlement at the Fuegian steppe (Argentina) based on diatom analysis. Lake Arturo as a case study. J. Archaeol. Sci. Rep. **18**, 775–781 (2018)

Franco, N.V.: Estrategias de utilización de recursos líticos en la cuenca superior del río Santa Cruz (Argentina). Unpublished Doctoral thesis, Facultad de Filosofía y Letras, Universidad de Buenos Aires (2002)

Grayson, D.: Quantitative Zooarchaeology. Topics in the Analysis of Archaeological Faunas. Academic Press, Department of Anthropology and Burke Memorial Museum. University of Washington (1984)

Hogg, A.G., Q Hua, P.G., Blackwell, C.E., Buck, T.P., Guilderson, T.J., Heaton, M., Niu, J.G., Palmer, P.J., Reimer, R.W., Reimer, C.S., Turney, M., Zimmerman, S.R.H.: Calibration data set: shcal13.$^{14}$c. Radiocarbon **55**(4) (2013). https://doi.org/10.2458/azu_js_rc.55.16783

Holliday, V.: Soil formation, time and archaeology. In: Holliday, V. (ed.) Soils in Archaeology. Landscape evolution and human occupation, pp. 101–117. Smithsonian Institution Press, Washington (1992)

Kilian, R., Hohner, M., Biester, H., Wallrabe-Adams, H.J., Stern, C.R.: Holocene peat and lake sediment tephra record from the southernmost Chilean Andes (53–55° S). Revista Geológica de Chile **30**(2), 47–64 (2003)

Lyman, R.L.: Bone density and differential survivorship of fossil classes. J. Anthropol. Archaeol. **3**, 259–299 (1984)

Lyman, R.L.: Quantitative Paleozoology. Cambridge University Press, New York (2008)

Marsh, W., Marsh, B.: Wind erosion and sand dune formation on high Lake Superior bluffs. Geographiska Annaler **69A**(3–4), 379–391 (1987)

Martin, F., Borella, F.: Tafonomía de Tierra del Fuego: reevaluación de la arqueología de Cabeza de León. In: Soplando en el viento…Actas de las III Jornadas de Arqueología de la Patagonia, Neuquén and Buenos Aires, pp. 439–450 (1999)

Mengoni Goñalons, G.L.: Zooarqueología en la práctica. Algunos temas metodológicos. Xama **19–23**, 83–113 (2010)

Montes, A., Santiago, F., Salemme, M., López, R., Magneres, I.: Geoarqueología en la duna de Laguna Las Vueltas, Tierra del Fuego, Argentina. In: XIX Congreso Nacional de Arqueología Argentina, digital abstracts, San Miguel de Tucumán (2016)

Munsell, A.H.: Munsell soil colour charts. Kollmorgen Corporation, Baltimore (1973)

Musotto, L.L., Borromei, A.M., Candel, M.S., Mehl, A., Bianchinotti, M.V., Coronato, A., Ponce, J.F.: Condiciones ambientales durante el Holoceno medio-tardío registradas en paleosuelos de Laguna Arturo, norte de Tierra del Fuego, en base al análisis palinológico. Libro de Resúmenes Séptimo Congreso Argentino de Cuaternario y Geomorfología (Puerto Madryn, 18–21 septiembre 2018). Naturalia Patagónica **10**, 70–71 (2018)

Orgeira, M.J., Vásquez, C.A., Coronato, A., Ponce, J.F., Moreto, A., Osterrieth, M., Egli, R., Onorato, R.: Magnetic properties of Holocene edaphized silty eolian sediments from Tierra del Fuego (Argentina). Revista de la Sociedad Geológica de España **25**(1–2), 45–56 (2012)

Oría, J.: Patrones de movilidad pre-europeos en el norte de Tierra del Fuego. Una aproximación geoarqueológica. Unpublished Doctoral Thesis, Facultad de Ciencias Naturales y Museo, Universidad Nacional de La Plata (2012)

Oría, J.: Tierra adentro. Distribuciones artefactuales y movilidad en la estepa fueguina, Cazadores de mar y tierra. In: Oría, J., Tívoli, A. (eds.) Estudios recientes en arqueología fueguina, pp. 289–312. Editoral Cultural Tierra del Fuego. Ushuaia, Tierra del Fuego, Argentina (2014)

Oría, J.: Movilidad y asentamiento en el interior de la estepa fueguina: la localidad Amalia. Arqueología de la Patagonia, Ediciones CIEP, pp. 267–276 (2016)

Oría, J., Mari, F.: Cronología y resolución temporal y espacial del sitio Yowen ko. Laguna Amalia (Tierra del Fuego). Arqueología de la Patagonia: el pasado en las arenas. IDEAus-CONICET, Puerto Madryn (in press)

Oría, J., Salemme, M.: Visibilidad y preservación en Laguna Arturo, norte de Tierra del Fuego (Argentina). Un análisis geoarqueológico. Intersecciones en Antropología, La Geoarqueología en la Argentina, aportes y avances, Volumen Especial **4**, 89–100 (2016)

Oría, J., Salemme, M., Santiago, F., Montes, A.: Localidad Tres Marías: un caso para evaluar la pérdida de información arqueológica por erosión eólica en la estepa fueguina. Cazadores Recolectores Del Cono Sur. Revista De Arqueología **4**, 159–173 (2011)

Oría, J., Villarreal, M.L., Coronato, A., Salemme, M., López, R.: Geoarqueología en la estepa fueguina, Argentina. Efectos de los procesos geomorfológicos sobre material arqueológico de superficie. Revista de la Sociedad Española de Geología **27**(2), 51–68 (2014)

Oría, J., Salemme, M., Coronato, A., Vázquez, M.: Procesos de formación del registro arqueológico en dunas de la estepa fueguina. In: 19° Congreso Nacional de Arqueología Argentina, Actas, San Miguel de Tucumán, pp. 1631–1637 (2016a)

Oría, J., Salemme, M., Vázquez, M.: Site formation processes in relation to surface bone assemblages in the Fuegian steppe (Tierra del Fuego, Argentina). J. Archaeol. Anthropol. Sci. **8**(2), 291–304 (2016b)

Oría, J., Bártoli, M.V., Fernández, M., Salemme, M.: Fuentes secundarias intermitentes. El uso de materias primas líticas y del espacio en Laguna Arturo, Tierra del Fuego. X Jornadas de Arqueología de la Patagonia, Abstracts, Puerto Madryn, p. 115 (2017)

Oría, J., Salemme, M., Coronato, A., Montes, A., Bártoli, M.: Lagunas estacionales como fuente secundaria de materias primas líticas. Norte de Tierra del Fuego, Argentina. In: GEOCUAR 2018, VII Congreso de Cuaternario y Geomorfología, Naturalia Patagonia, Puerto Madryn, vol. 10, pp. 101–102 (2018)

Riches, J.R.: Optical Dating Studies of Southeastern Patagonian Sand wedges in Argentina and Chile. Unpublished dissertation for the Degree of Master of Science, Simon Fraser University, 156 p. (2016)

Santiago, F.: La ocupación humana en el norte de Tierra del Fuego durante el Holoceno medio y tardío. Editora Cultural Tierra del Fuego, Ushuaia (2013)

Santiago, F., Salemme, M.: A guanaco kill site in Tierra del Fuego, Argentina. The case of Las Vueltas 1. Before Farming, 2010/2 (2010)

Santiago, F., Oría, J.: Lo que el viento no se llevó. Análisis de sitios de superficie en la estepa fueguina. Magallania **35**, 117–128 (2007)

Santiago, F., Salemme, M.: Guanaco hunting strategies in the northern plains of Tierra del Fuego, Argentina. J. Anthropol. Archaeol. **43**, 110–127 (2016). http://dx.doi.org/10.1016/j.jaa.2016.07.002

Schiffer, M.: Formation Processes of the Archaeological Record. University of New Mexico Press (1987)

Stein, J.: Organic matter in archaeological contexts. In: Holliday, V. (ed.) Soils in Archaeology. Landscape Evolution and Human Occupation, pp. 193–216. Smithsonian Institution Press, Washington (1992)

Stern, C.: Holocene tephrochronology record of large explosive eruptions in the southernmost Patagonian Andes. Bull. Volcanol. **70**(4), 435–454 (2008)

Tuhkanen, S.: The climate of Tierra del Fuego from a vegetation geographical point of view and its ecoclimatic counterparts elsewhere. Acta Botanica Fennica **145**, 1–64 (1992)

Vázquez, M., Oría, J., Salemme, M.: Acumulaciones óseas en estratigrafía en la estepa fueguina. Resolución e integridad en el sitio Yowen Ko. In: IV Congreso Nacional de Zooarqueología, Abstracts, Ushuaia, p. 141 (2016)

Villarreal, M.L., Coronato, A.: Characteristics and nature of pans in the semiarid temperate-cold steppe of Tierra del Fuego. In: Rabassa, J. (ed.) Advances in Geomorphology and Quaternary Studies in Argentina, pp. 203–224. Springer, Heidelberg (2017). https://doi.org/10.1007/978-3-319-54371-0

# Morphotectonic and Gravimetric Analysis of the La Burra Valley, Central Precordillera, San Juan, Argentina

L. P. Perucca[1,2(✉)], L. M. Rothis[1,2], J. Alcacer[1,2], N. Vargas[1], G. Lara[1], F. A. Audemard[3], F. H. Bezerra[4], and D. L. Vasconcelos[4]

[1] Gabinete de Neotectónica y Geomorfología, INGEO-DEPARTAMENTO GEOLOGÍA, Facultad de Ciencias Exactas, Físicas y Naturales, Universidad Nacional de San Juan, Av. Ignacio de La Roza y Meglioli, 5400 San Juan, Argentina

lperucca@unsj-cuim.edu.ar

[2] CIGEOBIO - CONICET, San Juan, Argentina

[3] Funvisis Earth Sciences Department, Venezuelan Foundation for Seismological Research, FUNVISIS, Caracas, Venezuela

[4] Departamento de Geologia, Universidade Federal do Rio Grande do Norte - UFRN, Campus Universitário, Natal, RN 59072-970, Brazil

**Abstract.** This work describes new evidence of Quaternary tectonic activity in the La Burra intermountain valley in the Central Precordillera, San Juan Province, Argentina. Quaternary structures are located in the eastern piedmont of the Sierra de la Crucecita, and western piedmont of the sierra de Talacasto (30° 54'–31° S and 68° 47'–68° 55' W), about 70 km northwest of the city of San Juan. By means of morphotectonic analysis, the main deformation features have been identified and located in both piedmonts, named from north to south: Las Crucecitas and Vertientes faults (in the eastern piedmont of the Sierra de La Crucecita) and the Western Talacasto fault system (in the western piedmont of the homonymous range). The faults affect alluvial deposits from Late Pleistocene to Holocene ages. They are defined by a series of features, typical of compressive environments such as: slope scarps (eastern piedmont of the Sierra de La Crucecita) and counter-slope scarps (western piedmont of the Sierra de Talacasto), fluvial strath terraces restricted to the hanging block of faults, sag-ponds, aligned springs, diverted and obstructed drainage, deeply incised and sinuous streams in the hanging blocks, alluvial fans with anomalous slopes, among other features that evidence their Quaternary activity. Six Quaternary alluvial levels were identified, from the most recent $Q_1$ to the oldest $Q_6$, according to their relative elevation, presence or absence of desert varnish and pavement and degree of incision. The faults affect $Q_6$ to $Q_2$ alluvial levels in coincidence with the east-verging thin-skinned structural style typical of the Central Precordillera. This deformation is related to the regional tectonics that have been active in the Precordillera area.


M. M. Collantes et al. (Eds.): *Advances in Geomorphology and Quaternary Studies in Argentina*, SPRINGEREARTH, pp. 92–113, 2020.
https://doi.org/10.1007/978-3-030-22621-3_4

Besides, the main geological structures were identified by the analysis and interpretation of gravimetric anomalies, employing data obtained from WGM12 (World Gravity Map 2012), which includes earth and satellite gravity data derived from the EGM2008 model. The gravimetrical response of the crustal blocks that compose the area under study, shows a correlation with the most superficial crustal structure and could be associated to the different morphotectonic domains recognized in the region.

**Keywords:** Morphotectonics · Faults · Gravimetry · Analytical signal · Precordillera · Andean ranges · San Juan · Argentina

# 1 Introduction

The Andean Precordillera of western Argentina is a first-order morphotectonic unit of the Pampean Segment of the flat subduction zone between 27° S and 33° S, being one of the most seismically active zones of thrust tectonics in the world. Nevertheless, up to now, papers describing Quaternary deformations in this portion of the Central Precordillera are scarce and isolated (Casa et al. 2011, 2014; Perucca and Vargas 2014).

The study area is located in a tectonic valley between the Sierras de La Crucecita and Talacasto, in the Central Precordillera geological province (30° 54' and 31° 00' S and 68° 55' and 68° 45' W), northwest of San Juan city, Argentina (Fig. 1). Sierra de La Crucecita is a mountain range that trends NNE, with a maximum elevation of 2,082 m a.s.l. Sierra de Talacasto has N-S to NW trend and a maximum altitude of 2,756 m a.s.l. The rivers located in the piedmonts of both mountain ranges drain by the de La Burra river, the main tributary of the Talacasto river, from north to south. Its headwaters are located to the east of the Pampa de Gualilán (Fig. 1). The Quaternary deformations identified in both foothills modify this drainage network, since small tributary streams are susceptible to variations, even subtle topographic modifications, depending on their erosion rate (Audemard 1999). There are very few previous papers in the area referring to neotectonic structures in this depression. Casa et al. (2011, 2014) mentioned some structures in the western foothills of the Talacasto mountain range, although they neither indicate their kinematics nor the alluvial levels affected. Perucca and Vargas (2014) described a natural exposure located in the northern portion of the eastern foothills of the La Crucecita mountain range, where they identified a reverse fault verging to the east that has been called the Las Crucecitas fault. Levina et al. (2014) evaluated the Cenozoic basin fill of the frontal sector of the Precordillera foreland basin in three valleys located along the San Juan river (one of them, the La Burra valley) concluding that the Neogene successions located in these depressions showed the internal construction of the Cordillera Frontal, the regional arc volcanism and the initial exhumation of the Precordillera. The changes of provenance recorded in the age of the detrital zircons, suggested an initial shortening in the Frontal Cordillera in coincidence with the change of aeolian to fluvial accumulation in the adjacent foreland basin, during the Early Miocene (24–17 Ma). They considered as the probable age of exhumation induced by the shortening and rise of the Precordillera to be between 12 and 9 Ma (Levina et al. 2014).

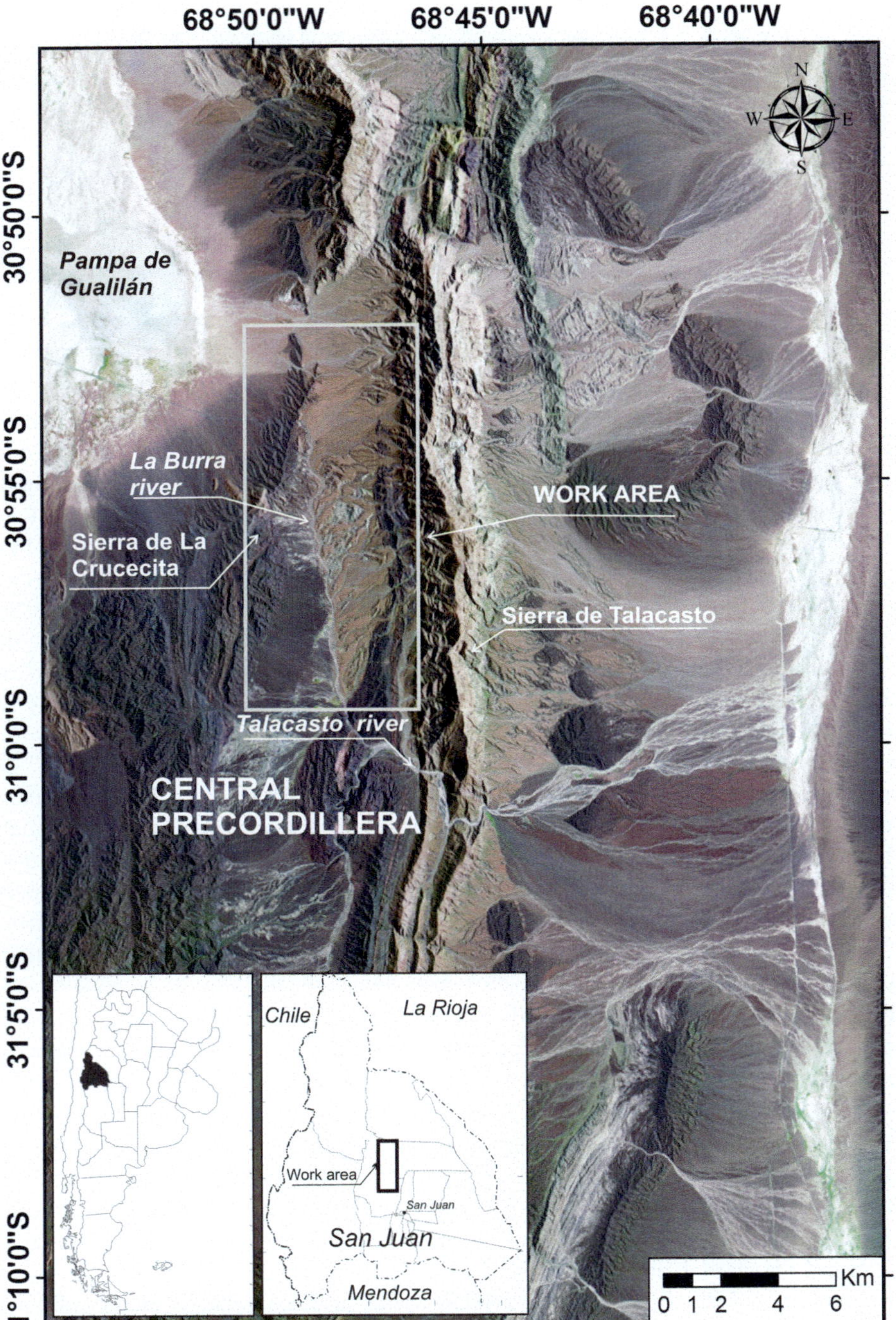


**Fig. 1.** Landsat satellite image with the location of the study area.

In this work, a morphotectonic analysis in the piedmonts located in the La Burra river valley is carried out, in which the main faults with Quaternary tectonic activity were identified. They were named from north to south: the Las Crucecitas and Vertientes (eastern piedmont of Sierra de La Crucecita) and the Western Talacasto fault system (western piedmont of Sierra de Talacasto). These faults disturbed deposits of Late Pleistocene to Holocene age and are defined by a series of typical landforms of compressive environments such as slope escarpments (western piedmont) and counter slope (eastern piedmont), strath terraces restricted to the hanging block, sag ponds, aligned slopes, diverted or disrupted streams, strongly incised and sinuous rivers in the hanging blocks of the faults, alluvial fans with anomalous slopes, among other features as evidence of Quaternary tectonic activity.

From the geomorphological analysis of the valley, the following morpho-structural units were recognized and described: Mountainous (high mountains and hills) and Transition (piedmonts). In addition, six Quaternary alluvial levels were identified (from the most recent to the oldest: $Q_1$, $Q_2$, $Q_3$, $Q_4$, $Q_5$) and a sixth level ($Q_6$) in the western piedmont of the Sierra de Talacasto. They were distinguished by their relative position, incision degree, presence (or not) of pavement and desert varnish. Quaternary faults affect almost all the alluvial levels (except the current channels) and they are consistent with structural style of the Central Precordillera, which are east-verging low-angle faults. This deformation is linked to the tectonism that disturbed the area of Precordillera at a regional scale.

On the other hand, our research tries to define the geological structures of the area based upon satellite gravimetric data processing and analysis. Gravimetric data obtained from the satellite missions have the advantage that allows viewing underground density variations within a high spatial stability and without disturbances by mathematical approximations.

This paper also tries to determine if there is a correspondence between the morphotectonic units recognized in the region and its spatial continuity based on satellite gravimetric data processing and analysis. In order to achieve a geophysical characterization, Free-air, Bouguer and Isostatic anomaly maps were calculated and analyzed.

Bouguer anomalies were filtered to separate shallow and deep gravity sources. Filters such as Upward Continuation and Band Pass can be used to separate a regional gravity anomaly resulting from deep sources from the observed gravity. When the estimated regional anomaly obtained from the Bouguer anomaly is subtracted, a residual anomaly may be obtained that clearly shows the location of the upper crustal structures.

Finally, it is intended to improve the knowledge about the active structures located in an intermountain depression located at scarcely 60 km northwest of the city of San Juan (>700,000 inhabitants) and near the main hydroelectric dams in the region.

## 2 Tectonic Setting

In the present geotectonic setting of convergence between the Nazca and South American plates (southern portion of the Central Andes), the Pampean flat-slab occurs between 27° and 34° S and had existed from 20 Ma ago (Jordan et al. 1983; Yáñez et al. 2001; Ramos et al. 2002). This is characterized by the absence of active volcanoes and by an intense seismic activity (Cahill and Isacks 1992) attributed to the presence of relevant active faults that thrusts the Neogene sequences over the Quaternary alluvial deposits. In this way, evidence of Quaternary deformations has been described at these latitudes (Perucca et al. 1999; Costa et al. 2000; Perucca et al. 2012, 2014a, 2014b, 2015; Perucca and Vargas 2014, Audemard et al. 2016, among others).

The Precordillera is a Paleozoic orogenic unit that has been subdivided, according to its structural style in Western, Central and Eastern Precordillera (Ortiz and Zambrano 1981). The study area is located in the area of Central Precordillera, which constitutes a E-verging thin-skinned fold and thrust belt, and imbricated structures rooting down towards a 10–15 km deep main decollement (Zapata and Allmendinger 1996, Ramos 1988), with mountain ranges trending N–S.

Cenozoic rocks are preserved in the footwalls of major thrusts (Jordan et al. 1993; Alonso et al. 2005) and they were differentially eroded to form N-S trending inter-mountain valleys between the thrust-generated ranges (Levina et al. 2014).

# 3 Methodology

**Structural and geomorphic analysis**

Both analyses were made using topographic data, fieldwork and digital satellite imagery (Landsat 7-TM and SPOT 5). A geological sheet published by the Argentina Mining Geological Service (Furque et al. 1998) on a 1:250,000 scale was used to determine lithology that was mapped on screen in a GIS environment (in vector format) and later was verified by fieldwork.

Besides, three swath averaged topographic-profiles (north, center and south), were built across the valley to report the main landscape characteristics of the area, where the mean, maximum and minimum elevations of the DEM (SRTM 30) are projected into cross section planes and the strip width was set to 0.5 km with a length of $\sim$10 km in order to examine trends in slope and local relief at this scale of work. These orthogonal (W-E) swath profiles crossing the La Burra Valley were selected to represent the asymmetry of the ranges, the main morpho-structural units, the differences in their heights, slopes and extension and the influence of the faults. In addition, a longitudinal swath profile trending N-S along the La Burra river was made. The elaboration of the geomorphological map allowed the integration of all these data, and the characterization of the geomorphology of the sector and the structures that disturb both piedmonts. The data processing of a Digital Terrain Elevation Model (DEM), (ASTER 30 m), was also carried out through the ArcGIS software, in order to highlight the main topographic characteristics of the La Burra valley. The fieldwork consisted in an analysis of the natural exposures identified in the channels that cross the Las Crucecitas and Western Talacasto faults, in order to determine the kinematics of these faults and the age of the disturbed deposits. From these exposures, two samples were taken and analyzed using Optically Stimulated Luminescence (OSL) Method carried out at the Laboratory of Datação, Comércio e Prestação de Serviços LTDA de Sao Paulo (Brazil).

**Data and Gravity Anomalies**

The area under study was covered with a gravity database obtained from the WGM12 (World Gravity Map 2012). The WGM is the first set of global gravity anomaly maps that take into account a realistic Earth model and that consider the contribution of most surface masses (atmosphere, land, oceans, inland seas, lakes, ice caps and ice shelves). The WGM denotes a set of high-resolution gravity anomaly maps and digital grids computed at global scale from available reference Earth's gravity and elevation models. The Release 1.0 (WGM12) includes a set of three anomaly maps (surface free-air, complete Bouguer and isostatic anomalies) derived from the EGM2008 Geopotential model and the ETOPO1 Global Relief Model.

Gravity information used is derived from the EGM2008 model, which was developed in spherical harmonics up to 2160 grade by the National Geospatial-Intelligence Agency (NGA). EGM2008 model includes surface gravity measurements (from terrestrial, marine or aerial studies), satellite altimetry and satellite gravimetry (GRACE Mission). Gravity anomalies are used to investigate the mass-density distribution of the Earth's interior to provide constraints on the geological structures from subsurface, crustal to upper mantle depths.

The surface free-air gravity anomaly was computed at the Earth's surface in the context of Molodensky theory (Heiskanen and Moritz 1967) and includes corrections for the mass of the atmosphere. The resulting complete Bouguer anomaly thus reflects the disturbance between the observed gravity and that computed for a given reference Earth model at a particular point P. It is also defined from the free-air anomaly ($\Delta g_{FA}$). The complete spherical Bouguer anomaly is determined over the whole Earth by computing in a single step the gravity contribution of all surface masses above or below the mean sea surface. The spherical Bouguer gravity anomaly at a given point P of the Earth's surface is given by:

$$\begin{aligned}\Delta \mathbf{g_B} &= \Delta \mathbf{g_{surface\ free-air}} - \Delta \mathbf{g_{surface\ masses}} \\ &= \mathbf{g_{obs}}(\mathbf{P}) - \gamma(\mathbf{Q}) + \delta \mathbf{g_{FC}} - \delta \mathbf{g_{TOP}} + \delta \mathbf{g_{Atm}}\end{aligned}$$

In the same way, the contribution of their compensation at the crustal-mantle boundary is also computed in spherical geometry on the base of isostatic equilibrium (Airy-Heiskanen model) to determine the corresponding isostatic anomaly. A spherical harmonic approach has been used to provide homogeneous and accurate global computations of gravity corrections and anomalies up to degree 10800 (1' × 1' half-wavelength equivalent spatial resolution).

Free-air, Bouguer and Isostatic Anomaly charts were gridded according to the minimum curvature method with a 1.5 km grid pass, using Oasis Montaj software. A density of 2.67 g/cm$^3$ was considered for the upper crust rocks according to the proposal by Hinze. Gravity data covers the analyzed region and its boundary sectors. In order to eliminate edge effects a data expansion was performed.

**Anomalies Filtering**
It is well known that the Bouguer anomaly contains the sum of gravimetric effects of various sources and there are many techniques that allow for a proper separation of such. The Bouguer anomaly map contains two effects; one regional (attributed to the lower crust and mantle) and another residual (attributed to upper crust and intra-basement gravity sources). For separating regional gravity anomalies from residual ones, it is assumed that a regional anomaly is caused by sources that have wide horizontal extent, are deep in terms of the depth to the top of the sources, and have similar depth. Likewise, it is assumed that a residual anomaly is produced by sources that are limited in horizontal extent, are shallow, and have depths different from the deep anomalies but similar to one another.

Different frequency techniques as Upward Continuation, Band Pass filter, were applied. Nevertheless, due to the powerful influence of the Andean root, the filter that proved to be most efficient, was obtained by a Band Pass Filter.

In order to obtain the regional Bouguer anomaly map, grid data was filtered with a Band pass, rejecting all wavelength upper than 11 km and lower than 3 km. A residual anomaly chart was obtained subtracting the estimated regional anomaly acquired from the Complete Bouguer anomaly.

# 4 Geological Setting

Detailed mapping of the study area allowed the identification of several geological units (Fig. 2). From the oldest to the most recent, the stratigraphic units are: (1) Cambrian–Ordovician sedimentary rocks, mainly consisting of limestones, dolomites and shales; (2) Silurian–Devonian greywackes and shales, of marine origin; (3) Carboniferous continental sandstones and shales; (4) Neogene sedimentary rocks (conglomerates, sandstones and shales); (5) Quaternary colluvial–alluvial consolidated deposits and (6) unconsolidated modern deposits (gravel, sand, silt and clay), in the piedmont and in river channels.

The Neogene deposits are located in the central portion of the valley with an estimated thickness of 1000 m. Levina et al. (2014) described clastic depositional patterns in these intermountain exposures of Central Precordillera assigned to the Late Oligocene-Miocene retroarc foreland basin system. The oldest strata recorded a ~24 Ma transition from ephemeral playa lake/incipient dune field conditions to a regional eolian system. About 17 Ma ago, the eolian system replaced a fine-grained floodplain deposition. Upward coarsening fluvial and alluvial fan deposits reflect a 17–8 Ma shift from distal to proximal foreland basin facies, locally influenced by syn-orogenic volcanic activity. However, in the area of the La Burra valley, the majority of A-He ages fall between 12 and 10 Ma (Levina et al. 2014). These authors pointed out that this pattern is consistent with an eastward (cratonward) progression of thin-skinned thrusting and resultant structural fragmentation of the foreland basin into isolated intermountain segments.

In the eastern piedmont of the Sierra de La Crucecita and from east to west, Neogene rocks outcrop, trending N 50° E and dipping from 10° W that later increases to

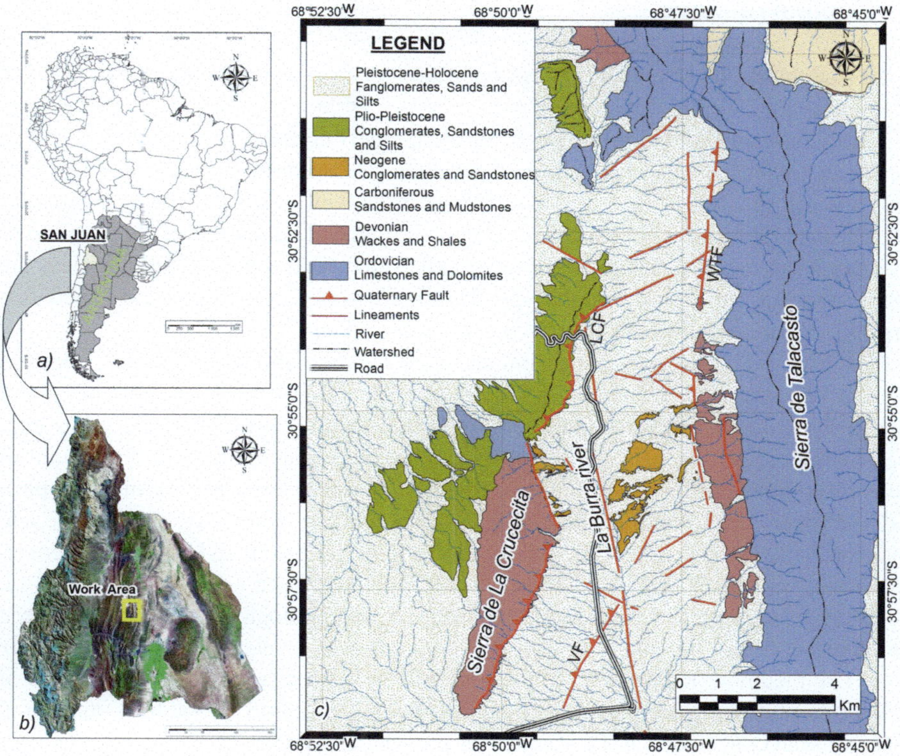


**Fig. 2.** Geological map of the La Burra river valley with Quaternary faults location, Las Crucecitas fault (LCF), Vertientes fault (VF) and Western Talacasto fault (WTF).

45° W westward. Near to the mountain front the strata dip 70° E, forming a synclinal that is faulted in its western flank. This reverse fault whose azimuth is 160° and dips from 47° to 67° W involves Devonian and Neogene rocks.

In the western piedmont of the Sierra de Talacasto, the Neogene strata have a trend of N 4° E, N 16° E and N-S and dip between 40° and 45° W. Towards the West, the trend of the Neogene strata is N 8° W and the dip increases to 70° W as far as the western flank of the Sierra de Talacasto, the Neogene deposits rest unconformably on rocks of Devonian age, with azimuth 355° and dip 60° W. This would suggest that the rise of the Sierra de Talacasto by a reverse fault verging to the east and located on its eastern flank, would have generated the tilt of the entire Neogene basin. The Pleistocene and Holocene alluvial deposits, consisting of poorly selected fanglomerates and conglomerates, with angular to subangular clasts of different sizes from the Paleozoic and Cenozoic sediments, are placed onto the Neogene outcrops by an erosion unconformity. Its matrix is a coarse-grained sand with frequent lenticular intercalations of coarse sandstones. Finally, modern deposits in the fluvial channels tributaries of the La Burra river and similar to those previously described, are mentioned.

## 5 Geomorphology

In the study area, the Mountainous and Transition morpho-structural units are identified as follows (Fig. 3):

The Mountain Unit is formed by the positive reliefs uplifted by faults. The Sierra de La Crucecita trends NNE with its highest elevations in the central sector (2,024 m a.s.l.), with a width ranging from 1 to 5 km. It is an important positive relief that provides detrital material towards the east. This mountain range is composed of Paleozoic sedimentary rocks and it is characterized by a gentle slope on its western flank, whereas the eastern one has an abrupt morphology due to the presence of the reverse fault that thrusted Devonian sequences over Neogene rocks.

The Sierra de Talacasto is a mountain range trending almost N-S with a width of about 3 km, and average altitude of 2,500 m a.s.l. Its W-E profile shows an abrupt eastern flank due to a reverse fault verging to the east that uplifts the calcareous unit.

The hill relief is represented by those positive units that contrast from the point of view of altitude with the high mountains, being less outstanding and forming a typical badland relief. They comprise sandstones, shales and conglomerates of Neogene and Pliocene-Pleistocene age, in some sectors with a thin Pleistocene to Holocene (?) cover. They show smooth and rounded slopes, with elevations between 1,650 and 1,700 m a.s.l.

The transition unit or piedmont establishes the morphological connection between the aforementioned mountain ranges and the local base level, which is the La Burra river. The piedmont located east of the Sierra de la Crucecita has an average slope of 18% in its northern area, with width that increases from north to south. In its central portion the average slope is 12.5%, and in the southern sector the slope reaches only 7% in average, with a greater extension than the northern and central portions. On the other hand, the western piedmont of the Sierra de Talacasto has an average width of 2 km, which is reduced to less than 500 m at its southern end, with a slope that varies between 10% in the northern part and 5% in the central and southern sector. Both piedmonts are covered by alluvial deposits that are formed by the accumulation of debris coming from the erosion of the structural highs (mountainous and hills units). Climate plays an essential role in its formation, since these deposits are rare in humid climates, whereas in areas of arid climate, scarcity of vegetation on the relief allows the disintegration of it into smaller fragments that accumulate and extend at the foot of the ranges. The alluvial levels overlie the erosion surface carved in Neogene outcrops, with the top surfaces of the oldest Quaternary units standing out.

Six alluvial levels ($Q_6$, the oldest, to $Q_1$ the present fluvial channels) were preliminary identified in both piedmonts according to their relative topographic position, grade of incision and presence or not of desert varnish (Fig. 3). However, these alluvial levels do not develop along the entire piedmonts and are only clearly identifiable where the Neogene rocks outcrop.

The drainage pattern is dendritic (divergent) in both piedmonts, typical of alluvial fans. However, several anomalies in the drainage occur when the streams cross the fault traces, increasing for example incision and sinuosity and showing diverted, obstructed, hanging and beheaded channels and showing aligned springs and unpaired terraces (Fig. 4a–d).

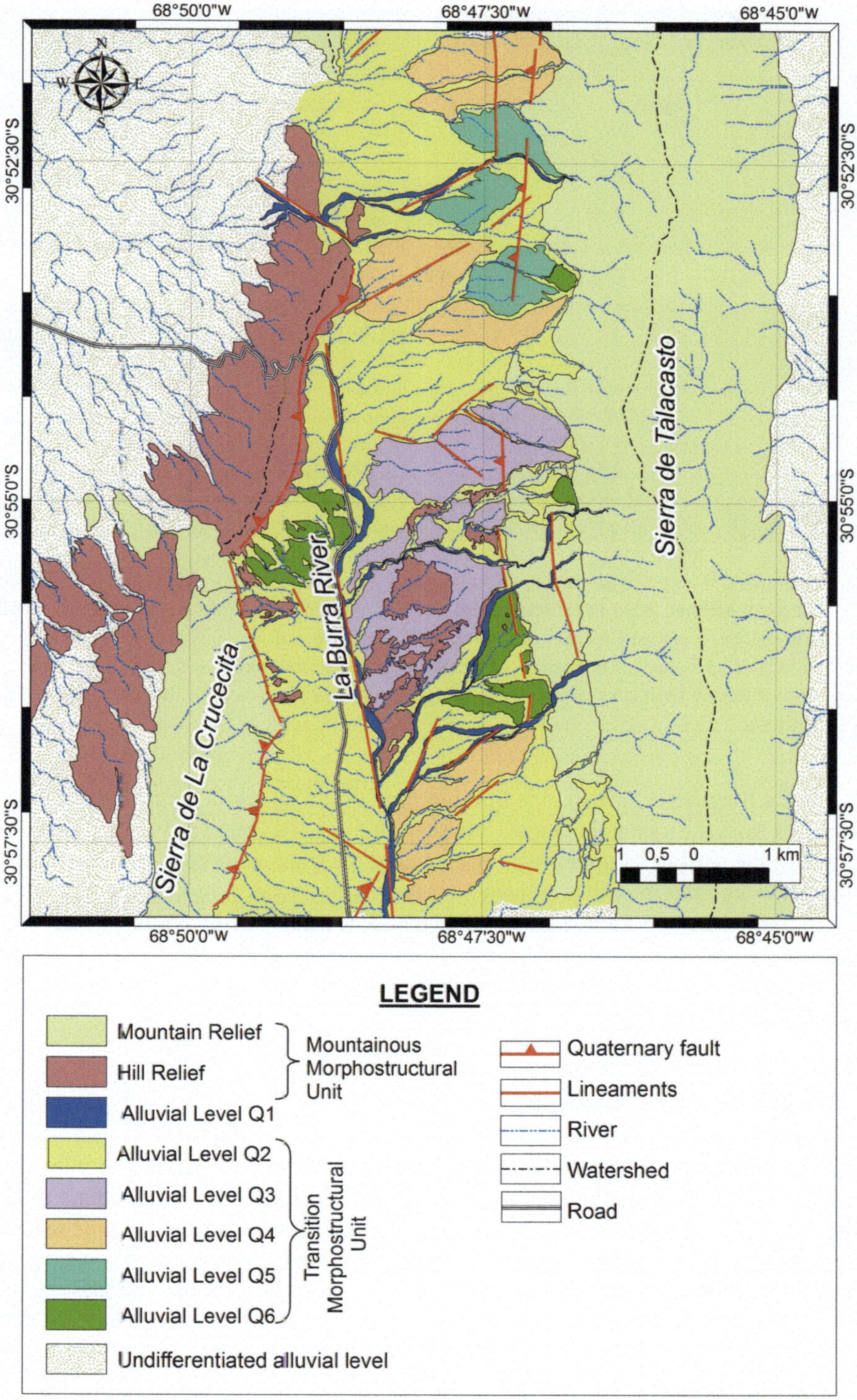


**Fig. 3.** Geomorphological map.

**Fig. 4.** Detailed Google Earth images showing (a) counter-slope fault scarp with river diversion and beheaded streams. Also the drainage pattern becomes more sinuous in the hanging block, (b) counter-slope scarp and natural exposure of Western Talacasto fault where Neogene rocks overlies Quaternary alluvial deposits, (c) shape-broom and diverted rivers, (d) aligned springs along the Vertientes fault trace and diverted rivers and unpaired terraces in the left margin of the La Burra river.

The La Burra river, which runs from north to south, becomes the main collector of tributaries from the mountain ranges. It has N-S trend and a length of 9.8 km. This river is tilted to the west in its upper river basin and to the south is located to the east of the valley showing a basin with a remarkable asymmetry. Besides, only at the left margin of this river there are three terraces (the two oldest are strath terraces being topographically higher and the youngest one, a fill-in terrace). That asymmetric distribution may be because of the variation of the local base level due to the tectonic uplift of the eastern portion of the valley.

## 6 Swath Profiles Analysis

As it was mentioned in the methodological section, three cross-sectional profiles were drawn up across the study area (North, Center and South) and one longitudinal profile along the La Burra river, in order to highlight the most distinctive topographic features of the valley, its piedmonts and the alluvial plain characteristics of the main river (Fig. 5a–e). With the standard deviation value between the extreme altitudes, slopes, relative heights, incision and the influence of the structures on the relief could be distinguished. The northern profile shows the most significant elevation difference between the Sierra de La Crucecita to the west ($\sim$1,850 m a.s.l.) and the Sierra de

Talacasto to the east (~2,510 m a.s.l.), the latter with an extensive piedmont (~3 km) showing low incision towards the distal portion and which increases towards the mountainous front of the calcareous range. On the other hand, the eastern piedmont of the Sierra de La Crucecita is very narrow (<300 m) and the alluvial levels are much incised (Fig. 5b). The alluvial plain of La Burra river lies to the western portion of the intermountain valley. In this upper section, the river has a gradient of 4.5%, and an average width of 87 m, with a dominant vertical incision. The central profile shows the decrease in elevation difference between both mountain ranges, even though the Sierra de La Crucecita does not exceed 2,000 m a.s.l. in this sector, whereas the Sierra de Talacasto has an elevation of 2,460 m a.s.l. Extensive piedmonts develop in both flanks of the valley, but the eastern one is strongly incised. The La Burra river is located symmetrically in the central portion of the valley (Fig. 5c). In this middle section, the river gradient diminishes, being 3%, whereas the river floodplain width decreases to 42 m. In the southern profile, the difference between the two mountain ranges is less than 200 m, with both piedmonts having a similar extension, although the western piedmont has a steeper slope, whereas the eastern one has a gentler slope, but it shows a greater incision. The La Burra river is located towards the east of the depression. The lower river section gradient diminishes to 1.5% and the river floodplain width increases to 114 m, where lateral erosion predominates showing a flat bottom (Fig. 5d).

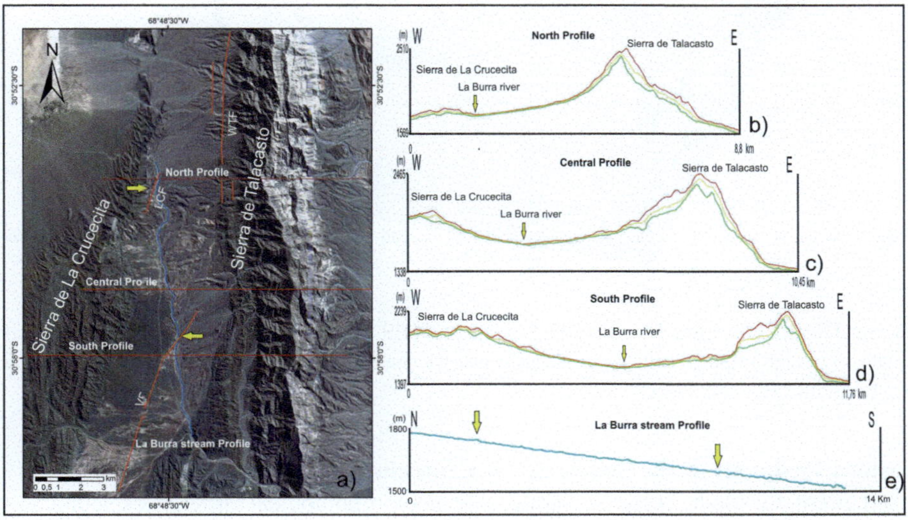


**Fig. 5.** (a) Satellite image showing main Quaternary faults: Las Crucecitas fault (LCF), Verientes fault (VF) and Western Talacasto fault (WTF) and topographic profiles location; (b) Northern, (c) Central and (d) Southern swath topography profiles (with 0.5 km window); performed across the river valley. Each profile shows focal statistics of the highest, mean and the lowest elevation, respectively, (e) Longitudinal profile of the La Burra River. Yellow arrows point knickpoints related to the Las Crucecitas and Vertientes faults, suggesting a tectonic origin.

Finally, in the N-S La Burra river profile it is possible to appreciate that this profile is not in equilibrium and has two main slope breaks or knickpoints, located in its middle section. Both knickpoints would coincide with the location of the NNE Quaternary structures analyzed in this work; these are the Las Crucecitas and Vertientes faults (Fig. 5e).

## 7 Neotectonic Analysis

Faults with evidence of Quaternary tectonic activity were identified in both piedmonts of the La Burra river valley. They are reverse faults that trend N-S to NE, with eastern vergence, consistent with the structural style of the Central Precordillera (Fig. 6).

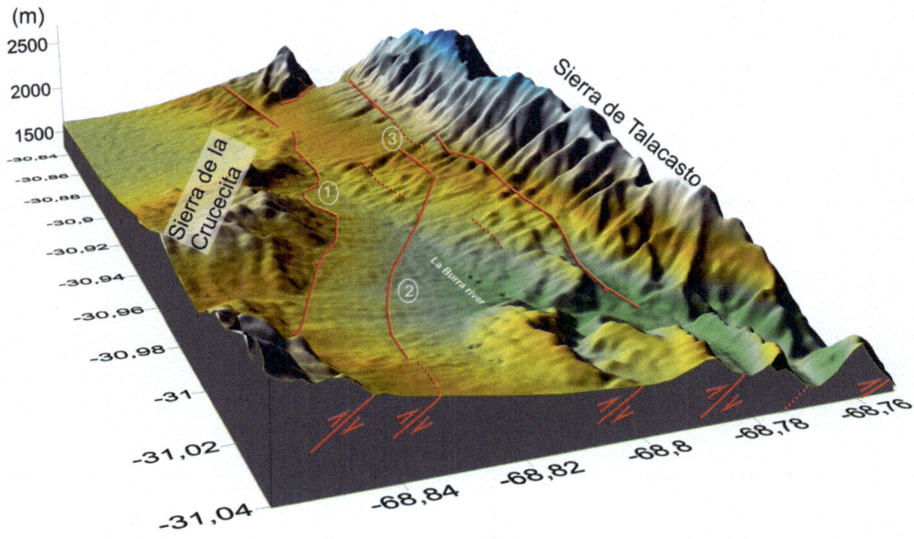


**Fig. 6.** Digital Elevation Model (DEM) map (ASTER GDEM) of the La Burra river valley showing its topographic features and location of the main Quaternary structures, Las Crucecitas (1), Vertientes (2) and Talacasto (3).

### Eastern Piedmont of the Sierra de La Crucecita

From north to south, there are noticeable geomorphological differences along this piedmont. For example, four strongly incised alluvial levels ($Q_4$ to $Q_2$ and the present river channels $Q_1$) have been recognized in its northern portion. These Quaternary deposits cover Neogene rocks and have evidence of glacio-planation processes (Fig. 7a). However, to the south, only the present ($Q_1$) and $Q_2$ alluvial levels were identified, although Neogene rocks outcrop in the sectors where the Vertientes fault is recognized.

**Fig. 7.** (a) View to the northwest (northern end of the Sierra de La Crucecita) and location of natural exposures T1 and T2. (b) T1 trench in a strath terrace on the left margin of a small tributary of the La Burra river. The reverse fault that dips to the west places Neogene rocks over alluvial deposits assigned to the Holocene. Behind the terrace there is a favor-slope scarp corresponding to $Q_3$ alluvial level. The scarp is rounded and $\sim$3 m height. (c) View to the south of the natural exposure T2 showing the Las Crucecitas fault, which dips to the west and disturbs the alluvial level Q3. (d) Fault detail in trench T2. The person in the photograph points the sample CRUZ 1 location.

On the northern end of the eastern piedmont of the Sierra de la Crucecita (Fig. 7b), Las Crucecitas reverse fault is located (Perucca and Vargas 2014). This fault, with a 210° azimuth and 40° W dip, was identified in a natural exposure (T1) as a strath terrace of a tributary of the La Burra river. The structure places Neogene rocks over alluvial deposits that have been assigned to the Holocene (Fig. 7b).

Further north, in the natural exposure T2, the fault affects the $Q_4$ alluvial level and shows a $\sim$1.70 m high, slightly rounded, east-looking flexural scarp. The fanglomerates of level $Q_4$ present lenses of fine materials as well as clasts oriented and flexured as a result of deformation (Fig. 7c, d). Striae were measured on the fault plane, whose pitch is 57° NW, indicating thrust with a horizontal sinistral component. A yellowish, clayish lime deposit (Sample Cruz 1), located below the fault plane (Fig. 7d) was dated by OSL, yielding an age of 17 $\pm$ 1.75 kyr.

In the central and southern portion of the eastern piedmont of the Sierra de La Crucecita, there is another fault running N 24° in average with a soft inflection in its southern trace where it changes to N 11° (Fig. 8a), exhibiting an east-looking favor-slope scarp. This fault, named as the Vertientes Fault affects the $Q_2$ alluvial level and even the present $Q_1$ alluvial level, modifying the drainage pattern, which presents hanging, diverted and obstructed channels with spring alignment (Figs. 4d and 8b, c). The fault length is ~6 km which crosses La Burra river course and crosses also the distal portion of the western piedmont of the Sierra de Talacasto. Here, this thrust trends N 10°, dips 45° W, and uplifts Neogene rocks.

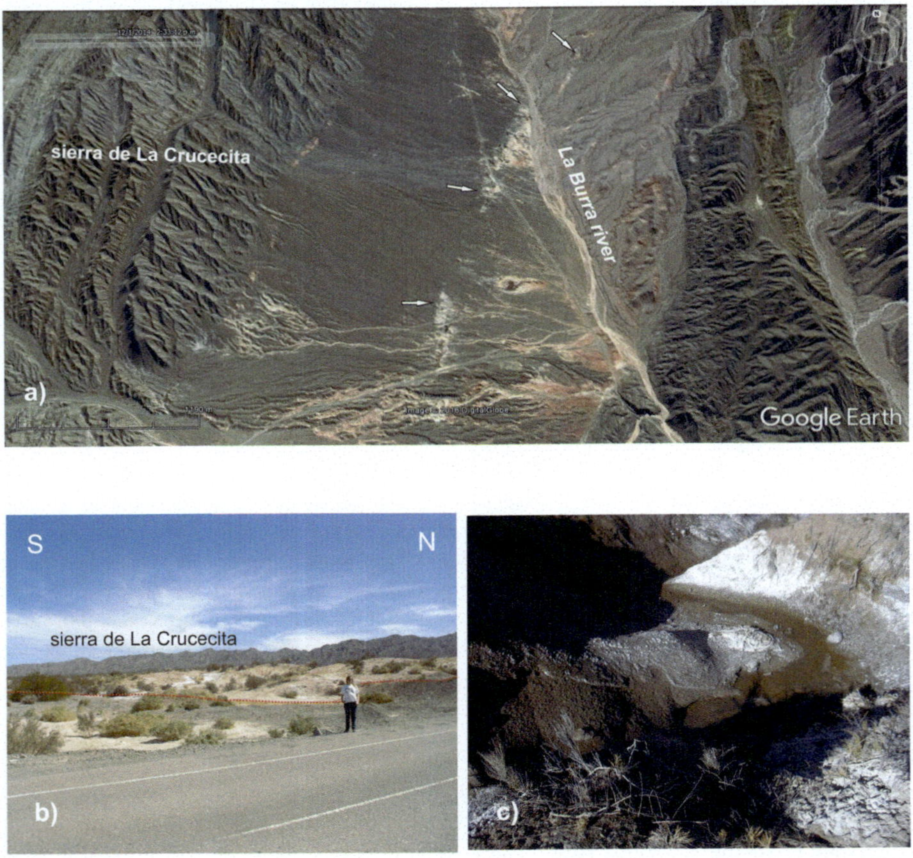


**Fig. 8.** (a) Google Earth image of the study area. White arrows point the location of the Vertientes fault. (b) View to the west of the silty deposits and aligned springs along the fault. (c) When the aquifer encounters the fault, it emerges as a surface spring.

## Western Piedmont of Sierra de Talacasto

In the western piedmont of the Sierra de Talacasto, several discontinuous and subparallel fault sections and alignments were identified. These structures show simple and compound counter-slope and flexural scarps, strath terraces in the hanging blocks and stream anomalies like diverted, obstructed and shape-broom streams (Figs. 4a, b and 9a, b). In addition, an increase in the incision and sinuosity of the channels in the hanging block of the fault is observed, where Neogene rocks outcrop (Fig. 4a, b, c). The scarp height ranges between 0.30 m and 10 m, being the most pronounced ones in the northern portion of the piedmont.

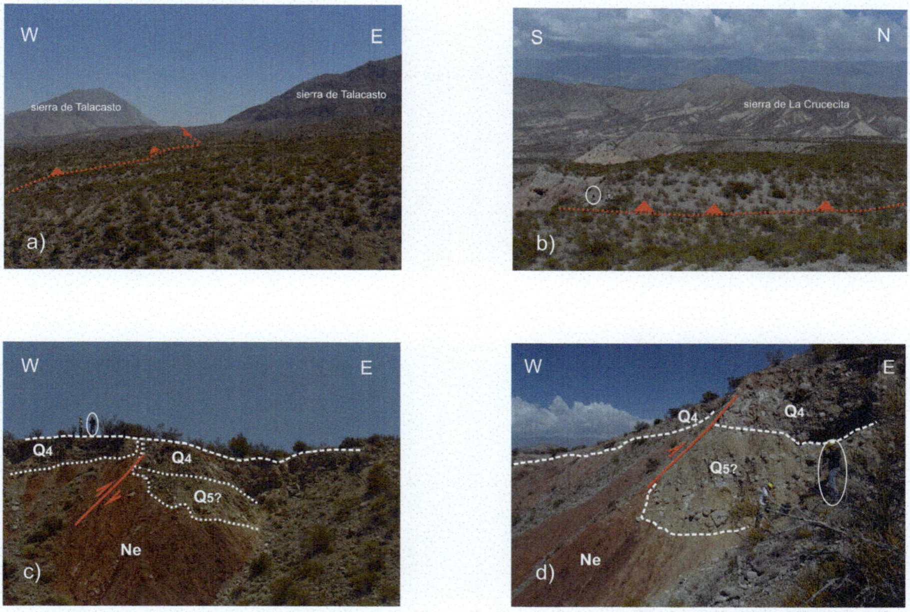


**Fig. 9.** (a) View to the north of the Western Talacasto fault. The counter-slope scarp affects $Q_3$ and $Q_4$ alluvial levels. (b) View to the west of the scarp. (c) Natural exposure showing a reverse fault that places Neogene rocks over alluvial deposits assigned to $Q_4$ level. (d) NW view showing the eastern vergence of the fault affecting $Q_4$ and $Q_5$ (?) alluvial levels. White circles point to a person for scale, where the sample CRUZ 2 was obtained.

In a natural exposure the fault thrusted Neogene rocks over alluvial deposits of level $Q_4$ and $Q_5$. It dips 74° W and trends 25° (Fig. 9c and d). Deposits assigned to the $Q_5$ level, represented by the sample Cruz 2, yielded an age of 52.3 ± 4.5 kyr (Fig. 9d).

## 8 Gravimetric Analysis

The anomalies charts obtained for la Burra area reflect high-density contrasts to a crustal and lithospheric scale, where positive and negative values are apparently associated to the different units exposed in the study area. It is important to highlight the strong influence of the Andean root on the observed anomalies.

The Free-air anomaly map (Fig. 10) shows correspondence with the thrust fronts that uplift the mountain ranges of Central Precordillera, that is, the Sierra de Talacasto to the east and the Sierra of the Crucecita to the west. In addition, changes of gradient observed between the elevated morphotectonic units to the east (Sierra de Talacasto) and to the west (Sierra de la Crucecita) would bound the depression corresponding to the La Burra valley.

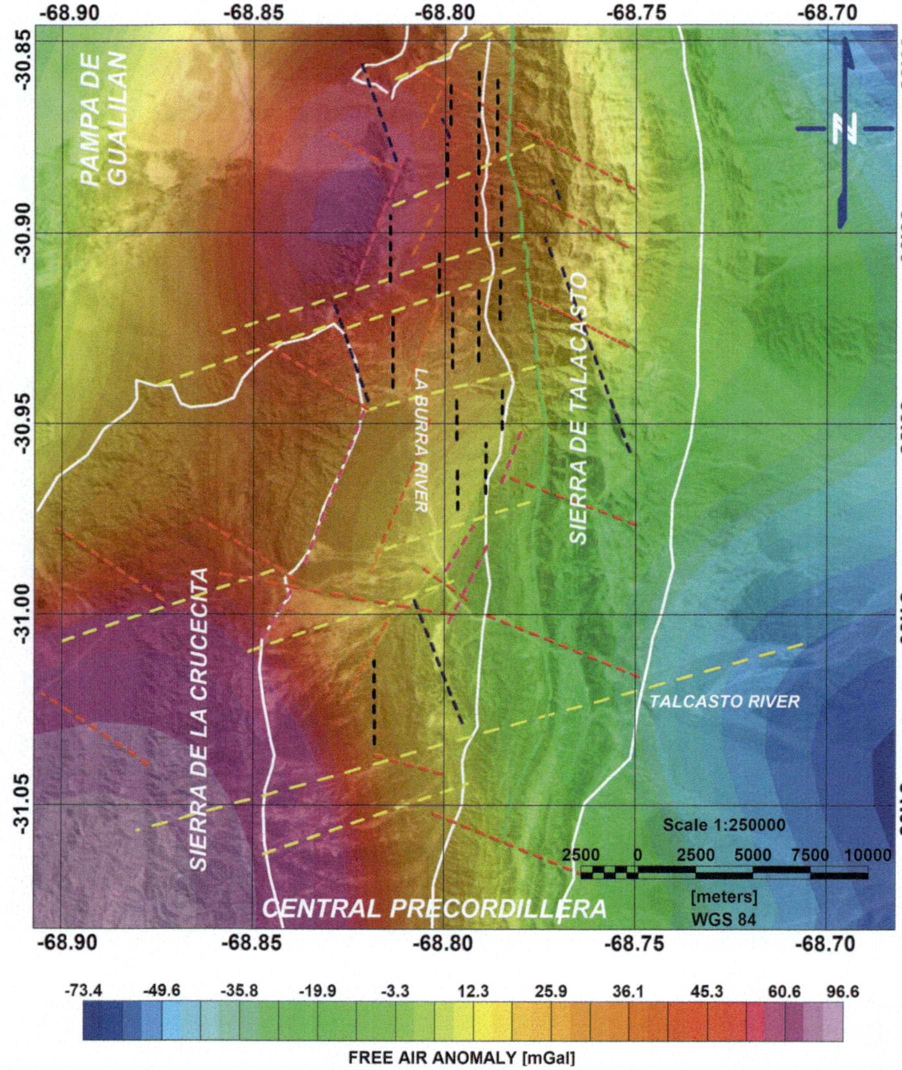


**Fig. 10.** Free-air gravity anomaly map.

In the complete Bouguer anomaly chart, negative values, associated with the mountain ranges that exhibit Paleozoic units and that were uplifted by N-S faults, were observed (Fig. 11). On the other hand, the positive values would seem to show the existence of intermountain depocenters (La Burra valley) where Neogene-Quaternary deposits predominate. These valleys seem to be controlled by NE and NW structures.

The isostatic anomaly chart preliminarily suggests a strong disequilibrium for the region under study, evidenced by negative values along the main mountain ranges as in Sierra de Talacasto to the east. On the other hand, the Cenozoic depocenters show positive values which also suggest an isostatic imbalance, so that tectonic adjustments would be expected for the region analyzed in order to reach the isostatic equilibrium (Fig. 12).

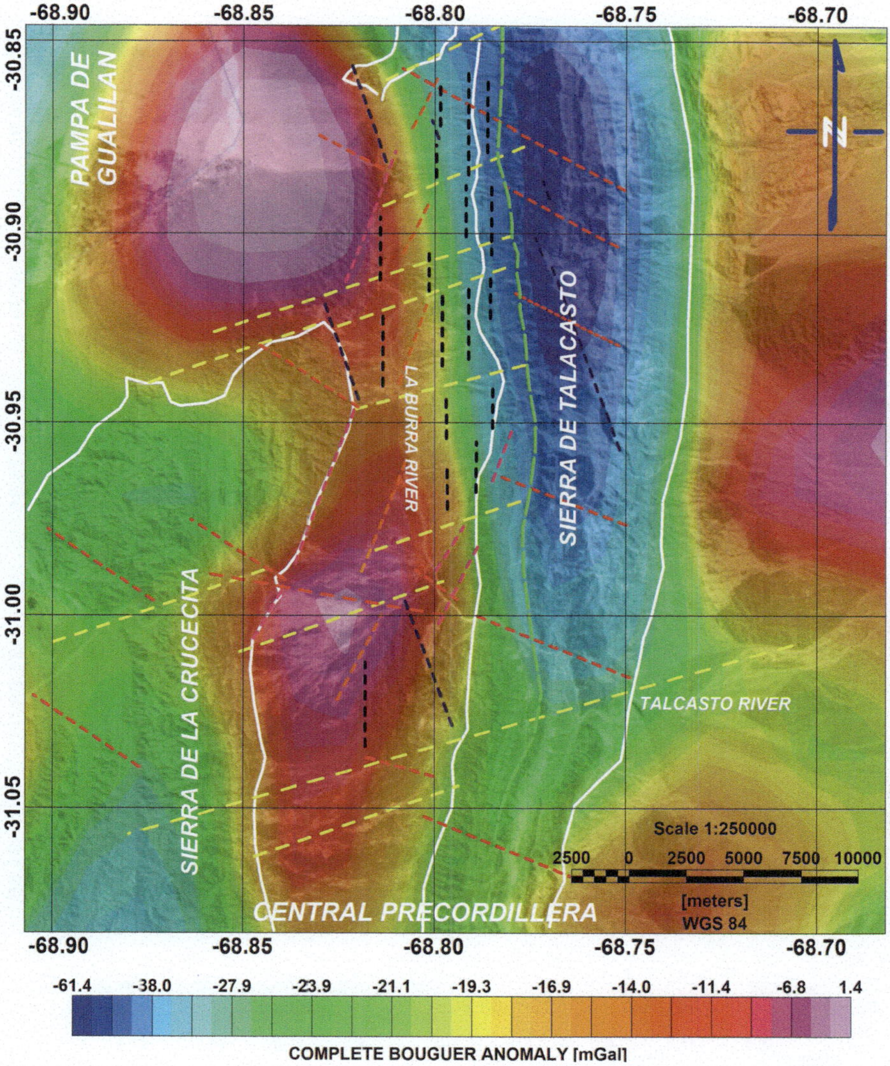


**Fig. 11.** Complete Bouguer anomaly map.

The analysis of the residual anomaly chart shows the existence of structural controls (thrust fronts), trending N-S, that exposed the Paleozoic units, reaching the highest topographical elevations. These structures bound both the east and west of the intermountain valleys filled by Neogene to Quaternary deposits. Besides, the residual anomaly chart reveals lateral density contrasts in a north-south direction associated with NE and NW structures that would segment the intermountain valleys (Fig. 12) (Fig. 13).

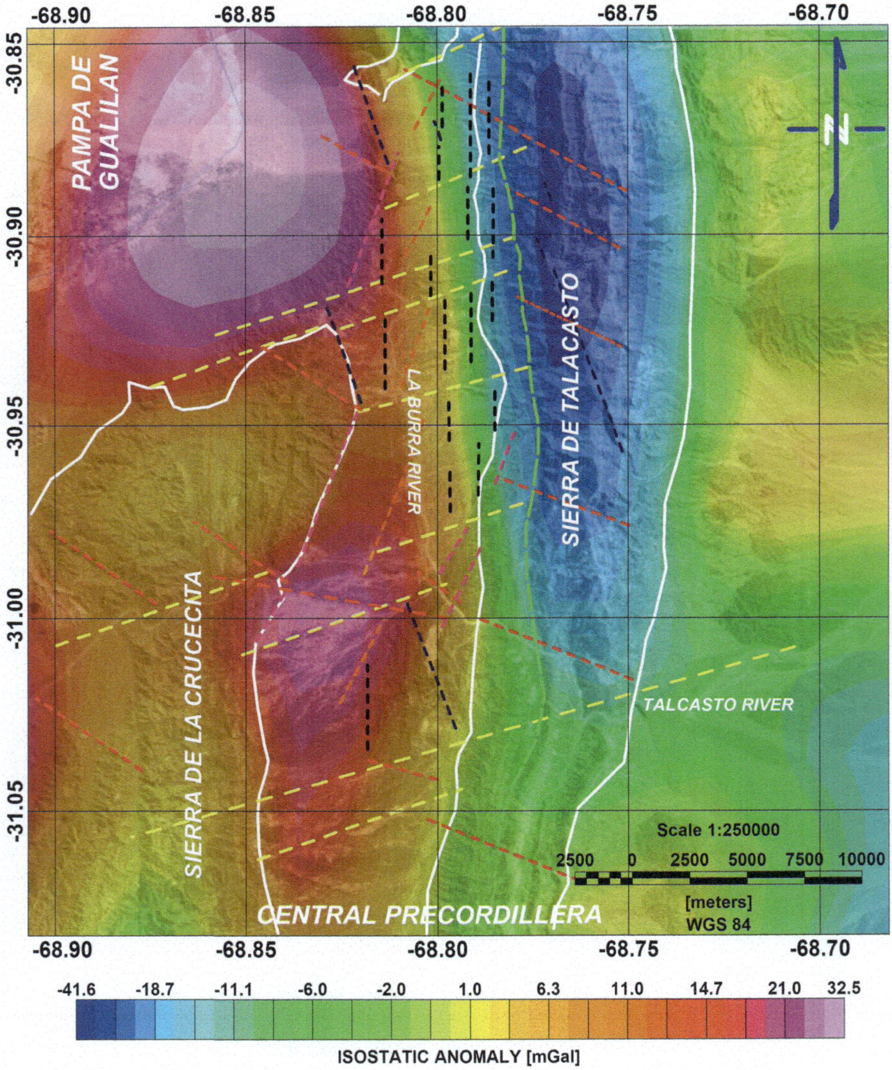


**Fig. 12.** Isostatic anomaly map. The isostatic anomaly chart shows the influence of the main superficial structures in the region under study.

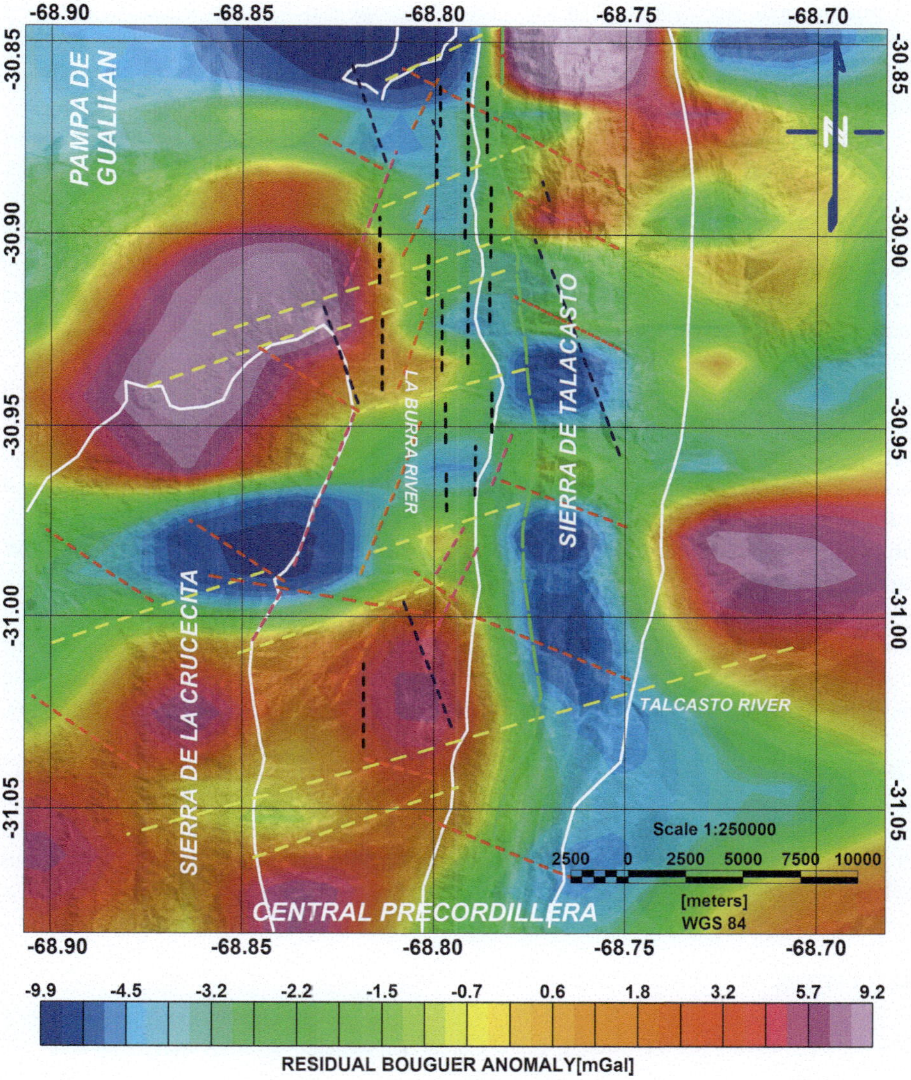


**Fig. 13.** Residual Bouguer anomaly map.

## 9 Conclusions

A neotectonic analysis in the longitudinal La Burra valley was performed between the Sierra de La Crucecita at the west and the Sierra de Talacasto at the east. In this depression, evidence of Late Pleistocene to Holocene tectonic activity was identified, consistent with the fold-and-thrust belt area of the Central Precordillera.

Identification of several geomorphological markers showed the existence of active deformation along both piedmonts. These active structures have N-S preferential orientation, with the exception of the Vertientes fault which trends N 30° E, affecting the $Q_4$ to $Q_2$ alluvial levels. Linear, sinuous, obstructed, diverted and incised streams, aligned springs, favor and counter-slope and flexural scarps, are common geomorphological anomalies, associated to all fault sections.

From the analysis of the two samples dated by OSL method, as well as the topographic position, pavement and varnish desert, it is suggested that the age of the alluvial levels would oscillate between the Late Pleistocene for the $Q_6$–$Q_3$ levels to the Holocene for the $Q_1$ and $Q_2$ level. From the analysis of those morphostructures identified in both piedmonts, it is possible to determine the structural control exerted by these structures and their Holocene tectonic activity. From the morphotectonic markers and natural exposures analyzed thrusts have been identified trending almost N-S, exhibiting average dips of 40° W.

The Bouguer anomalies recognized in the study area show high positive and negative values, reflecting high-density contrasts to a crustal and lithospheric scale, as well as possible isostatic disequilibrium. On the other hand, from the analysis and processing of the gravimetric data, a remarkable correlation with the most superficial crustal structure is noted. The gravimetric charts reveal lateral density contrasts in N-S direction in coincidence with main mountainous ranges, associated with NE and NW structures that cross the depression filled by Holocene deposits.

Finally, it is considered that these structures may act as potential seismogenic sources in the area of the Precordillera and that they should not be underestimated for seismological purposes, since these faults are close to four important dams located along the San Juan river and close to several of the main cities of central–western Argentina.

**Acknowledgements.** The present authors acknowledge funding from PID CICITCA 2016-17 (Universidad Nacional de San Juan), CAPES- MINCYT Br1201 and IPGH (GEOF 02-18) for supporting this research project. We are thankful to CIGEOBIO for providing funds for ArcGIS 10.3 software license and to Dr. Fernando Hongn and Dr. Jorge Rabassa for reviewing a previous draft of the manuscript.

## References

Alonso, J.L., Rodríguez Fernández, L.R., García-Sansegundo, J., Here-dia, N., Farías, P., Gallastegui, J.: Gondwanic and Andean structure in the Argentine Central Precordillera: the Río San Juan section revisited. In: 6th International Symposium on Andean Geodynamics. IRD Editions (Institut de Recherche pour le Développement), Extended Abstracts, Paris, pp. 36–39 (2005)

Audemard, F.A.: Morpho-structural expression of active thrust fault systems in the humid tropical foothills of Colombia and Venezuela. Zeitschrift fur Geomorphologie **118**, 1–18 (1999)

Audemard, F.A., Perucca, L.P., Pantano, A., Avila, C., Onorato, M.R., Vargas, H.N., Alvarado, P., Viete, H.: Holocene compression in the Acequión valley (Andes Precordillera, San Juan Province, Argentina): geomorphic, tectonic, and paleoseismic evidence. J. S. Am. Earth Sci. **67**, 140–157 (2016)

Cahill, T., Isacks, B.L.: Seismicity and the shape of the subducted Nazca plate. J. Geophys. Res. **97**, 17503–17529 (1992)

Casa, A., Yamín, M., Wright, E., Costa, C., Coppolecchia, M., Cegarra, M.: Deformaciones Cuaternarias de la República Argentina, Sistema de Información Geográfica: Instituto de Geología y Recursos Minerales, Servicio Geológico Minero Argentino, Publicación 171, v1.0 DVD, Buenos Aires (2011)

Casa, A., Yamín, M., Wright, E., Costa, C., Coppolecchia, M., Cegarra, M., Hongn, F. (eds.) Deformaciones cuaternarias de la República. Argentina. Sistema de Información Geográfica. Instituto de Geología y Recursos Minerales. Servicio Geológico Minero Argentino, Publicación N°171, v.1.0 DVD, Buenos Aires (2014)

Costa, C., Machette, M.N., Dart, R.L., Bastías, H.E., Paredes, J.D., Perucca, L.P., Tello, G.E., Haller, K.M.: Map and database of Quaternary faults and folds in Argentina. U.S. Geological Survey Open-File Report 00-0108, 81 p. (2000)

Furque, G., González, P., Caballé, M.: Descripción de la hoja geológica 3169-II, San José de Jáchal (Provincias de San Juan y La Rioja). Servicio Geológico y Minero Argentino 259, 150 p., Buenos Aires (1998)

Heiskanen, W.A., Moritz, H.: Physical Geodesy, p. 364. Freeman and Co., San Francisco (1967)

Jordan, T.E., Allmendinger, R.W., Brewer, J.A., Ramos, V.A., Ando, C.J.: Andean tectonics related to geometry of subducted Nazca plate. Geol. Soc. Am. Bull. **94**, 341–361 (1983)

Jordan, T.E., Allmendinger, R.W., Damanti, J.F., Drake, R.E.: Chronology of motion in a complex thrust belt: the Precordillera. J. Geol. **101**, 135–156 (1993)

Levina, M., Horton, B., Fuentes, F., Stockli, D.: Cenozoic sedimentation and exhumation of the foreland basin system preserved in the Precordillera thrust belt (31–32° S), southern central Andes, Argentina. Tectonics **33**, 1659–1680 (2014). https://doi.org/10.1002/2013TC003424

Ortiz, A., Zambrano, J.: La provincia geológica de Precordillera Oriental. Proceedings, 8° Congreso Geológico Argentino. San Luis **3**, 59–74 (1981)

Perucca, L.P., Vargas, H.N.: Neotectónica de la provincia de San Juan, centro-oeste de Argentina. Boletín Soc. Geológica Mex. **66**, 291–304 (2014)

Perucca, L.P., Paredes, J., Tello, G., Bastías, H.: Fallamiento activo en el área norte del sistema de fallamiento El Tigre, San Juan-La Rioja. Rev. Asoc. Geológica Argent. **54**, 206–214 (1999)

Perucca, L.P., Lara, G., Vargas, N.: Nueva evidencia de actividad tectónica cuaternaria en la depresión Zonda-Maradona, provincia de San Juan. Rev. Asoc. Geológica Argent. **69**, 97–105 (2012)

Perucca, L.P., Rothis, M., Vargas, H.N.: Morphotectonic and neotectonic control on river pattern in the Sierra de la Cantera piedmont, Central Precordillera, province of San Juan, Argentina. Geomorphology **204**, 673–682 (2014a)

Perucca, L.P., Onorato, M., Millán, J., Bustos, A., Vargas, H.N.: Variación del estilo de deformación a lo largo del sistema de falla La Cantera, Precordillera Central, San Juan, Argentina. Rev. Soc. Geológica España **27**(2), 69–79 (2014b)

Perucca, L.P., Rothis, M., Bezerra, F., Vargas, N., Lima, J.: Late Quaternary evolution of the La Cantera Fault System (Central Precordillera, Argentina): a morphotectonic and paleoseismic analysis. Tectonophysics (2015). https://doi.org/10.1016/j.tecto.2015.08.041

Ramos, V.A.: The tectonics of the central andes; 30° to 33° S latitude. Geol. Soc. Am. Spec. Pap. **218**, 31–54 (1988)

Ramos, V.A., Cristallini, E.O., Pérez, D.J.: The pampean flat-slab of the central andes. J. S. Am. Earth Sci. **15**, 59–78 (2002)

Yáñez, G.A., Ranero, C.R., von Huene, R., Díaz, J.: Magnetic anomaly interpretation across the southern central Andes (32°–34° S): the role of the Juan Fernández Ridge in the late tertiary evolution of the margin. J. Geophys. Res.-Solid Earth **106**, 6325–6345 (2001)

Zapata, T.R., Allmendinger, R.W.: Thrust-front zone of the Precordillera, Argentina: a thick-skinned triangle zone. Am. Assoc. Pet. Geol. Bull. **80**, 359–381 (1996)

# Morphotectonic Analysis in the Eastern Piedmont of the Sierra de La Huerta, Western Sierras Pampeanas, San Juan, Argentina

Luis Martín Rothis[1,2(✉)], Federico Miguel Haro[1], Laura Patricia Perucca[1,2], Paula Santi Malnis[1,3], Juan Manuel Alcacer[1], and Nicolás Vargas[2]

[1] CIGEOBIO - Consejo Nacional de Investigaciones Científicas y Tecnológicas (CONICET), Gabinete de Neotectónica y Geomorfología, INGEO, Facultad de Ciencias Exactas, Físicas y Naturales, UNSJ, Av. Ignacio de La Roza y Meglioli s/n (5400), San Juan, Argentina
mrothis@unsj-cuim.edu.ar

[2] Departamento de Geología, Facultad de Ciencias Exactas, Físicas y Naturales, Universidad Nacional de San Juan, San Juan, Argentina

[3] Instituto y Museo de Ciencias Naturales, Universidad Nacional de San Juan, San Juan, Argentina

**Abstract.** Tectonic geomorphology is a useful discipline to investigate the effects of active tectonic processes upon landforms. The Andean back arc between 27° S and 33° S is characterized by intense seismicity, being the epicentre of the most damaging earthquakes in Argentina. Those events produced deformation, causing changes on the land surface morphology. A geomorphological and a morphotectonic analysis were carried out in the eastern flank of the Sierra de la Huerta to characterize the available evidence of tectonic activity. Three morphostructural units (named as Mountain, Transition and Depression) were recognized. The transition unit shows Quaternary active faulting where eight fault traces with a NW trend and associated scarps were identified. These evidences of Quaternary faulting were assigned to the Marayes Fault System. Detailed topographic profiles perpendicular to the strike of fault scarps allowed to measure main deformation to the north of the study area. Neotectonic deformation in the Arroyo Los Papagayos (arroyo = creek, sometimes an ephemeral one) produced anomalies in its pattern and longitudinal profile. Thus, this study allowed the consideration of the Marayes Fault System as a potential seismogenic source.

**Keywords:** Neotectonics · Topographic profiles · River anomalies · Marayes Fault System · Seismic hazard

## 1 Introduction

Tectonically active landforms result from a complex integration of the effects of vertical and horizontal motions of crustal rocks and erosion or deposition by surface processes (Burbank and Anderson 2001). Tectonic geomorphology is an excellent integration field that investigates the effects of active tectonic processes – faulting,


M. M. Collantes et al. (Eds.): *Advances in Geomorphology and Quaternary Studies in Argentina*, SPRINGEREARTH, pp. 114–128, 2020.
https://doi.org/10.1007/978-3-030-22621-3_5

tilting, folding, uplifting and subsidence – upon landforms (Hugget 2007). The use of GIS (Geographic Information System) techniques is relatively recent in the area of Tectonic Geomorphology. This methodology and the availability of DEM (Digital Elevation Models) provide valuable tools to evaluate the tectonic activity of a certain region.

Up to now, papers describing the relationship between the geomorphology and the tectonic process in the Sierras Pampeanas Occidentales (Western Sierras Pampeanas) are very scarce. The Eastern Sierras Pampeanas (between 27° and 33° S) involve basement uplifts in the Andean foreland (Fig. 1a and b). Associated to these ranges several marginal faults with evidence of Quaternary deformation have been documented. It is commonly understood that these typical intraplate faults have long-term earthquake recurrence so they do not present remarkable geomorphic evidence (Costa et al. 2006). Bastías et al. (1990) recognized clear evidence of Quaternary deformation in the westernmost Western Sierras Pampeanas, defining two main fault systems: the Pie de Palo Fault and the Las Chacras Fault System (Fig. 1c). The last system is located in the western flank of the Sierra de la Huerta where numerous features suggesting neotectonic activity were descripted by different authors (Bossi 1976; Bastías et al. 1990; Rothis et al. 2012). Rothis et al. (2018) analysed morphotectonic features associated with the Las Chacras Fault System. The most conspicuous markers (fault scarp, wind gap, river diversion, springs, change of incision depth and sinuosity along river courses, among others) correspond with the Pan de Azúcar fault. Besides, they carried out the analysis of drainage basins in the western piedmont of Sierra de la Huerta suggesting that the present asymmetric morphology of these river basins reveal their tilt by the fault activity. On the contrary, scarce morphotectonic markers have been observed in the eastern piedmont of the Sierra de La Huerta. Nevertheless, the analysis of satellite imagery allowed the identification of alignments with a NW direction disturbing the Quaternary deposits.

In order to characterize the evidence of tectonic activity in the eastern piedmont of the Sierra de la Huerta, the first results are herein presented from detailed geomorphological, morphotectonic and geological observations. Besides, this information allowed the establishment of this region as a potential seismogenic source.

## 2 General Tectonic Features of the Western Sierras Pampeanas

In the western margin of the South American continent, between 27° and 33° S, the Nazca plate is subducting with a flat geometry beneath the South American plate (Fig. 2a) at a rate of 6.3 cm/year. This flat geometry is attributed to the oblique subduction of the Juan Fernández ridge (Fig. 1b) (Pilger 1981; Martinod et al. 2010; Rosenbaum and Mo 2011). This tectonic setting produced strong Quaternary deformation and a gap in the arc volcanism in the back arc and foreland regions. Compression produced crustal shortening and uplift of the thin-skinned Precordillera and the thick-skinned Pampean Ranges of Argentina (Sierras Pampeanas) since the Late Miocene-Quaternary. The Sierras Pampeanas are roughly elongated following a NW direction and they show an asymmetric morphology in a W-E cross section, with the

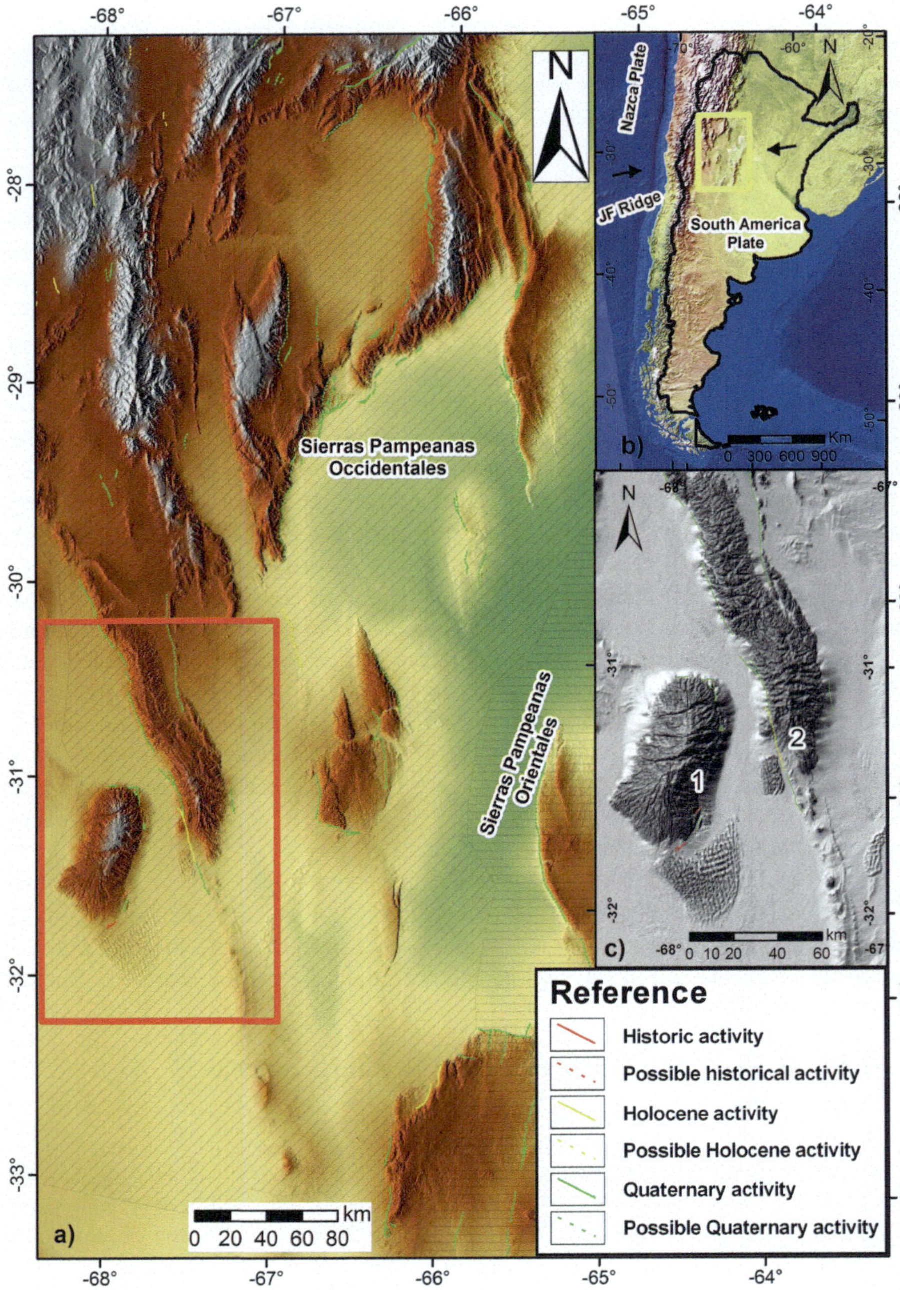


**Fig. 1.** (a) Regional satellite image with the location of the structures with neotectonic activity. Red box shows the Western Sierras Pampeanas; (b) Fig. 1a, location in Argentina; (c) Satellite image showing Pie de Palo (1) and Las Chacras (2) Fault Systems.

western flank steeper than the gentler eastern slope. These are bounded to the west by listric marginal reverse faults dipping to the east (González Bonorino 1950; Rosello et al. 1996) which uplifted the basement core that is composed of Late Proterozoic-to Early Paleozoic metamorphic rocks. In addition, strike-slip structures affect the ranges oblique to the main strain (Massabié 1987; Costa et al. 2014). Deformation reported in this region consists mainly in piedmont faults, characterized by discontinuous fault traces affecting Quaternary alluvial surfaces. Paleozoic and Mesozoic structural and lithological anisotropies present in the crystalline basement control these structures (Ramos et al. 2002; Costa et al. 2018). In this region, the instrumental seismicity show a bimodal distribution (Fig. 2a and b), where interplate events present focal depths between 90–120 km and crustal seismicity has focal depths shallower than 35 km (Alvarado and Beck 2006; Costa et al. 2014). More than 90% of the total continental seismic-moment is released in this region (Costa et al. 2001), clustered in the Pie de Palo range (Western Sierras Pampeanas area), and surroundings, as it is shown in Fig. 2a. Three destructive historical crustal earthquakes (focal depths less than 30 km) in Argentina occurred in this area during the past century (Alvarado and Beck 2006). The first and the most damaging event, with a magnitude Mw = 7.0 (Alvarado and Beck 2006) occurred in 1944 and produced a surface rupture of 6 to 8 km long and a scarp of 0.6 m height. The second earthquake in this list took place in 1952, with a magnitude Mw = 6.8, which did not show any surface rupture evidence, whereas the

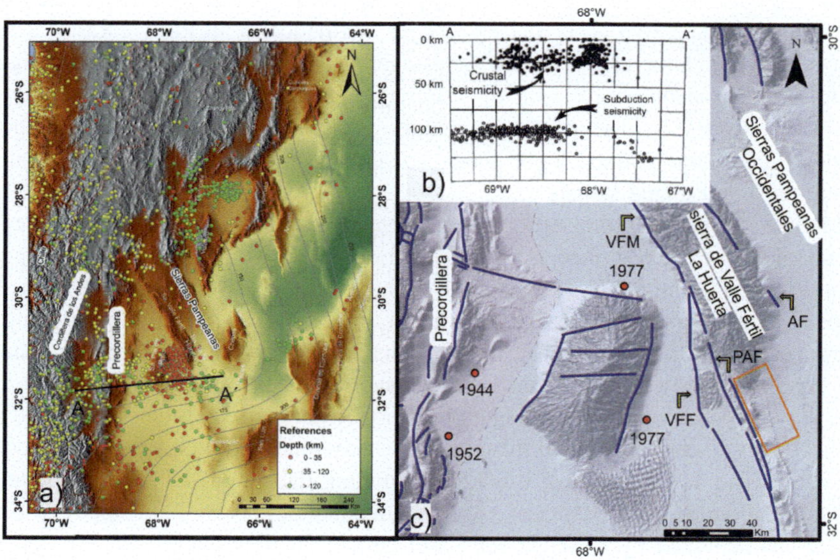


**Fig. 2.** (a) Regional image of the central-west portion of Argentina with the instrumental seismicity. The solid grey curves show the slab contour of the top of the Wadati-Benioff zone. The solid black line represents the location of the transect A-A′. (b) Bimodal instrumental seismicity in the study area (modified from Siame et al. 2006). (c) Regional image showing Precordillera and Western Sierras Pampeanas domains. Red points represent four historical earthquakes; VFM (Valle Fértil Megafault), PAF (Pan de Azúcar Fault), AF (Astica Fault). The orange rectangle indicates the location of detailed fieldwork.

1977 earthquake (Mw = 7.4) produced a surface rupture with a vertical slip of 0.3 m, with epicentre in the eastern piedmont of Sierra de Pie de Palo, very close to the study zone (Fig. 2c).

## 3 Local Geology

The Sierra de Valle Fértil-La Huerta (in the domain of the Western Sierras Pampeanas) is located in central-western Argentina, 130 km east of the city of San Juan (Fig. 3). This mountain range is a tectonic block, uplifted along its western flank by a regional reverse fault with western vergence (the Valle Fértil Megafault System), showing an asymmetrical profile, deeper in the west and slightly extended in the east. A Neoproterozoic-Early Paleozoic complex of igneous-metamorphic rocks (the Valle Fértil Complex, according to Cuerda et al. 1984), with medium to high grade of metamorphism (Castro de Machuca et al. 2007), in outcrops of this area. These rocks present a NNW foliation with dips to the ENE. Sedimentary Mesozoic rocks outcrop in the eastern piedmont of the Sierra de La Huerta (Fig. 3). These units correspond with the Carrizal, Quebrada del Barro and Balde de Leyes formations and the Gigante Group. The Carrizal Formation is a continental Triassic sequence composed of light grey conglomerates and sandstones that overlie the crystalline basement-core (Bossi 1976). The Quebrada del Barro Formation is formed by a sedimentary succession from thin to medium conglomerates, interspersed with mica bearing sandstones and abundant shales (Bossi 1976). The Balde de Leyes Formation is an Early Jurassic association of red conglomerates and fine sandstones (Colombi et al. 2015). The Gigante Group is a Cretaceous succession of red conglomerates and sandstones (Flores and Criado Roque 1972). The piedmont mainly presents Quaternary alluvial deposits composed of coarse-grained loose sandy-gravel layers, with clasts of metatonalites, amphibolites and quartz schists.

Modern deposits of the Arroyo Los Papagayos (arroyo = creek, sometimes an ephemeral one) were described as a seasonal arid distributive fluvial system, composed of three zones (Santi Malnis et al. 2018). The first zone is characterized by a low sinuosity, gravelly-sandy bed, braided fluvial system. The second zone present the divergence of the main channel in two forming a fan shape deposit with a medial to high sinuosity sandy-gravelly anastomosed-anabranching fluvial system. In the third zone, the main channel converged again in a mono-channelized system interpreted as a high sinuosity, flashy ephemeral meandering, sandy-muddy fluvial system with eolian interaction. The development of this high aggrading Papagayos DFS is interpreted as the combination of many factors as a low piedmont slope (below 1° in slope), seasonal climate characterized by a steppe semi-arid climate in the upland and seasonal arid climate in the lowland, drainage basin morphometry and accommodation space availably related to a high subsidence basin (Santi Malnis et al. 2018).

The few Mesozoic outcrops recognized in the eastern piedmont of the Sierra de la Huerta show an east-dipping homoclinal structure and they are impacted by oblique structures, which produce inflections in the strike of the sedimentary beds.

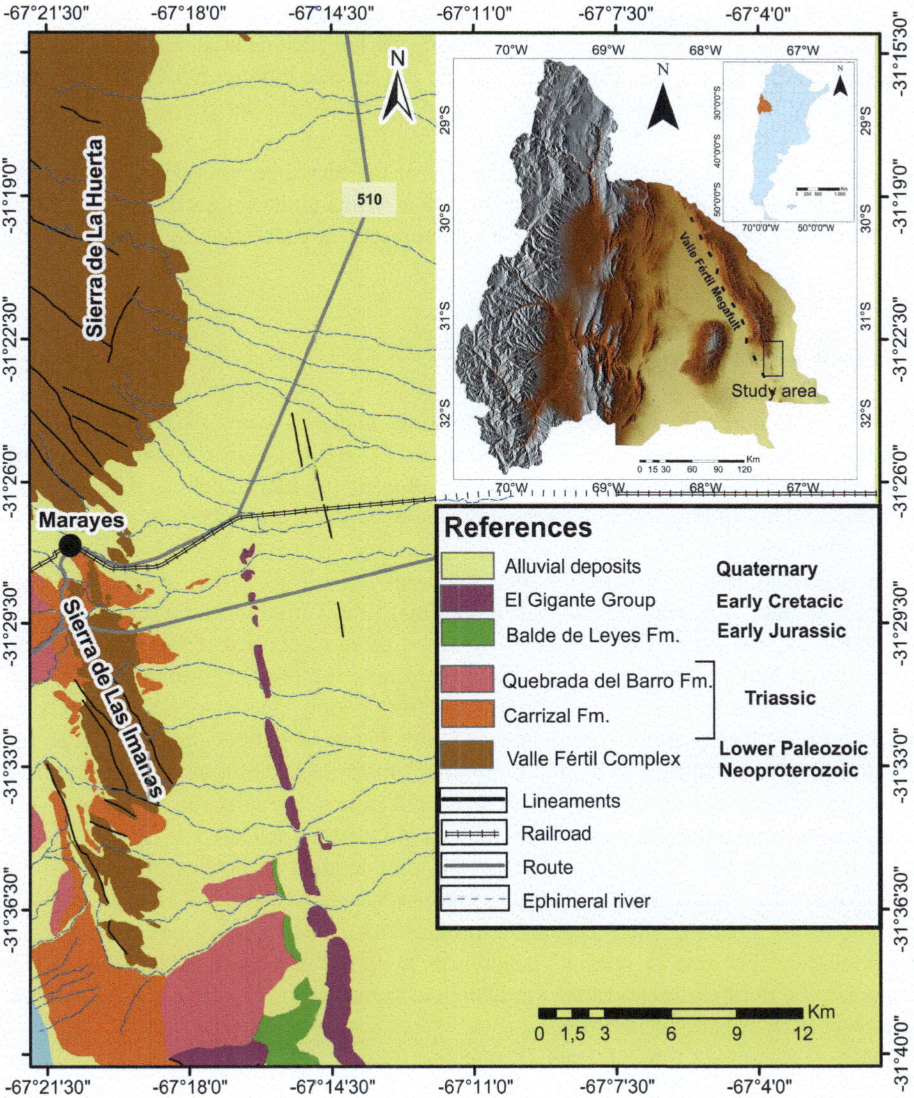


**Fig. 3.** Geological map of the study area and its location in San Juan province. De la Huerta and de Las Imanas ranges are basement core blocks. Mesozoic rocks are continental clastic sedimentary sequences.

## 4 Neotectonic Setting

In the western piedmont of the Sierra de Valle Fértil there is evidence of Quaternary structures grouped into the Las Chacras Fault System (Bastías et al. 1990), a subsidiary branch of the Valle Fértil Megafault System (Rothis 2016). The Las Chacras Fault System is composed of several sections, being the master trace the Pan de Azúcar fault

(Fig. 2c), which shows clear evidence of Pleistocene to Holocene tectonic activity. This fault had impacted different alluvial levels along 26 km (Rothis et al. 2018), producing free face scarps to the west. The Pan de Azúcar is a reverse fault with 325° azimuth that dips 25° NE. Natural exposures show rocks of the Quebrada del Barro thrusting over Quaternary deposits.

On the other hand, in the eastern piedmont of the Sierra de Valle Fértil, Casa et al. (2011) mentioned the presence of a structure (the Astica fault) suggesting the existence of Quaternary activity, although they do not present any description about the fault (Fig. 2c).

## 5 Data and Methods

Both the geomorphological and morphotectonic analyses were made using digital satellite imagery (Landsat 8-TM and Sentinel 2), topographic data and fieldwork studies. A detailed assessment of the region allowed the identification of fault traces and different morphotectonic markers associated with the most conspicuous neotectonic structures.

In order to improve, simplify and standardize fault data into potentially fault seismic sources for seismic hazard modelling, the Christophersen et al. (2015) methodology was used. These authors identified three categories of fault data: traces, sections and faults. Traces are locations where faults rupture the ground surface (e.g. fault scarp) or where its presence is interpreted from geophysical evidence. Sections are traces with attributes such as slip rate and dip. Faults are those structures that can rupture the ground surface in a single maximum magnitude earthquake. Faults consist of one or more sections, and they are represented in map view by one or more traces.

Detailed topographic profiles have been acquired with a DGPS in W-E transects (perpendicular to the fault scarps) selected through detailed photo-interpretation of the area. These data were improved with a post-processing DGPS calculation with an uncertainty of $\pm 0.1$ m. The topographic profiles made possible to highlight particular morphological patterns of terrain. In addition, a digital elevation model (DEM) was created in the northern sector of the study area to represent the topography of the terrain disturbed by two parallel fault traces.

Many researchers (Ouchi 1985; Keller and Pinter 1996; Holbrook and Schumm 1999; Zámolyi et al. 2010, among others) have documented the influence of vertical movements on a river channel pattern. Because of this, a morphotectonic analysis of the Arroyo Los Papagayos, an alluvial stream that crosses the Marayes Fault System, was carried out. A topographic profile along the Arroyo Los Papagayos was created to observe its shape in order to determine whether the river is in geomorphological equilibrium. Hack (1973) proposed the SL Index or stream length gradient index to identify morphometric anomalies along the longitudinal profile of a river. The longitudinal profile of a river under steady conditions show an ideally straight line on a semi-logarithmic graph. Conversely, if a stream runs across active structure systems, a convex profile is shown (Chen et al. 2006). The theoretical graded profile of the Arroyo Los Papagayos was calculated based on the gradient index concept taking the base level and the headwater as reference points. The G index (Jiménez-Cantizano et al.

2017) was used to compare it with the real longitudinal profile and illustrate how far the Arroyo Los Papagayos is from theoretical equilibrium.

G = AR/AT

AR (real area) is the difference between the area below the line that link the headwaters and the river mouth point with the area below of the longitudinal profile.

AT (theoretical area) is the difference between the area below the line that links the headwater and the river mouth point with the area below of the graded profile.

In order to determine the potential seismic hazard of the Marayes fault system, the empirical relationships of Wells and Coppersmith (1994), between moment magnitude and surface rupture length to estimate the earthquake moment magnitude (Mw) were applied.

## 6 Results

The Sierra de la Huerta is a basement core uplifted by the Valle Fértil fault System. In this region, three morphostructural units: mountain unit, transition unit and depression unit were recognized. The mountain unit is represented by the higher elevations like the Sierras de La Huerta and the Las Imanas, and by a hill sub-unit with elevations lower than 700 m a.s.l. Those ranges are elongated in a N-S direction with an elevation lower than 2060 m a.s.l. The ranges are composed of Neoproterozoic and Early Paleozoic rocks belonging to the Valle Fértil Complex. The northern portion of the ranges is higher than the southern ones, as Fig. 4a illustrates. The Sierra de la Huerta shows an asymmetric shape in the E-W topographic profile, with steeper slopes on the western hillside in comparison with the smoothened eastern flank. To the east of the Sierra de las Imanas the "hills sub-unit" is observed. These are low hills (<525 m a.s.l.) with an elongated shape in N-S trend (Fig. 4a and b). They are composed mainly of continental conglomerates of Cretaceous age (El Gigante Group). The transition unit represents the eastern piedmont of the Sierra de la Huerta that extends from the Sierra de la Huerta up to the Macasín Valley. This unit is elongated in a N-S trend, parallel to the Sierra de la Huerta and up to 13 km wide. In this unit, seven alluvial levels were identified, according to the degree of fluvial incision, their relative height and the presence of varnish desert on the clast surface. These alluvial levels are named $Q_1$ (the older level) to $Q_7$ (the younger level). In the northern sector of the study area $Q_1$, $Q_2$, $Q_3$ and $Q_7$ levels were recognized. The $Q_4$, $Q_5$ and $Q_7$ levels were identified south of the Arroyo Los Papagayos (Fig. 4b). On the other hand, the surfaces of $Q_1$, $Q_2$ and $Q_3$ alluvial levels located in the northern sector show darker shades in comparison to the alluvial levels located in the southern area. These darker colours are attributed to the occurrence of sediments from Sierra de la Huerta in the northern area and from Sierra de las Imanas in the southern sector. The $Q_1$, $Q_2$ and $Q_3$ alluvial levels are composed of angular clasts of gneiss, schist, amphibolite and basic intrusive rocks which long axes sizes vary between 40 cm and 10 cm, bearing desert varnish on their surfaces. However, $Q_4$, $Q_5$ and $Q_6$ alluvial levels show angular clasts of quartz, schist, amphibolite and granodiorite, with diameters ranging between 2 cm and 20 cm. The present piedmont level is $Q_7$, comprising gravels with angular clasts of amphibolite, quartz schists and granodiorite. The Salinas de Mascasín represents the depression unit; this is

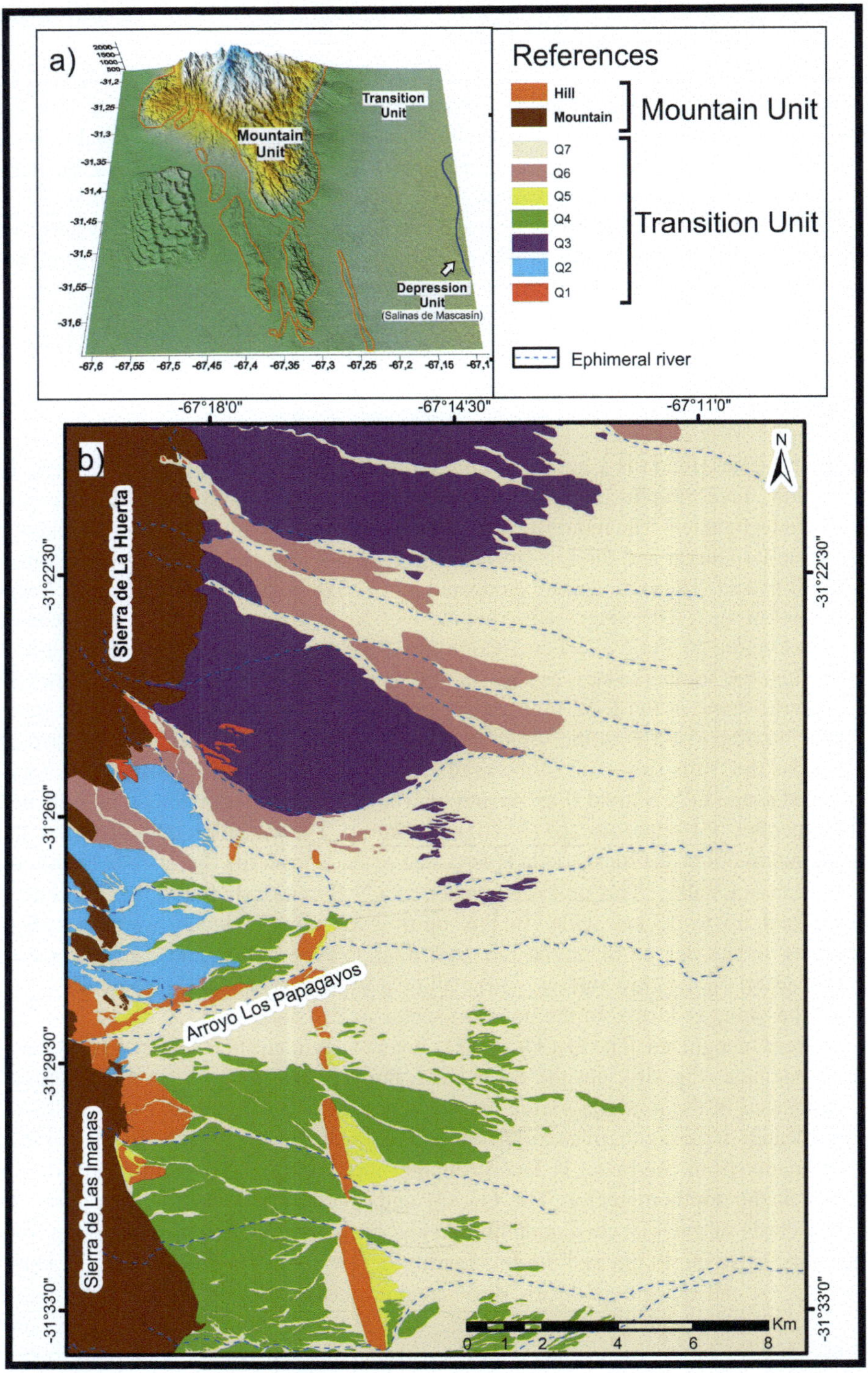


**Fig. 4.** (a) Regional DEM where the three-morphostructural units (Mountain, Transition and Depression units) are shown; (b) detailed geomorphological map of the mountain and eastern transition units.

a N-S elongated playa-lake, 29 km long and 8 km width, with dominant silt, clay, sand and evaporite rocks.

Almost the entire eastern piedmont of Sierra de la Huerta shows clear evidence of Quaternary tectonic activity, as fault scarps, drainage anomalies and alluvial levels offset. Those morphotectonic markers are associated with NW lineaments that were grouped into the Marayes Fault System (Rothis et al. 2017).

According to the methodology proposed by Christophersen et al. (2015), the Arroyo Los Papagayos shows eight fault traces (Fig. 5). Farther north of the Arroyo Los Papagayos, traces 1, 2, 3, 4 and 5 were recognized. They have smooth scarps with their free face to the west, with the exception of trace 3, which presents a scarp with free face towards the east.

South of the Arroyo Los Papagayos, traces 6 and 7 have scarps with the free face to the west, whereas the trace 8 has its free face to the east. In order to characterize scarp morphology, several high-resolution topographic profiles perpendicular to the different traces of the Marayes fault System using DGPS were measured. The topographic profiles exhibit small asymmetric bulges with decreasing elevations towards the south. The trace 2 is 0.5 m high and the trace 3 is 1.4 m (profile 1 and 2, Fig. 5b) high but the trace 6 reaches only 0.3 m (profile 3 and 4, Fig. 5b). Besides, a DEM made in the

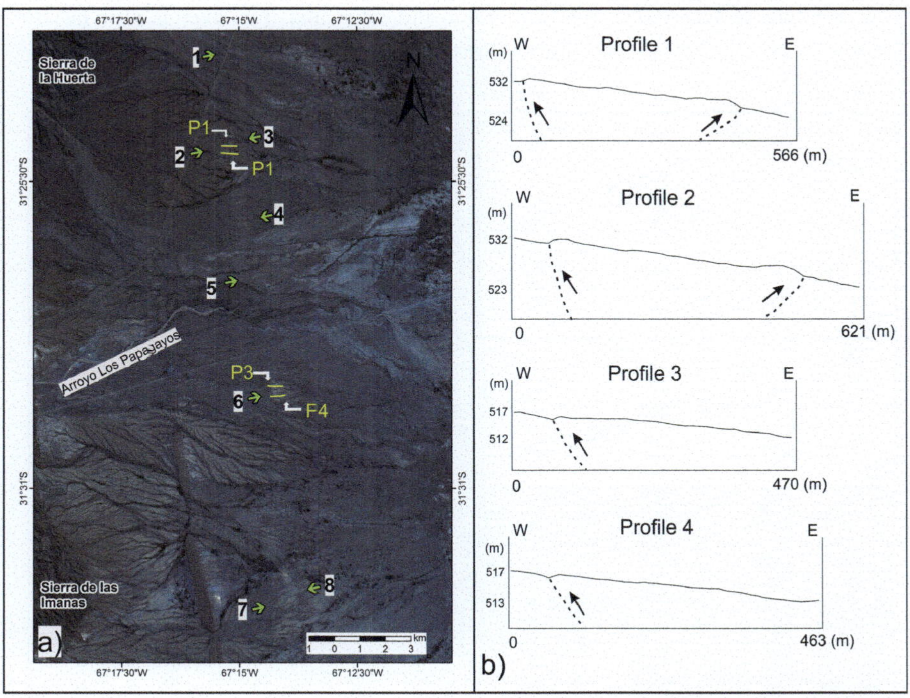


**Fig. 5.** (a) Satellite image of the study area showing the eight traces of the Marayes Fault System. P1, P2, P3, P4, and solid yellow lines depict the location of four high-resolution topographic profiles carried out perpendicular to the scarps; (b) Topographic profiles showing higher scarps in the northern section of the study area (Profiles 1 and 2).

northern sector suggested a small pop up structure, which presents an elongated form trending N-S trend and 450 m wide. Its surface is flat and it is limited by traces 2 and 3 (Fig. 6). This geomorphological feature suggests two faults dipping in opposite directions. Schumm et al. (2002) mentioned that riverbeds are sensitive to modifications in the longitudinal slope of the channel. Those modifications could be related to lithological contrasts or tectonic activity. Alluvial rivers respond to active tectonics in various ways, the main observable effect seems to appear in the channel pattern (Ouchi 1985). In the case of Arroyo Papagayos, the analysis of satellite imagery shows variations in its channel pattern. In the upper and middle courses, the Arroyo Los Papagayos shows a braided multi-trenched low sinuosity channel; however, in the lower course, the morphology of the river changes to mid to high sinuosity channels with a multi-trenched design (Fig. 7a). This pattern change is close to traces 4 and 5, where the river crosses the hanging block of the Marayes Fault System. As it was previously mentioned, the flat geometry of the eastern piedmont of the Sierra de la Huerta is only interrupted by the hills sub-unit, and no lithological contrast is observed. The longitudinal profile of the Arroyo Los Papagayos shows a slightly convex geometry, indicating that this river is not under stable condition (Fig. 7b). The semi-logarithmic plot in Fig. 7c shows a convex Hack profile for the Arroyo Los Papagayos and a straight line corresponding to a simulated graded profile. The G index used to compare how far is the real longitudinal profile to the ideal profile of a river (graded), shows a value of G = 0.07. This low value indicates that this river is still far from reaching its equilibrium profile, thus suggesting structural controls.

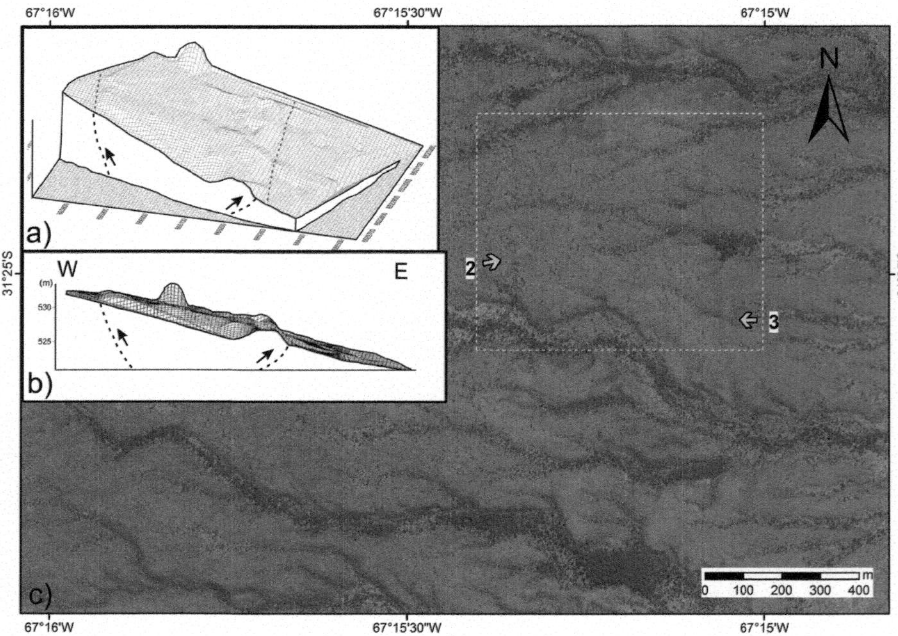


**Fig. 6.** (a) and (b) Block diagram showing the pop up structure between traces 2 and 3; (c) Detailed scale satellite Image showing traces 2 and 3. The yellow rectangle points the location of the DEM.

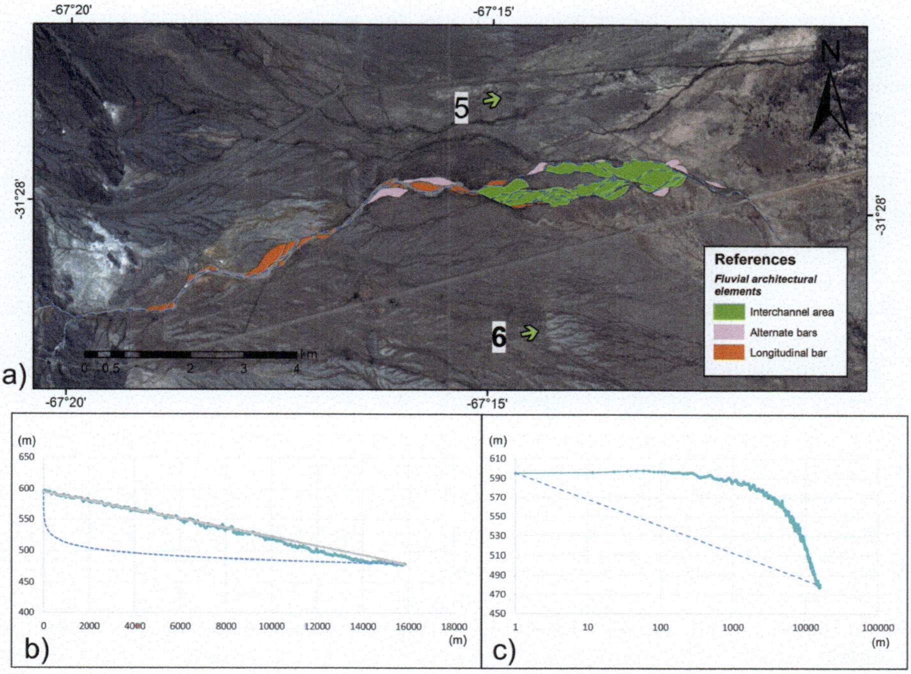


**Fig. 7.** (a) Satellite image showing the change in the fluvial architectural elements along the Arroyo Los Papagayos channel, near traces 5 and 6; (b) The solid light blue line represents the longitudinal profile of the Arroyo Los Papagayos. The dashed light blue line depict the simulated graded profile of the Arroyo Los Papagayos. The solid grey line is a straight line that link the headwater and the river mouth point; (c) Hack profile of the Arroyo Los Papagayos. D ashed line represents the graded profile.

In order to analyse the Marayes Fault System as a potential seismogenic source, the empirical relationships derived from Wells and Coppersmith (1994) were applied. A moderate to large earthquake of Mw = 6.6 was estimated for this fault system. This is in agreement with the tectonic context of an intraplate area like the Western Sierras Pampeanas.

## 7 Discussion and Conclusions

Three main morphostructural units called mountain, transition and depression were determined in the study area. The origin of these units are related to crustal shortening and propagation of a deformation wave into the Andean foreland since the Late Neogene, that is, when the Nazca plate swallowed due to the incorporation of the Juan Fernández Ridge into the subduction zone at ~14 Ma (Gutscher et al. 2000; Ramos et al. 2002; Costa et al. 2018).

Morphostructural units are elongated in a N-S trend with elevations that decrease from north to south. North of the Arroyo Los Papagayos four well-developed alluvial levels were identified. Nevertheless, towards the south three alluvial levels were observed but only the youngest two are extensive and well developed ($Q_6$ and $Q_7$). This could be related to a greater amount of uplift of the mountains unit towards the north. Moreover, in the transition unit there is geomorphological evidence of Quaternary tectonic activity. At least seven traces with NW strike were identified, which are related with discontinuous lineaments and smoothened scarps with free face to the west, except for traces 2 and 7 with their free faces to the east. In order to quantify the scarps located in the eastern piedmont of the Sierra de la Huerta, several high-resolution topographic profiles perpendicular to the Marayes fault System using DGPS were elaborated. The higher scarp is related to trace 3, with 1.4 m in the northern area. The height of the scarps decreases southwards, with values around 0.3 m.

From the analysis of the drainage pattern anomalies, several changes in the Arroyo Los Papagayos were recorded. Upstream of the Marayes Fault System the river shows a braided multi-trenched low sinuosity channel whereas downstream of the fault, the river changes to mid-high sinuosity multi-trenched channels.

Comparison of the distribution of the pattern design of the river in the sector analysed, under certain lithological and geomorphological features, reveals that changes in the pattern design could reflect neotectonic activity. The study of the longitudinal profile of the Arroyo Papagayos that crosses the eastern piedmont of the Sierra de la Huerta shows a significant difference with respect to the theoretical equilibrium profile. It exhibits a convex Hack profile in a semi-logarithmic graph. Taking into account this, if a Hack profile is convex this could be due to stronger uplift with relatively weaker denudation, as proposed by Brookfield (1998). Moreover, the obtained value of G index = 0.07 is so far from 1.0 that it should be considered as the value for a river under steady state conditions.

Deterministic formula that takes into account the complete length of the Marayes Fault System resulted in earthquakes of magnitude Mw = 6.6. This value is in agreement with moderate to large magnitudes defined for intraplate events in the Precordillera and Sierras Pampeanas regions. In addition, Costa et al. (2018) mentioned that the empirical relationships would underestimate the potential of the structures present in Sierras Pampeanas. For these reasons and considering the profuse geomorphological evidence of active tectonics described and discussed in this contribution, the seismogenic potential for this fault system could be greater than expected and should not be underestimated.

## References

Alvarado, P., Beck, S.: Source characterization of the San Juan (Argentina) crustal earthquakes of 15 January 1944 (Mw 7.0) and 11 June 1952 (Mw 6.8). Earth Planet. Sci. Lett. **243**, 615–631 (2006)

Bastías, H., Uliarte, E., Paredes, J.D., Bastías, J.A., Perucca, P.: Neotectónica de la provincia de San Juan. In: 11° Congreso Geológico Argentino, San Juan, pp. 228–244 (1990)

Bossi, G.: Geología de la Cuenca de Marayes-El Carrizal, provincia de San Juan, República Argentina. In: 6° Congreso Geológico Argentino, Buenos Aires, pp. 23–38 (1976)

Brookfield, M.E.: The evolution of the great river systems of southern Asia during the Cenozoic India-Asia collision: Rivers draining southwards. Geomorphology **22**, 285–312 (1998)
Burbank, D.W., Anderson, R.S.: Tectonic Geomorphology, 274 p. Blackwell Scientific, Oxford (2001)
Casa, A., Yamin, M., Wright, E., Costa, C., Coppolecchia, M., Cegarra, M.: Deformaciones Cuaternarias de la República Argentina, Sistema de Información Geográfica. Instituto de Geología y Recursos Minerales. Servicio Geológico Minero Argentino, Publicación 171, DVD-ROM, 1 (2011)
Castro de Machuca, B., Conte-Grand, A., Meissl, E., Pontoriero, S., Sumay, C., Morata, D.: El magmatismo neopaleozoico en la Sierra de la Huerta, Sierras Pampeanas Occidentales, provincia de San Juan: Los pórfidos Marayes Viejo y El Arriero. Rev. Asoc. Geológica Argent. **62**(3), 447–459 (2007)
Chen, Y.C., Sung, Q., Chen, C.N., Jean, J.S.: Variations in tectonic activities of the central and Southwestern Foothills, Taiwan, Inferred from River Hack Profiles. Terr. Atmos. Ocean. Sci. **17**(3), 563–578 (2006)
Christophersen, A., Berryman, K., Litchfield, N.: The GEM Faulted Earth Project (2015). https://doi.org/10.13117/gem.gegd.tr2015.02
Colombi, C.E., Santi Malnis, P., Correa, G., Martínez, R.N., Apaldetti, C.G., Alcober, O., Drovandi, J.: La Formación Balde de Leyes, una nueva unidad estratigráfica de la cuenca Triásica de Marayes-El Carrizal, San Juan. Rev. Asoc. Geológica Argent. **72**(4), 445–455 (2015)
Costa, C.H., Audemard, M.F., Becerra, A., Lavenu, M.N., Machete, G.: An overview of the main quaternary deformation of South America. Rev. Asoc. Geológica Argent. **4**(61), 461–479 (2006)
Costa, C.H., Massabie, A.C., Sagripanti, G.L., Brunetto, E., Coppolecchia, M.: Neotectónica. In: Relatorio XIX Congreso Geológico Argentino, pp. 725–746, Córdoba (2014)
Costa, C.H., Murillo, M.V., Sagripanti, G.L., Gardini, C.E.: Quaternary intraplate deformation in the southeastern Sierras Pampeanas, Argentina. J. Seism. **5**, 399–409 (2001)
Costa, C.H., Owen, L.A., Ricci, W.R., Johnson, W.J., Halperin, A.D.: Holocene activity and seismogenic capability of intraplate thrusts: Insights from the Pampean ranges, Argentina. Tectonophysics **737**, 57–70 (2018)
Cuerda, A.J., Cingolani, C.A., Varela, R., Schauer, O.C.: Descripción Geológica de la hoja 19 d, Mogna, provincia de San Juan. 86 p. (1984)
Flores, M., Criado Roque, P.: Cuenca de San Luis. In: 1° Simposio de Geologia Regional Argentina, Córdoba, pp. 567–580 (1972)
González Bonorino, F.: Algunos problemas geológicos de las Sierras Pampeanas. Rev. Asoc. Geológica Argent. **5**(3), 81–110 (1950)
Gutscher, M., Spakman, W., Bijwaard, H., Engdahl, E.: Geodynamics of flat subduction: seismicity and tomographic constraints from the Andean margin. Tectonics **19**, 814–833 (2000)
Hack, J.: Stream profile analysis and stream gradient index. US Geol. Surv. J. Res. **1**, 421–429 (1973)
Holbrook, J., Schumm, S.A.: Geomorphic and sedimentary response of rivers to tectonic deformation: a brief review and critique of a tool for recognizing subtle epeirogenic deformation in modern and ancient settings. Tectonophysics **305**, 287–306 (1999)
Hugget, R.J.: Introducing landforms and landscape. In: Hugget, R.J. (ed.) Fundamentals of Geomorphology, Abingdon, pp. 1–49 (2007)
Jiménez-Cantizano, F., Antón, L., Soria-Jáuregui, A., Pastor-Martín, C.: Cálculo del perfil teórico de equilibrio de un río en función del índice de gradiente. Geogaceta **62**, 51–54 (2017)

Keller, E.A., Pinter, N.: Active tectonics: Earthquakes Uplift and Landscape. 338 p. New Jersey (1996)

Martinod, J., Husson, L., Roperch, P., Guillaume, B., Espurt, N.: Horizontal subduction zones, convergence velocity and the building of the Andes. Earth Planet. Sci. Lett. **299**, 299–309 (2010)

Massabie, A.: Neotectónica y sismicidad en la región de las Sierras Pampeanas Orientales, Sierras de Córdoba, Argentina. X Congreso Geológico Argentino, Tucumán, pp. 271–274 (1987)

Ouchi, S.: Response of alluvial rivers to slow active tectonic movement. Geol. Soc. Am. Bull. **96**, 504–515 (1985)

Pilger, R.H.: Plate reconstructions, aseismic ridges, and low angle subduction beneath the Andes. Geol. Soc. Am. Bull. **92**, 448–456 (1981)

Ramos, V.A., Cristallini, E.O., Pérez, D.J.: The Pampean flat-slab of the central Andes. J. S. Am. Earth Sci. **15**(1), 59–78 (2002)

Rosello, E.A., Mozetic, M.E., Cobbold, P.R., de Urreiztieta, M., Gapais, D., López-Gamundi, O.: The Valle Fértil flower structure and its relationships with the Precordillera and Pampean Ranges, (30–32° S, Argentina). Third ISAG, St. Malo, pp. 481–484 (1996)

Rosenbaum, G., Mo, W.: Tectonic and magmatic responses to the subduction of high bathymetric relief. Gondwana Res. **19**, 571–582 (2011)

Rothis, L.M.: Análisis de la actividad tectónica cuaternaria en el sector sur de la falla Las Chacras y evaluación de la peligrosidad sísmica. Sierras Pampeanas Occidentales. Provincia de San Juan. Universidad Nacional de San Juan, unpublished Doctoral Thesis (2016)

Rothis, L.M., Santi Malnis, P., Pantano Zuñiga, A., Perucca, L.P.: Evidencias de reactivación de la falla maestra de la cuenca triásica de Marayes, al norte del Cerro Pan de Azúcar, Sierras Pampeanas Occidentales, San Juan, Argentina. XV Reunión de Tectónica, San Juan, p. 131 (2012)

Rothis, L.M., Haro, F.M., Perucca, L.P.: Nueva evidencia morfotectónica en el faldeo oriental de la Sierra de la Huerta: el sistema de falla Marayes, Sierras Pampeanas Occidentales. XX Congreso Geológico Argentino. San Miguel de Tucumán (2017)

Rothis, L.M., Perucca, L.P., Santi Malnis, P., Pantano Zuñiga, A., Alcacer, J.M., Haro, F.M., Vargas, N.: Análisis morfotectónico en el piedemonte suroccidental del antepaís andino, Sierras Pampeanas Occidentales, San Juan, Argentina. Geociências **37**(1), 55–74 (2018)

Santi Malnis, P., Colombi, C.E., Rodríguez Posatini, N., Rothis, L.M., Limarino, C.O.: Caracterización sedimentológica de un Sistema Fluvial Distributivo de clima árido, Arroyo Papagayos, en el piedemonte oriental de las sierras de la Huerta-Imanas, San Juan, Argentina. Andean Geol. **45**(2), 186–228 (2018)

Schumm, S.A., Dumont, J.F., Holbrook, J.M.: Active tectonics and alluvial rivers. J. Quat. Sci. **17**(8), 800–801 (2002)

Wells, D.L., Coppersmith, K.J.: New empirical relationship among magnitude, rupture length, rupture width, rupture area, and surface displacement. Bolletin Seism. Soc. Am. **84**(3), 974–1002 (1994)

Yáñez, G., Ranero, C., von Huene, R., Díaz, J.: Magnetic anomaly interpretation across the southern Central Andes (32°–33.5° S): the role of the Juan Fernández ridge in the late Tertiary evolution of the margin. J. Geophys. Res. **106**, 6325–6345 (2001)

Zámolyi, A., Székely, B., Draganits, E., Timár, G.: Neotectonic control on river sinuosity at the western margin of the Little Hungarian Plain. Geomorphology (2010). https://doi.org/10.1016/j.geomorph.2009.06.028

# Morphometric Analysis of River Basins Using GIS as a Basis for Peak Discharge Calculation. Case of Study: Pre-Andean Section of National Route 150, Argentina

María Yanina Esper Angillieri[1,2(✉)], Oscar Mario Fernández[2], Miguel Pereyra[2], Carla Ginesta Torcivia[1], and Natalia Ríos[1]

[1] CONICET-CIGEOBIO, San Juan, Argentina
yaninaesper@gmail.com
[2] Facultad de Ingeniería, EICAM, Universidad Nacional de San Juan, San Juan, Argentina
mfernandez@eicam.unsj.edu.ar

**Abstract.** This work analyzes various morphometric characteristics of river basins in order to evaluate debris flood or flash flood hazards. The chosen area includes a section of National Route 150 up to the International Border Pass to Chile, access to the future international tunnel of Agua Negra, between Argentina and Chile in a sector of the Western and Central Pre-Andean. One of the most important tools in hydrological analysis is the morphometric survey of a river basin, which allows establishing evaluation parameters on the behavior of the hydrological system of the river basin areas. Geographical information system (GIS) has emerged as an efficient tool in delineation of drainage pattern and ground water potential and its planning. A total of 122 river basins were identified and delineated, but only 3 (those of larger surface area) were selected for morphometric analysis and as a suitable case to evaluate the potential magnitude of flash floods in relation to their associated hazard level in the region. In addition, a peak discharge analysis was performed to predict approximately the behavior of the river basin in the presence of a series of theoretical rainstorms that may generate unusual runoff volumes that make up such flash floods. This study, when properly combined with geomorphology and geology analyses, helps to elaborate a primary hydrological diagnosis to predict the approximate behavior of the river basin at the time of heavy rainstorms.

**Keywords:** Morphometry · Peak discharge · Flash flood · GIS · Pre-Andean · San juan · Argentina

## 1 Introduction

The hydrological response of a drainage basin is defined by the production of runoff against a given rainfall, which in turn is characterized by basin morphometric properties, soil characteristics and land use pattern. While the soil characteristics and land use pattern control the infiltration loss, the distribution of the remaining excess rainfall is governed by basin morphometric properties (Ajibade et al. 2010).


M. M. Collantes et al. (Eds.): *Advances in Geomorphology and Quaternary Studies in Argentina*, SPRINGEREARTH, pp. 129–135, 2020.
https://doi.org/10.1007/978-3-030-22621-3_6

Discharge estimation of hydrological river basins based on rainfall-runoff analysis is a very frequent practice in Hydrology Engineering. But, also quite often, basin discharge measurements are not readily available for statistical analysis. This lack of information can be satisfactorily resolved in most instances by using basin delineation and morphometric analyses, which provide resources for describing the hydrological behavior of a basin and are a prerequisite for runoff modeling (Magesh et al. 2011; Thomas et al. 2012). Morphometric analysis of a river basin provides a quantitative description of the drainage system, which is an important aspect of the characterization of watersheds (Strahler 1964).

Morphometric analyses have been widely used for characterizing stream basins. Some recent examples are the following: Sreedevi et al. (2013), Mesa (2006), Esper Angillieri (2007, 2008, 2012), Esper Angillieri and Fernández (2017), Esper Angillieri and Perucca (2014a, 2014b), Esper Angillieri et al. (2016, 2017), Perucca and Esper Angillieri (2011), Magesh and Chandrasekar (2014), Magesh et al. (2012) and Vieceli et al. (2015), among many others.

The estimation of the maximum theoretical discharge is of critical importance because it controls the maximum flow velocities, their probable extent, their impact, and the possibility of building contention and/or mitigation works (Hungr 2000).

In this work, the quantitative analysis of morphometric parameters, using GIS, is carried out for a section of the Pre-Andean region of Argentina, focusing on peak discharge calculations. The method presented here is easy to reproduce and applicable to other mountain ranges of comparable topography and environment and the result from this investigation will aid in suggesting management techniques that can be adopted to mitigate the problems being caused by processes operating in the studied river basins.

## 2 Study Area

The present study area is located in a sector of the Pre-Andean ranges, in the county of Iglesia, province of San Juan, at about 206 km NW of the city of San Juan, Argentina. The chosen area includes a section of the National Road No. 150 up to the Agua Negra pass between Argentina and Chile. It also includes the road access to the future international tunnel (Fig. 1). This region has a general arid climate. The total annual average precipitation is very scarce: only about 121 mm. Winter temperatures are generally mild, ranging between −9 °C and 8 °C, whereas summers are very dry and hot, with maximum temperatures of 44 °C (Minetti et al. 1986). In the period between December and March, heavy rainfalls of short duration but torrential character are responsible for the generation of large flash floods. The steep gradient of this area rapidly drains these precipitated volumes of water, funneling them towards the Río Jáchal.

The Pre-Andean ranges are part of a fold-and-thrust belt that could be divided in the Western, Central and Eastern sections, based on the stratigraphy, timing and style of deformation (Ortiz and Zambrano 1981). The study area supports a wide range of geologic units. The sequence initiates with Cambrian sedimentary rocks mainly composed of limestone. Most of the outcrops consist of marine, clastic sedimentary rocks of Silurian-Devonian ages, continental Carboniferous sandstones and Neogene conglomerates, sandstones and claystones covered by Pleistocene/Holocene deposits. The Quaternary deposits are composed of gravels, sands, marls, and clays and are restricted to narrow river channels and valleys. Data were acquired from geological sheets published by the Servicio Geológico Minero Argentino (SEGEMAR, Argentine Mining Geological Service) on a 1:250,000 scale.

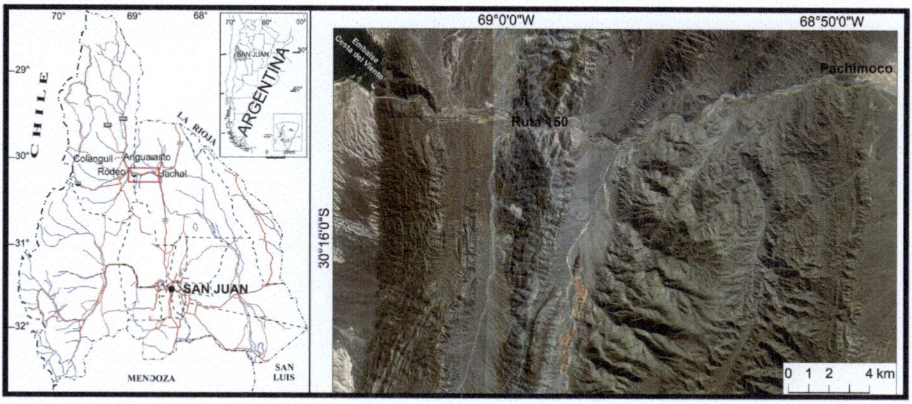


**Fig. 1.** Location of the study area. Image@2015CNES (SPOT 5) from Google Earth TM.

## 3 Materials and Methods

In the present research project, the river basins were identified from digital elevation models (ASTER GDEM V2 with a 30 m spatial resolution; NASA ASTER 2011), satellite images from Google Earth™ (SPOT 5 image), field work observations and GIS technology (ESRI's ArcGis 10.3). The topographic basin boundaries can be calculated using the standard terrain analysis methods based on a DEM, which are implemented with most GIS software based on the algorithms of flow-direction and flow-accumulation functions. Flow directions were then calculated with the eight-direction (D8) flow model which assigns the flow direction from each grid cell to one of its eight adjacent cells, on the direction with the steepest downward slope. The D8 method was introduced by O'Callaghan and Mark (1984). Several morphometric characteristics of the basins are analyzed and some basic parameters (area, perimeter, length, mean width, maximum and minimum heights, basin relief, main channel length and slope of the main channel) are calculated (Table 1). The main channel length, or longest flow path, was identified as the distance measured along the stream channel from source to outlet points (Schumm 1956). In GIS software, the accumulated cost of travelling from the source grid cell (the outlet in this case) to each other grid cell in the raster dataset can be easily calculated. The source of the longest flow path can then be selected as the cell with the highest cost accumulation value. The distance to the outflow profile was calculated using the previously calculated flow direction grid (Zlatanovic and Gavric 2013). The choice of the morphometric parameters have been found to correlate highly with peak discharge.

The Generalized Rational method (Ruhle 1966) for estimating direct runoff from storm rainfall was chosen to estimate the peak discharge. This method relates peak discharge to contributing drainage area, rainfall intensity for duration equal to a basin response time (time of concentration by California Culvert Practice (Department of Public Works 1995), and a coefficient that represents hydrological abstractions and hydrograph attenuation.

**Table 1.** List of derived parameters, equations and references. Basin 1: La Tranca river; basin 2: Caracol river and basin 3: Zanja Honda river.

| Basin | A [km$^2$] | P [km] | L [m] | H [m a.s.l] | h [m a.s.l] | | Lcp [m] |
|---|---|---|---|---|---|---|---|
| 1 | 141.297 | 99.151 | 27455.716 | 3406 | 1519 | | 32147.170 |
| 2 | 102.888 | 65.901 | 17996.786 | 3017 | 1399 | | 21966.260 |
| 3 | 55.623 | 59.461 | 16655.442 | 2579 | 1372 | | 25104.580 |
| | Elongation ratio (Schumm 1956) | Form factor (Horton 1931) | Sinuosity index (Schumm 1977) | Mean width | Basin relief (Hadley and Schumm 1961) | Relief ratio (Schumm 1956) | Melton ratio (Melton 1957) |
| | $(Re = \sqrt{4A/\pi})/L$ | $Ff = A/L^2$ | S = Lcp/L | Wm = A/L | Hr = H − h | Rr = Hr/L | MR = Hr/A$^{0.5}$ |
| 1 | 0.489 | 0.187 | 1.171 | 5146.378 | 1887 | 0.069 | 0.159 |
| 2 | 0.636 | 0.318 | 1.221 | 5717.014 | 1618 | 0.090 | 0.160 |
| 3 | 0.505 | 0.201 | 1.507 | 3339.653 | 1207 | 0.072 | 0.162 |

## 4 Results and Discussion

A total of 122 river basins were identified in the area, but only 3 (those of larger surface area) were selected for morphometric analysis and as a suitable case to evaluate the potential magnitude of flash floods in relation to their associated hazard level in the region (Fig. 2).

It is appropriate to take into account that the automatic extraction of basins was conducted using DEM of 30 m resolution, which subsequently restricts the accuracy of the estimated drainage areas. However, this methodology turns out to be more precise than the conventional manual methodology (Esper Angillieri and Fernández 2017).

Elongation ratio (Re) and form factors (Ff) are important parameters in analyzing the basin shape. The estimated elongation ratios (Table 1) are 0.48, 0.63 and 0.50, which shows that the basin areas are characterized by high run off capacity along the stream flow path, a feature associated with high relief and steep surfaces (Strahler 1964). The form factor value was measured and it indicates that the basins have more elongated shapes in nature, with the characteristic of flatter peak flow for a longer duration. As such, elongated basins are highly vulnerable to flood flows with respect to circular-shaped basin areas. Likewise, the basin relief (Hr) and relief ratio (Rr) indicate the intensity of the erosion process operating on the slope of the basins, low infiltration, and high runoff conditions (Table 1). Similar results have been obtained in other basins of this sector of the Precordillera geological unit (Esper Angillieri and Fernández 2017, Esper Angillieri et al. 2017).

In general, the Melton ratio (MR; Melton 1957) values are > 0.3, which imply a moderate probability of experiencing frequent flooding with low material content

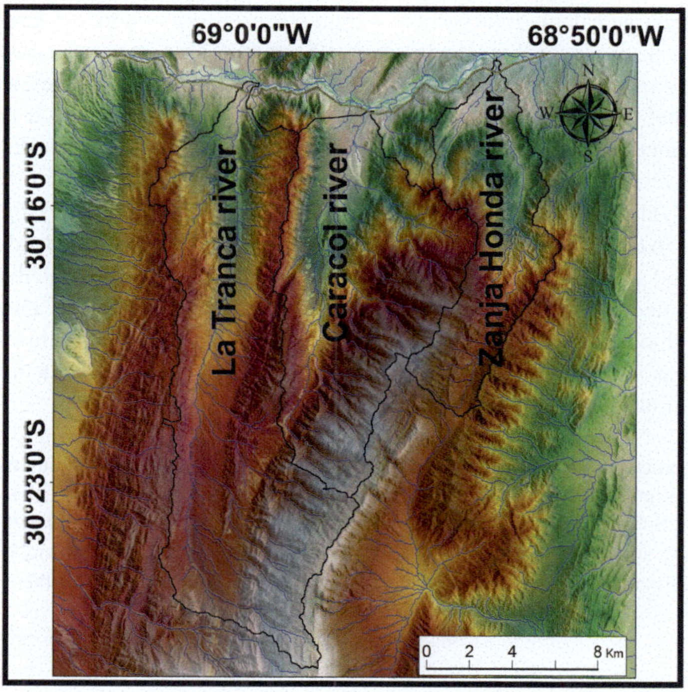


**Fig. 2.** The mayor surface basin area on the study area. Image@ digital topographic data (Aster GDEMV2).

(Table 1), which is not consistent with the field observations and analysis of events, which mostly try to debris flood or flash flood. Nevertheless, these statements can be interpreted ambiguously, because it will depend on the dimensions or extent of the storm as well as its duration and intensity. On the other hand, the probability for hydrograms of floods with pronounced peaks of short duration occurrence is high, considering the strong relief and recent events. Similar results were obtained and reported in Esper Angillieri and Perucca (2014a, 2014b).

The resulting peak discharges for the La Tranca, Caracol and Zanja Honda rivers, is 215.38 $m^3/s$, 200.93 $m^3/s$ and 93.03 $m^3/s$, respectively. These values point to a high probability for serious flood hazard, with peak discharges greater than 100 $m^3/s$.

## 5 Conclusions

The conventional methods of morphometric analysis are time consuming, tiresome and error prone, while use of GIS technique allows for more reliable and accurate estimation of similar parameters of basins. The computed morphometric characteristics of basins in the study area revealed that the basins are elongated, with high relief and steep slope. The Melton ruggedness number (MR; Melton 1957) suggests high susceptibility to flash flooding, with low amounts of sediment transported. The values obtained from the peak discharge calculations point to a high occurrence probability for serious flood hazard, with peak discharges greater than 100 $m^3$/s. These measurements permit to predict, approximately, the behavior of the basin in the presence of a series of theoretical rainstorms that may generate unusual runoff volumes that make up such flash floods. The morphometric analysis provides a basis for estimating subsequent peak flood flows for different return periods and rain duration, as a direct function of its morphometric parameters. It also provides important elements to the hydraulic and hydrological design of structures to contain flood and sediment, as well as for floods and landslides analysis, evaluation and susceptibility zoning by floods. The methodology presented here is easy to reproduce and it may be applied to other dry mountainous regions, as a useful tool to assess flash floods in relation to hazard. Finally, the results presented in this paper could be included as a tool when planning, designing and managing the roads through areas featuring similar conditions as those of the present work.

**Acknowledgements.** The authors express their acknowledgement for the funding received from CONICET (Argentine National Council of Scientific and Technological Research) and CICITCA (Secretary of Science and Technology, Universidad Nacional de San Juan) supporting this research. Same appreciation is given to CIGEOBIO for providing funds for ArcGIS 10.3 software license. Finally, the authors would also like to thank several reviewers for their constructive comments.

## References

Ajibade, L.T., Ifabiyi, I.P., Iroye, K.A., Ogunteru, S.: Morphometric analysis of Ogunpa and Ogbere drainage basins, Ibadan, Nigeria. Ethiop. J. Environ. Stud. Manag. **3**(1), 13–19 (2010)

Department of Public Works: California culvert practice, second. DPW, Division of Highways, Sacramento, 87 p. (1995)

Esper Angillieri, M.Y.: El aluvión del 13 de febrero de 1944 en la Quebrada del Carrizal, Departamento Iglesia, Provincia de San Juan. Rev. Asociación Geológica Argentina **62**, 283–288 (2007)

Esper Angillieri, M.Y.: Morphometric analysis of Colangüil river basin and flash flood hazard, San Juan, Argentina. Environ. Geol. **55**, 107–111 (2008)

Esper Angillieri, M.Y.: Morphometric characterization of the Carrizal basin applied to the evaluation of flash floods hazard, San Juan, Argentina. Quatern. Int. **253**, 74–79 (2012)

Esper Angillieri, M.Y., Fernández, O.M.: Morphometric analysis of river basins using GIS and remote sensing of an Andean section of Route 150, Argentina. A comparison between manual and automated delineation of basins. Rev. Mex. Cienc. Geol. **34**(2), 150–156 (2017)

Esper Angillieri, M.Y., Perucca, L.P.: Geomorphology and morphometry of the de La Flecha river basin, San Juan, Argentina. Environ. Earth Sci. **72**, 3227–3237 (2014a)

Esper Angillieri, M.Y., Perucca, L.P.: Caracterización morfométrica de la Cuenca del Río Seco a propósito de las fuertes precipitaciones de enero de 2013, Departamento Sarmiento, San Juan, Argentina. Bol. Soc. Geol. Mex. **66**(2), 235–245 (2014b)

Esper Angillieri, M.Y., Espejo, K., Lara, G., Perucca, L., Rothis, M.: Morfometría de cuencas como base en la evaluación de la susceptibilidad/peligrosidad por aluviones, caso de estudio: cuenca del Río del Agua, San Juan, Argentina. Rev. Asoc. Geol. Argentina **73**(4), 513–519 (2016)

Esper Angillieri, M.Y., Perucca, P., Vargas, N.: Catastrophic flash flood triggered by an extreme rainfall event in El Rodeo village, January 2014. Northwestern Pampean Ranges of Argentina. Geogr. Ann. Ser. A Phys. Geogr. **99**(1), 72–84 (2017)

Hadley, R.F., Schumm, S.A.: Sediment sources and drainage basin characteristics in upper Cheyenne River Basin. U.S. Geological Survey Water-Supply Paper 1531-B, 198 p. (1961)

Horton, R.E.: Drainage basin characteristics. Am. Geophys. Union Trans. **13**, 350–361 (1931)

Hungr, O.: Analysis of debris flow surges using the theory of uniformly progressive flow. Earth Surface Process Land. **25**, 483–495 (2000)

Magesh, N.S., Chandrasekar, N.: GIS model-based morphometric evaluation of Tamiraparani subbasin, Tirunelveli district, Tamil Nadu, India. Arab. J. Geosci. **7**(1), 131–141 (2014)

Magesh, N.S., Chandrasekar, N., Soundranayagam, J.P.: Morphometric evaluation of Papanasam and Manimuthar watersheds, parts of Western Ghats, Tirunelveli district, Tamil Nadu, India: a GIS approach. Environ. Earth Sci. **64**, 373–381 (2011)

Magesh, N.S., Chandrasekar, N., Kaliraj, S.: A GIS based automated extraction tool for the analysis of basin morphometry. Bonfring Int. J. Ind. Eng. Manag. Sci. **2**(1), 32–35 (2012)

Melton, M.A.: An analysis of the relation among elements of climate, surface properties and geomorphology. Office of Naval Research Project NR389–042. Department of Geology Columbia University, New York, Technical report 11 (1957)

Mesa, L.M.: Morphometric analysis of a subtropical Andean basin (Tucumán, Argentina). Environ. Geol. **50**, 1235–1242 (2006)

Minetti, J.L., Barbieri, P.M., Carleto, M.C., Poblete, A.G., Sierra, E.M.: El régimen de precipitación de la provincia de San Juan. Informe técnico 8. CIRSAJ-CONICET, San Juan (1986)

NASA ASTER Global Digital ElevationMapV2 (2011). http://gdem.ersdac.jspacesystems.or.jp

Ortiz, A., Zambrano, J.: La provincia geológica Precordillera oriental. In: 8° Congreso Geológico Argentino, Actas, vol. 3, pp. 59–74 (1981)

O'Callaghan, J.F., Mark, D.M.: The extraction of drainage networks from digital elevation data, computer vision. Graph. Image Process. **28**, 328–344 (1984)

Perucca, L.P., Esper Angillieri, M.Y.: Morphometric characterization of the Molle Basin applied to the evaluation of flash floods hazard, Iglesia Department, San Juan, Argentina. Quatern. Int. **233**, 81–86 (2011)

Rühle, F.G.: Determinación del Derrame Máximo Superficial de las Cuencas Imbríferas. La Ingeniería, N° 987. Centro Argentino de Ingenieros. 2ª edición. Buenos Aires (1966)

Schumm, S.A.: Evolution of drainage systems and slopes in badlands at Perth Amboy, New Jersey. Geol. Soc. Am. Bull. **67**, 597–646 (1956)

Schumm, S.A.: The Fluvial System, 338 p. Wiley, New York (1977)

Sreedevi, P.D., Sreekanth, P.D., Khan, H.H., Ahmed, S.: Drainage morphometry and its influence on hydrology in a semi-arid region: using SRTM data and GIS. Environ. Earth Sci. **70**(2), 839–848 (2013)

Strahler, A.N.: Quantitative geomorphology of drainage basin and channel networks. In: Chow, V.T. (ed.) Handbook of Applied Hydrology, pp. 4–76. McGraw Hill, New York (1964)

Thomas, J., Joseph, S., Thrivikramji, K., Abe, G., Kannan, N.: Morphometrical analysis of two tropical mountain river basins of contrasting environmental settings, the southern Western Ghats, India. Environ. Earth Sci. **66**(8), 2353–2366 (2012)

Vieceli, N., Bortolin, T.A., Mendes, L.A., Bacarim, G., Cemin, G., Schneider, V.E.: Morphometric evaluation of watersheds in Caxias do Sul City, Brazil, using SRTM (DEM) data and GIS. Environ. Earth Sci. **73**(9), 5677–5685 (2015)

Zlatanović, N., Gavrić, S.: Comparison of an automated and manual method for calculating storm runoff response in ungauged catchments in Serbia. J. Hydrol. Hydromech. **61**(3), 195–201 (2013)

# Distributary Drainage Systems in La Huerta Range, Western Pampean Ranges, San Juan, Argentina


Paula Santi Malnis[1,2(✉)], L. Martin Rothis[1,3], and Carina E. Colombi[1,2]

[1] CONICET – CIGEOBIO, San Juan, Argentina
[2] Instituto y Museo de Ciencias Naturales, Facultad de Ciencias Exactas, Físicas y Naturales, Universidad Nacional de San Juan, Av. España 400 Norte, 5400, San Juan, Argentina
{pmalnis, ccolombi}@unsj.edu.ar
[3] Laboratorio de Geomorfología y Neotectónica, Facultad de Ciencias Exactas, Físicas y Naturales, Universidad Nacional de San Juan, Av. Ignacio de la Roza y Meglioli, Rivadavia, 5400, San Juan, Argentina
marothis@gmail.com



**Abstract.** Transversal drainage systems in the southern portion of the Bermejo intermontane basin -San Juan province, Argentina- are organized in five main rivers that reach the longitudinal Río Bermejo valley. Geomorphological and sedimentological survey of the western piedmont of the La Huerta range shows that there is a north-south trend change of drainage basin morphometry and facies architecture. Facies distribution of the Las Chacras river, in the northern portion of the piedmont, fits with the classical alluvial fan model, with a divergent net of channels from the fan apex. In the middle portion of the piedmont (the Arenal and Puquial rivers), also alluvial fan facies are present with eolian interaction, triggering disconnection of mid fan facies from distal facies. In the southern portion of the piedmont, drainage system flows parallel to axial drainage system (Bermejo River). Facies architecture is different from those observed in the northern portion. In the Cañada River, flows keep channelized until its distal portion in mid to high sinuosity channels that deposit alternate and scroll bars. In the distal portion, main channel diverge in a distributary net of channels that develops a fan-shaped deposit characterized by terminal splays and eolian dunes. Facies distribution allow considering the La Cañada River as a distributive fluvial system. The last system under study is the Los Palos River which frequently disappears because it has a permeable substratum and deposit mostly terminal splays. The Los Palos River is interpreted as a flood-out system.



**Keywords:** Alluvial · Distributive fluvial · Seasonal climate · Tectonics · Andean ranges




M. M. Collantes et al. (Eds.): *Advances in Geomorphology and Quaternary Studies in Argentina*, SPRINGEREARTH, pp. 136–157, 2020.
https://doi.org/10.1007/978-3-030-22621-3_7

## 1 Introduction

The definition of new sedimentary models, as the distributive fluvial systems (DFS), challenges sedimentation patterns in piedmonts and continental valleys, where they have been considered the most common modern sedimentary environment (Weissmann et al. 2010; Fielding et al. 2012; Davidson et al. 2013). With the aim of increasing our knowledge about sedimentation patterns in piedmonts and to get a model that could characterize the DFS systems, different transversal drainage systems that drain the western piedmont of the La Huerta range, Western Pampean Ranges, in the broken Andean foreland (Jordan and Allmendinger 1986), have been described (Fig. 1). All these fluvial systems are graded to the Bermejo basin as the local base level. This is a longitudinal basin that corresponds to the fore-deep basin of the Central Andes in these latitudes (30°–32° S) and it behaves as a high aggrading basin (Damanti 1993). Active tectonics of the Andes, plus the highly seasonal climate and climate changes occurred in the last 150,000 years, promote the renewal of the Bermejo stream profile, including its role as regional base level. This triggered changes in the stream equilibrium profile all over the Bermejo drainage basin, including in it transverse systems and consequently its tributaries, with different dilatation times depending of the distance to the Bermejo River. The combined allogenic factors that influenced the Bermejo stream profile produced episodes of avulsion and aggradation, turning this region in a great scenario to study alluvial sedimentation and DFS formation in continental settings.

## 2 Geological Setting

Western Pampean Ranges are basement blocks located in the foreland of the Andes, which exhumation is related to the flat subduction of the Nazca plate (Ramos et al. 2002). This region presents high deformation and seismicity related to active tectonics, becoming one of the greatest seismic hazard regions in Argentina (Costa et al. 2006; Rothis et al. 2018). The La Huerta Range is one of the Pampean ranges and is a NW-SE basement block composed of crystalline basement rock of mica and garnet-bearing, feldspar gneiss and schist, quartz-diorite and quartz-tonalite. Besides, this block includes sedimentary rocks of the Triassic Marayes Group and finally it is covered with alluvial terraces of the Late Pleistocene to Holocene deposits (Vujovich et al. 2007). The study area presents three morphostructural units (Fig. 1): (1) the Mountain region of La Huerta and Las Chacras Ranges; (2) the Transitional region, which includes many alluvial fans levels, and (3) the Valley region that embraces the Las Chacras Eolian field, small-scale eolian fields and the Bermejo River (Rothis et al. 2018). The mountain region is divided in three sub-units: (a) the higher mountain sub-unit, reaches 2,060 m a.s.l., composed mostly of crystalline rocks of the La Huerta Range, (b) the hill sub-unit presents elevations between 945–680 m a.s.l., composed of Mesozoic sedimentary rocks, (c) the lower mountain sub-unit reaches up to 750 m a.s.l., formed by crystalline basement rocks. The Transitional region has different levels of Quaternary deposits which are deeply disturbed by active tectonics (Rothis et al. 2018), being the most conspicuous the scarp of the Pan de Azúcar fault, a north-south oriented, strike inverse fault, considered as the southern expression of the Valle Fértil Mega-fault (Bossi 1976; Bastías et al. 1990; Rothis et al. 2018). Another prominent geomorphological unit is the Las Chacras eolian field of Pleistocene and Holocene age (Tripaldi and Forman 2007). It is a 100 $km^2$ eolian field located between the Bermejo flood-plain and the piedmont, triggering the change of central and southern piedmont river courses toward the south.

The climate in the study area is divided in two types according to its topographic position: a seasonal steppe climate (semi-arid), occupying the upland catchment area and a seasonal desert climate (arid) in the lowlands, which presents lower mean annual precipitations than the highlands (BW wha Köppen classification; Pereyra 1996).

Morphometric studies carried out by Rothis et al. (2018) in the Western piedmont of La Huerta range identified 12 drainage basins, which present elongated to very elongated morphologies, NE-SW orientation and have an asymmetric cross-section, with a broader northern side for Las Chacras, Arenal and tributaries of Cañada and Los Palos rivers (Rothis et al. 2018). Drainage basin morphologies are typical of a region with active tectonics, while asymmetry is related to differential tectonic offset in a north-south trend, controlled by the Pan de Azúcar fault and NE structures (Rothis et al. 2018).

## 3 Materials and Methods

Based on satellite imagery and remote sensing data analyzed through a GIS platform (QGIS 2.14.2, 2016) thematic maps were performed for the characterization of fluvial geomorphology and drainage patterns in the piedmont.

A GIS platform allowed to obtain drainage basin parameters and morphometric-derived parameters, for these field calculation allowed to obtain the perimeter, basin area, either of the polygons or each drainage basin. The drainage basin was digitalized using aerial photographs (23 × 23 cm) and high resolution satellite images from Google Earth and Bing Maps available from Openlayer plugin of QGIS 2.14.2, using a 1:2,000 scale. Multispectral Sentinel 2A satellite images were used to support mapping of different landforms as dunes and salt ponds and to identify the distributary network through vegetation concentration (Image: S2A 20160928T192805_A006628_T19JFF). For these purposes, some RGB combinations were used as: 12-8-2, 11-8-4 and 8-11-4, in order to identify different land cover as vegetation, sedimentary rocks, eolian dunes, gravels, sands, fine materials, salty areas, etc.

In this work, the twelve drainage basins defined by Rothis et al. (2018) where grouped in five, which are the five major order rivers that reach the local base level. Basic parameters are area (A), perimeter (P), length (L), mean width (W), basin order (from Horton 1945), basin relief (Hr, Hadley and Schumm, 1961), and total stream length (Tsl). Derived parameters are compactness index (Kc), circularity index (C), elongation ratio (Re), form factor (Ff), drainage density (Dd), Melton ruggedness number (MRN). These were obtained from the following equations:

| | |
|---|---|
| $Kc = 0,28(P/\sqrt{A})$ | Gravelius 1914 |
| $C = 4\pi A/P2$ | Miller 1953 |
| $Re = (\sqrt{4A/\pi})/L$ | Schumm 1956 |
| $Ff = A/L2$ | Horton 1932 |
| $Dd = Tsl/A$ | Horton 1945 |
| $MRN = Hr/A^{0.5}$ | Melton 1957 |

In a tributary channel networks stream order is classified in ascending order, where a new stream order is acquired when two channels of the same order join to form a new channel (Strahler 1952). However, in distributary networks the bifurcation order is used to classify distributary channels (Edmonds and Slingerland 2007). The bifurcation order is defined as "the number of times a channel splits upstream of a given bifurcation" and is notated with "n" in ordinal ascending order, meaning that higher bifurcation orders represent distributary channels farther down the trunk channel or the delta (Edmonds and Slingerland 2007). Terminal distributary channels forms through multiple successive splits from the main channel, in a reverse pattern of the drainage basin (Olariu and Bhattacharya 2006). Consequently, distributary networks lack new tributary channels, so the distributary channel would inherit the Strahler stream order of the trunk channel. In this work, the classification of distributary channels will be described as follows: adding a suffix to the Strahler stream order of the trunk or feeding channel that corresponds to the bifurcation order of that channel, this is: $S_n$: $7_3$. In this example, the number 7 means a seventh order stream and the suffix 3 means a third-order bifurcation order. The suffix increases when distributary channels bifurcate again and form another generation (order) of terminal distributary channels (Olariu and Bhattacharya 2006).

To describe each drainage system, different geomorphological elements were identified as longitudinal, alternate and lateral accretion bars, proximal splays, distal splays, saline lakes, ponds, eolian dunes and floodplains.

Fieldwork controls in points of interest allow the checking of the interpreted data through the GIS platform, using natural trenches. To better understand the geomorphological elements interpreted in the GIS platform of La Cañada River, a longitudinal transect upstream from the fan-shaped cone has been surveyed. In this transect, texture, color, sedimentary structures and internal organization (bounding surfaces) of the geomorphological elements were recognized.

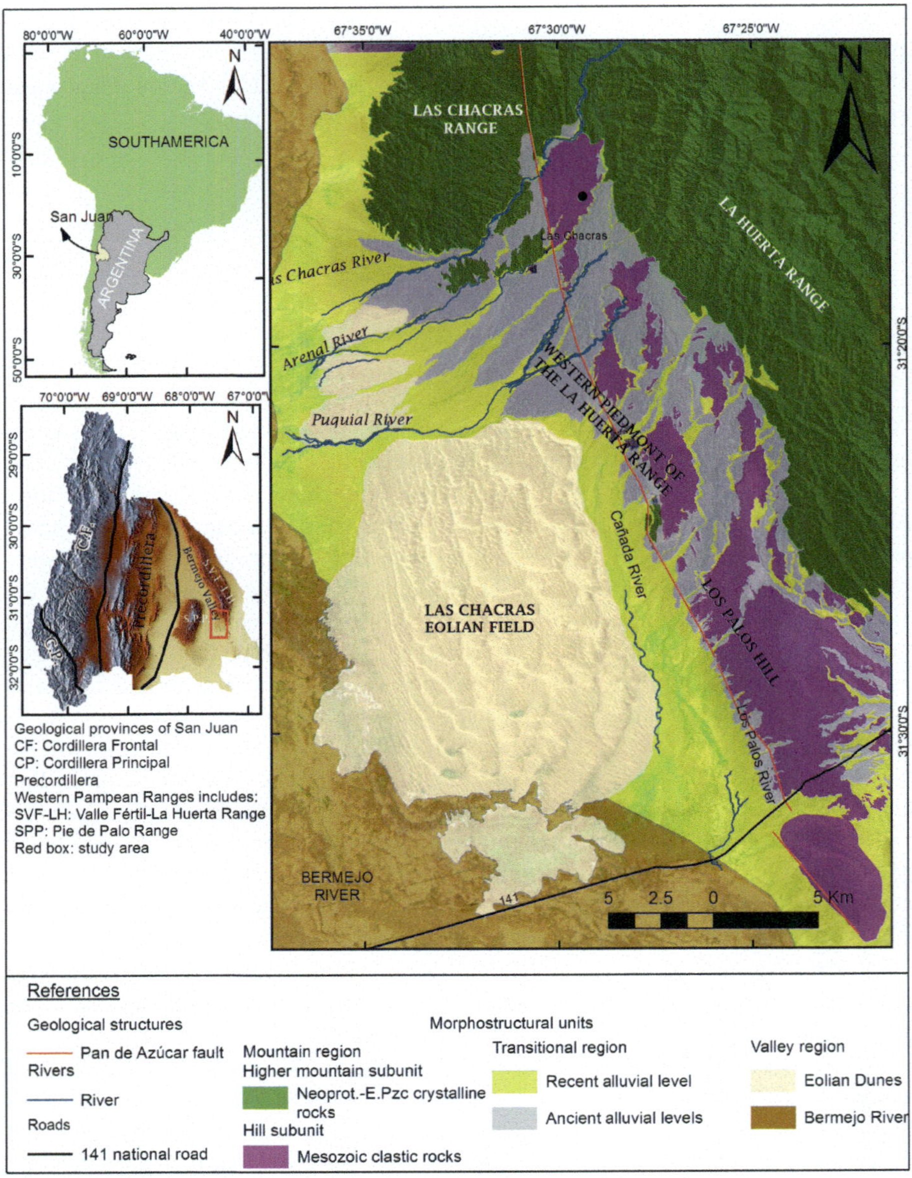


**Fig. 1.** Location map of San Juan provinces, geological provinces of San Juan and geological map of the study area, the western piedmont of the La Huerta Range in Western Pampean Ranges.

# 4 Results

## 4.1 Basin Morphometry

Five main rivers drain the western piedmont of the La Huerta Range to the Bermejo River valley. These rivers are, from north to south: (1) Las Chacras, (2) Arenal, (3) Puquial, (4) La Cañada and (5) Los Palos (Fig. 2).

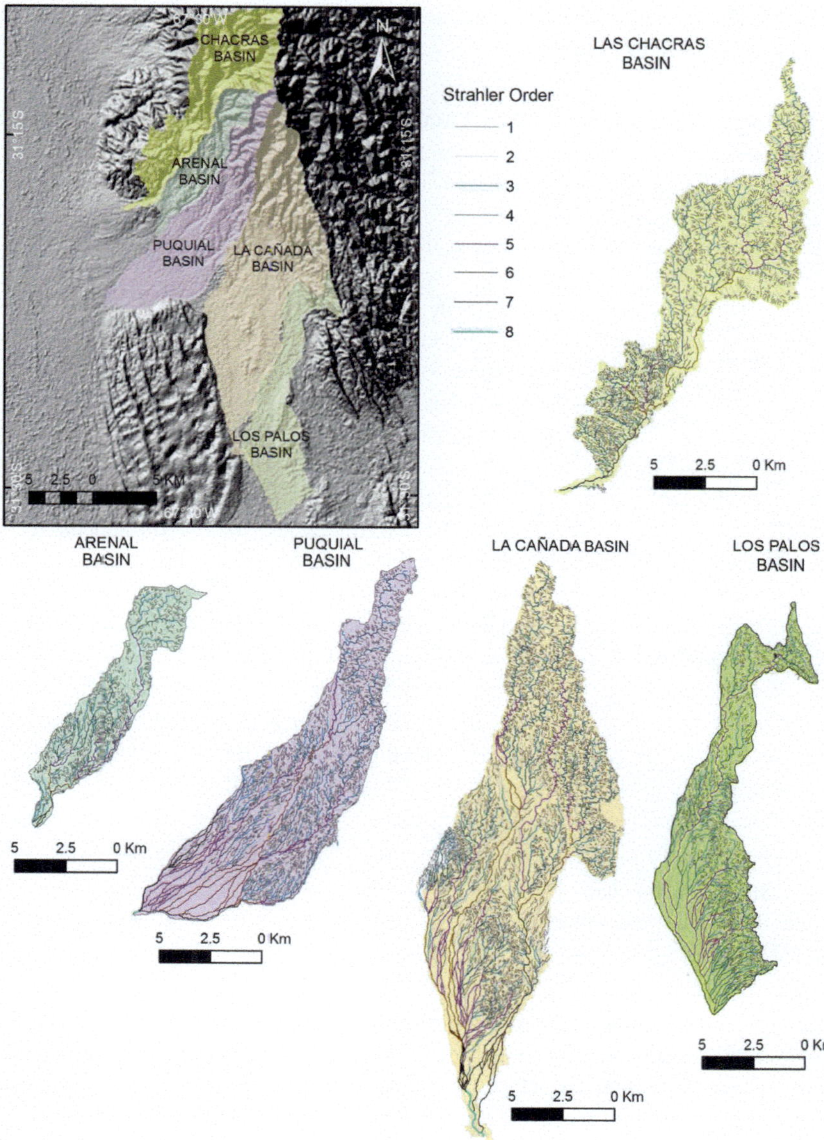


**Fig. 2.** Drainage basins of the Las Chacras, Arenal, Puquial, La Cañada and Los Palos rivers. Strahler stream order are shown in colors.

Drainage basins of these rivers were studied through basic morphometric analysis; the parameters are detailed in the Materials and methods section and listed in Table 1. The Las Chacras and Arenal basins belong to a seventh order, as the Puquial and La Cañada are eighth order basins and Los Palos is a sixth order basin. The drainage basins have different areas, being the smallest the Arenal basin with 29 km$^2$ and the largest, La Cañada basin with 122 km$^2$. Perimeter, length and mean width show a similar trend, with minimum values for the Arenal basin and maximum values for the La Cañada basin. Derived morphometric parameters (Kc, C, E, Ff) are similar to those obtained by Rothis et al. (2018), concluding that drainage basins present elongated forms with a NE-SW orientation.

**Table 1.** Morphometric parameters of the Las Chacras, Arenal, Puquial, Cañada and Los Palos drainage basins.

| Basic parameters | | | | | | | |
|---|---|---|---|---|---|---|---|
| Basin | A (km$^2$) | P (km) | L (km) | W (km) | On | Hr (km) | Tsl (km) |
| Chacras | 87 | 71 | 24.44 | 2.99 | 7 | 1.99 | 597.75 |
| Arenal | 29 | 36 | 14 | 2.06 | 7 | 1.34 | 271.58 |
| Puquial | 75 | 54 | 21.98 | 3.41 | 8 | 1.7 | 815.12 |
| Cañada | 122 | 71 | 28 | 4.36 | 8 | 1.62 | 1143.37 |
| Los Palos | 54 | 58 | 20 | 2.7 | 6 | 0.59 | 523.72 |
| Derived parameters | | | | | | | |
| Basin | Kc | C | E | Ff | Dd | MRN | |
| Chacras | 2.85 | 0.1212 | 0.3946 | 0.1223 | 8.19 | 0.23 | |
| Arenal | 1.86 | 0.28 | 0.43 | 0.15 | 9.44 | 0.25 | |
| Puquial | 1.75 | 0.32 | 0.44 | 0.16 | 10.87 | 0.19 | |
| Cañada | 1.8 | 0.3 | 0.45 | 0.16 | 9.37 | 0.14 | |
| Los Palos | 2.21 | 0.2 | 0.41 | 0.14 | 9.7 | 0.08 | |
| Rock composition | | | | | | | |
| Basin | | CBR% | | CSR% | | MS% | |
| Chacras | | 95 | | 1 | | 4 | |
| Arenal | | 47 | | 28 | | 25 | |
| Puquial | | 30 | | 10 | | 60 | |
| Cañada | | 52 | | 16 | | 32 | |
| Los Palos | | 9 | | 39 | | 52 | |

CBR: Crystalline basement rocks
CSR: Clastic sedimentary rocks
MS: Modern Sediments

Drainage basin pattern changes according to the resistance of the underlying rocks. Here, substrate composition is grouped in three types: (i) crystalline basement rocks of the Valle Fértil Complex, (ii) clastic Triassic sedimentary rocks, and (iv) gravelly sediments of older alluvial levels. When the substrate is composed of rocks of the Valle Fértil Complex, drainage basins exhibit a trellis-dendritic pattern. Instead, when the substrate are the clastic Triassic rocks, a dendritic pattern with a high drainage density is observed. Finally, when the substrate is gravelly sediments, the drainage presents a dendritic -locally divergent- pattern with lower drainage density. Drainage basins present drainage density values from 7 to 10 with a mean value of 9.51. A similar value is obtained from crystalline rocks of the Pampean ranges in Córdoba province (Doffo and González Bonorino 2005). This is a remarkably high value considering the classification by Delgadillo and Páez (2008); however, a different methodology used for mapping the drainage network could explain these differences. Drainage density values corresponds to a medium texture showing that the substrate composition presents more resistant rocks (igneous and metamorphic rocks; Smith 1950). Generally, high drainage density indicates impervious materials, dispersed vegetation and mountainous relief (Nag 1998). High basin orders and high drainage density favor rapid surface run-off and flood occurrence (Ruiz 2001).

A derived parameter that presents a closer relation between drainage morphometry and resultant sedimentary processes is the Melton ruggedness number (Melton 1957; Kostachuk et al. 1986). Threshold values of the Melton ruggedness number (MRN) separating fluvial from debris flow fans is between 0.25–0.30 (Marchi and Dalla Fontana 2005). In the study area, MRN presents a decreasing trend from north to south, with values close to 0.25 in the Las Chacras and Arenal drainage basins and 0.08 in Los Palos drainage basin (see Table 1). According to this, the Las Chacras and Arenal drainage basins could produce steep fans dominated by debris flow processes, whereas the Puquial, La Cañada and Los Palos basins could generate gently fans, dominated by fluvial processes.

## 4.2 Las Chacras River

Las Chacras River is a NE-SW oriented, seventh order trending, perennial stream located in the northern portion of the western piedmont of the La Huerta Range. Las Chacras Range and some other hills surround the narrow valley of Las Chacras river and they diverge 7 km away from the mountain front (point 3, Fig. 3). Once it diverges, its distributary network reaches a $7_6$ bifurcation order (Fig. 3).

Four sections were defined in the Las Chacras River course (Fig. 3). Along the section one, the river is still confine and develops a narrow valley of 120–250 m wide. The river presents mid to high sinuosity channels with a multi and mono-channelized design. Fluvial geomorphological elements which have been identified here are mostly alternate bars, scrolls bars and scarce longitudinal bars. In the section two, the river is unconfined and the valley widens (240–500 m), although the main channel is still surrounded by terraces of its own older deposits. At this point, the Las Chacras river presents multi-channelized low sinuosity channels, where longitudinal bars domain. In the section three, the river diverges in a distributary net characterized by large and wide longitudinal lobes. Then, in the fourth section the river deposited a fan-shaped cone opening toward distal area that reached up to 4 km wide. Distributary channels disappear between eolian deposits and saline ponds.

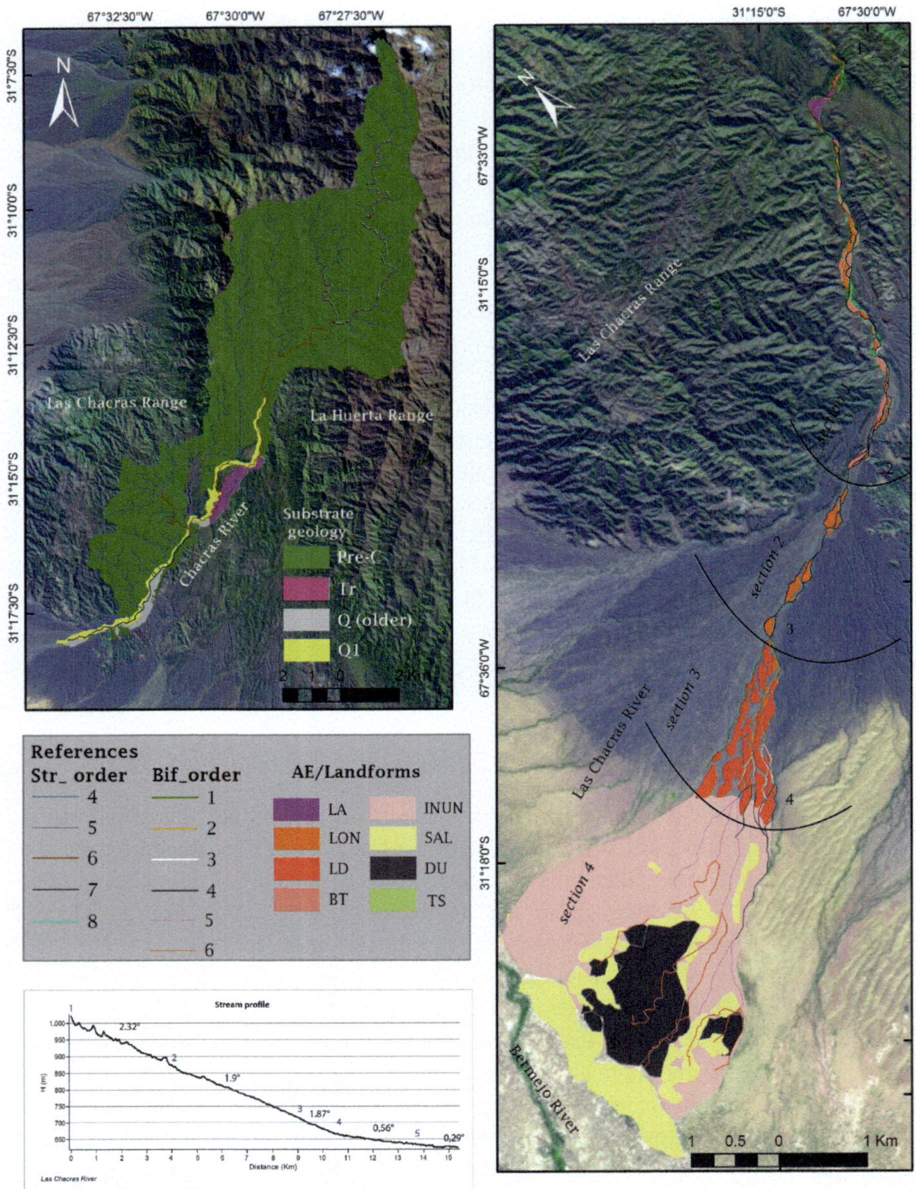


**Fig. 3.** Drainage basin rock composition, geomorphological map and stream profile of the Las Chacras River. Composition of drainage basin rocks are: Pre-C: crystalline Neoproterozoic-Early Paleozoic rocks, Tr: Triassic rocks, Q (older): ancient alluvial level, Q1: Recent alluvial level. Strahler order refers to stream order of tributary streams and bifurcation order refers to hierarchy order of distributary channels. Architectural elements or Landforms: LA: lateral accreted bars, LON: longitudinal bars, LD: gravel lobes, BT: alternate bar, INUN: floodplain, SAL: saline ponds, DU: eolian dunes, TS: terminal splays.

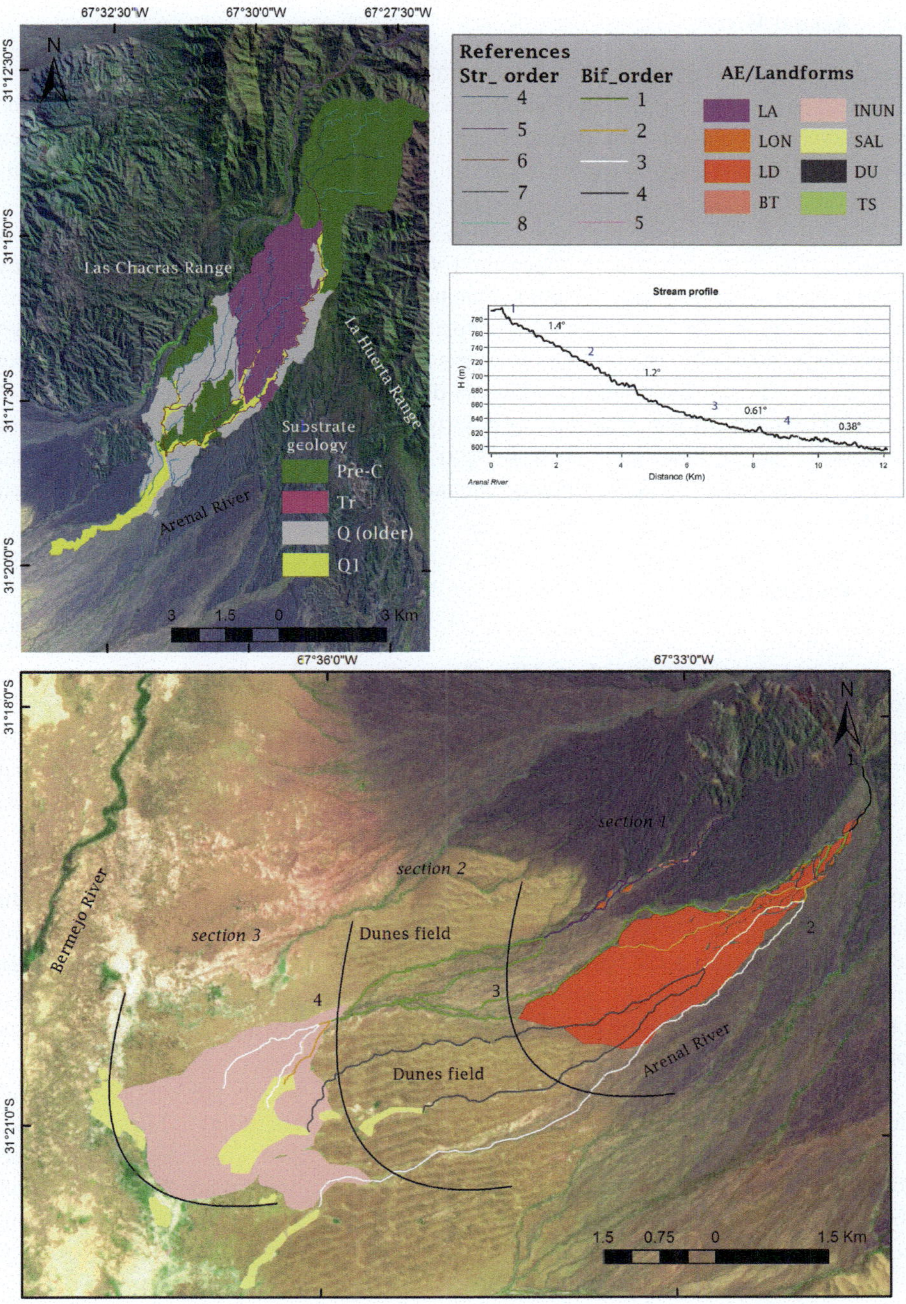


**Fig. 4.** Drainage basin rock composition, geomorphological map and stream profile of the Arenal River. Composition of drainage basin rocks are: Pre-C: crystalline Neoproterozoic-Early Paleozoic rocks, Tr: Triassic rocks, Q (older): ancient alluvial level, Q1: recent alluvial level. Strahler order refers to stream order of tributary streams and bifurcation order refers to hierarchy order of distributary channels. Architectural elements or Landforms: LA: lateral accreted bars, LON: longitudinal bars, LD: gravel lobes, BT: alternate bar, INUN: floodplain, SAL: saline ponds, DU: eolian dunes, TS: terminal splays.

## 4.3 Arenal River

The Arenal River is a seventh order, temporary river with a NE-SW trend, which joins an unnamed river to cross a dune field that borders the piedmont and finally reaches the Bermejo river floodplain. The Arenal River diverges 2.5 km away from the point when it loses confinement (point 2, Fig. 4) and it distributary network reaches a $7_4$ bifurcation order.

The Arenal River presents three section characterized by slope changes and their deposits. The first section starts when the river lose confinement (point 1), the valley widens to 350–500 m. From this point, channels deposit longitudinal bars downstream, the drainage gradually diverges until it deposits a 1.5 km wide cone, composed of large lobes (point 2, Fig. 4). The section two starts when the Arenal River crossed the eolian dune field and most distributaries channels disappear by infiltration. Although one of the main branches of the Arenal River diverts to the north and joins an unnamed river to cross the small dune field. Then, in section three, this main branch of the Arenal River diverges again in a distributary channel net that disappears, accumulating salt ponds.

## 4.4 Puquial River

The Puquial River is an eighth order river that drains the central portion of the piedmont. It is a temporary river located in the northern border of the Las Chacras dune field. Its tributaries are the Quebrada del Barro, El Carrizal River and a branch of El Arenal River, which are also temporary rivers that join the Puquial river 5 km away from the mountain front. Four kilometers away from the convergence point, the Puquial River finally spreads out in the Bermejo River and it drainage network reaches an $8_4$ bifurcation order.

Along the Puquial River three sections are recognized. The section one starts where the Quebrada del Barro and El Carrizal rivers lose confinement (point 1, Fig. 5). In point 1, their valley has a width of 200–450 m and they develop a multi-channelized network with longitudinal bars. Downstream both rivers converge and deposit large lobes in a cone-shaped deposit 2 km wide and 5 km long, where intense eolian interaction is observed in the top of these bars (point 2). Downstream, these tributaries join the Puquial River.

The section two is described by the Puquial River, which presents high to mid sinuosity channels with longitudinal, alternate and scroll bars (point 3, Fig. 5). Inactive bars are capped by salt deposits and vegetation; the plant cover increases as the abandonment time of the bars increase. In the section three (point 4, Fig. 5), main channel diverges in distributary channels that forms a cone shaped deposit. According to the satellite images, finer sediments and salt deposits compose the floodplain deposits in this area and increasing vegetation cover in sandier sediments close to the border of the Bermejo River. Main distributary channels disappear by infiltration in the distal portion of the cone.

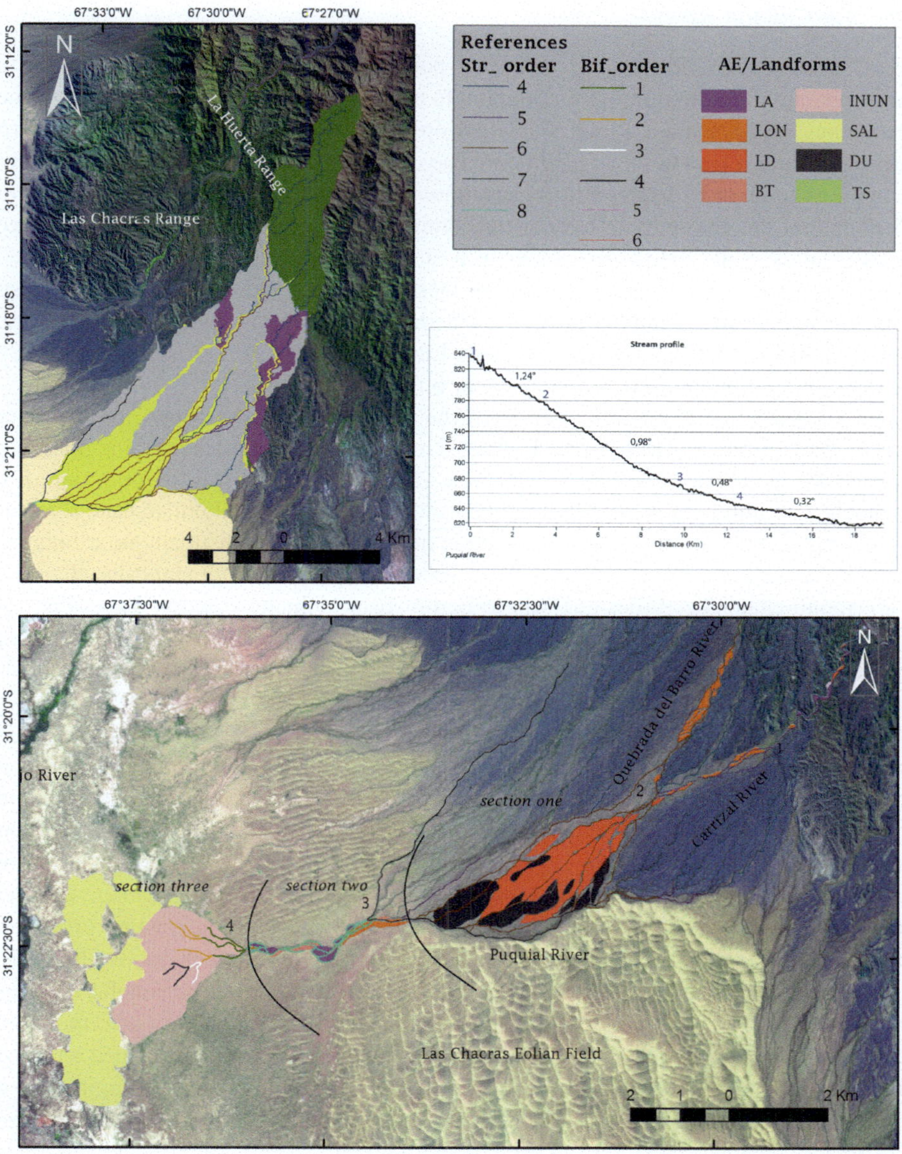


**Fig. 5.** Drainage basin rock composition, geomorphological map and stream profile of the Puquial River. Composition of drainage basin rocks are: Pre-C: crystalline Neoproterozoic-Early Paleozoic rocks, Tr: Triassic rocks, Q (older): ancient alluvial level, Q1: recent alluvial level. Strahler order refers to stream order of tributary streams and bifurcation order refers to hierarchy order of distributary channels. Architectural elements or Landforms: LA: lateral accreted bars, LON: longitudinal bars, LD: gravel lobes, BT: alternate bar, INUN: floodplain, SAL: saline ponds, DU: eolian dunes, TS: terminal splays.

## 4.5 La Cañada River

Las Chacras dune field occupies the distal portion of the central-southern La Huerta range piedmont causing drainage diversion to a SSE trend (162°) in this portion of the pediment. The La Cañada River is an eight order river that drains the central-southern portion of the La Huerta range piedmont. The catchment basin of the La Cañada River is one of the largest observed in the piedmont of the La Huerta Range, just slightly smaller than the Papagayos River which drains to the east of the La Huerta Range (Santi Malnis et al. 2018). The drainage basin of the La Cañada River is developed mostly over the hills sub-unit composed of Mesozoic clastic sedimentary rocks and also over the Valle Fértil Complex in the La Huerta Range. Many tributaries as Del Toro, Del Mercado and Las Cavaderas rivers join La Cañada River (Fig. 6). The distributary network of La Cañada River reaches an $8_4$ bifurcation order.

In a point located 7.5 km away from the mountain front, measured on the river course, most tributaries join the La Cañada River (point 1, Fig. 6). From this point, La Cañada River develops an up to 400 m wide valley, where low to mid sinuosity channels with multi-channelized design deposit longitudinal and alternate bars, some of them with laterally accreted deposits (section one, Fig. 6b). Downstream, no tributaries feed the system and the La Cañada River presents middle to high sinuosity channels with mono-channelized design, where scroll bars domain (section two, Fig. 6). Fieldwork in this portion of the system showed that the channelized facies have massive to laminated, coarse and very coarse sands and gravelly sands, ripple lamination are capped with laminated muds (Fig. 7a, b and c). There are important hollows that reach up to 1.5 m deep filled with bushes and branches (Fig. 7a). In addition, there is reverse grading (coarsening in upward trend) in laminated sands suggesting eolian reworking of channel deposits. Scroll bars deposits are compound of massive and ripple laminated middle to fine sand, capped with laminated muds. In this portion (section two, Fig. 6), the main channel of the La Cañada River reached up to 20 m wide. Towards the distal portion (point 3, section three Fig. 6), the main channel gradually decreased it width to 2 m and presents middle to fine sands with horizontal, through-cross and ripple lamination and scour marks preserved in muds (Fig. 7d). Eolian reworking is very common, developing reverse grading in channelized facies and low angle cross lamination in levees which are just 0.60 m thick. Once the main channels of the La Cañada River diverge, they developed a large cone shaped terminal deposit (5 km long and 4 km wide). Low-relief terminal splays appears composed of ripple laminated, massive fine sands and reverse grading in small-scale dunes. Distributary channels have 2.0 to 0.40 m width with laminated very fine sand and muds (silty-clays, Fig. 7e). Other frequent floodplain elements observed are sandy-silty ponds with weakly developed desiccation cracks and eolian dunes of up to 2.0 m high with low angle cross and horizontal lamination (Fig. 7f, g, h).

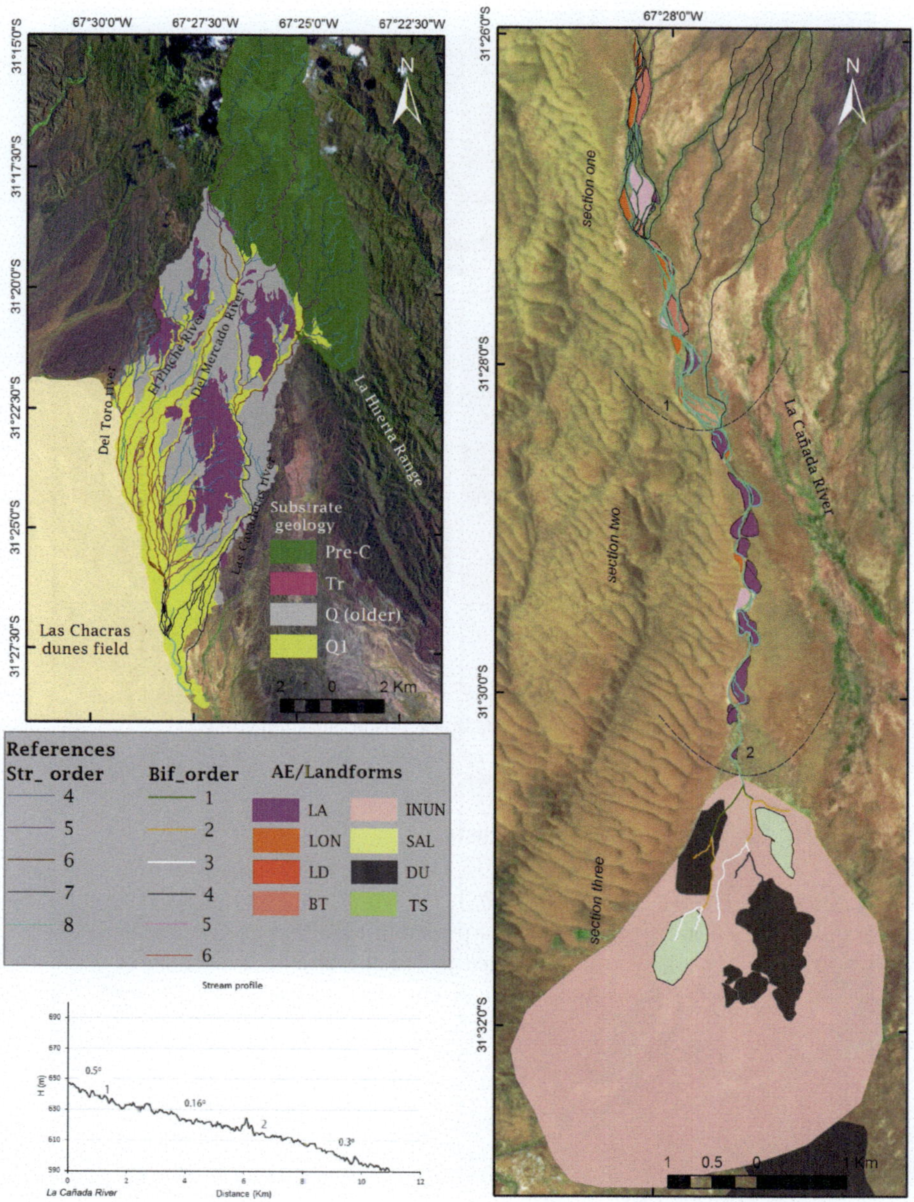


**Fig. 6.** Drainage basin rock composition, geomorphological map and stream profile of the La Cañada River. Composition of drainage basin rocks are: Pre-C: crystalline Neoproterozoic-Early Paleozoic rocks, Tr: Triassic rocks, Q (older): ancient alluvial level, Q1: recent alluvial level. Strahler order refers to stream order of tributary streams and bifurcation order refers to hierarchy order of distributary channels. Architectural elements or Landforms: LA: lateral accreted bars, LON: longitudinal bars, LD: gravel lobes, BT: alternate bar, INUN: floodplain, SAL: saline ponds, DU: eolian dunes, TS: terminal splays.

**Fig. 7.** Architectural elements of the La Cañada River. (A) Main channel of La Cañada River, HO: hollows, (B) Scroll bar deposit of the main channel, (C) Lithofacies deposits of channels, Sh: horizontal lamination, Sl: low-angle cross lamination, St: through-cross lamination, Sr: ripple lamination, Fl: fine lamination, very small ripples, (D) Lithofacies deposits of channels Sh: horizontal lamination, parting lineation, Sr: ripple lamination, (E) Distal distributary channels of the distal fan of the La Cañada River, Sm: massive or structuresless sands (F) Dunes and floodplain in the distal fan, (G) Ponds in distal fan of the La Cañada River with weakly developed desiccation cracks, (H) Eolian dunes in the distal fan of the La Cañada River.

### 4.6 Los Palos River

The Los Palos River is a temporary river that drains the southwestern portion of the piedmont and flows parallel to the La Cañada River. The drainage basin of the Los Palos River is developed over clastic sedimentary rocks of the hills sub-unit and its drainage network reaches a $5_I$ bifurcation order, since bifurcation occur upstream and the drainage network reaches its sixth stream order (point 1, Fig. 8).

Los Palos River presents narrow, discontinuous, low sinuosity channels that diverge downstream in two main channels. This river disappears by infiltration, probably due to a permeable substrate, and channelized zones are completely covered by vegetation. Some architectural element as interchannel areas are recognized, but it is mostly characterized by broad areas of fine sediments (sands and muds) and salt ponds with a fan morphology that includes terminal splays and floodplain. In its distal portion, it laterally overlaps the La Cañada distal cone. This depositional area is surrounded by eolian dunes, which become more important towards the south.

## 5 Discussion

Distributary drainage systems are common in piedmont regions, developed by rivers when they lose confinement after leaving a mountainous region and enter into a sedimentary basin (Blair and Mc Pherson 1994; Weissmann et al. 2010). Two main sedimentary environment models domain these regions: alluvial fans (AF) and distributive fluvial systems (DFS; Blair and Mc Pherson 1994; Weissmann et al. 2010; Hartley et al. 2010). To identify these sedimentary environments many geomorphological and sedimentological parameters have been considered (Nichols and Fisher 2007; Suriano and Limarino 2009; Colombo 2010; Davidson et al. 2013). According to the literature, alluvial fans are "conspicuous conical landform commonly developed where a channel emerges from a mountainous catchment to an adjoining valley" (Blair and Mc Pherson 1994). Alluvial fans are developed with piedmont slope that varies from maximum values of 35°–25° close to the mountain front, medial zone slope values are between 10° to 5°, and they decrease to up to 1° slope values at the toe (Blair and Mc Pherson 1994; Suriano and Limarino 2009; Colombo 2010). On the other hand, DFS are "the deposits of a fluvial system which, in plan view, displays a radial distributive channel pattern" (Hartley et al. 2010). Differing from alluvial fans, DFS occurred in piedmont regions of low to very low slope values, usually below 1° o even less far away from the mountain front (0.5°; Hartley et al. 2010). Besides slope, length is another criteria used for identifying alluvial fans from DFS, considering that DFS are larger, with an apex-toe distance variable from 30 to 50 km, and up to 700 km (Hartley et al. 2010; Davidson et al. 2013). However, Hartley et al. (2010) concluded that there is an overlap between alluvial fans and DFS, and that only slope and/or length parameters are not enough to distinguish them.

The AF models present three regions: proximal (apex), medial and distal (toe) zone. Sediment gravity-flows, non-aqueous flows and sheet-floods in unconfined flows are characteristic, especially in the proximal zone; confined and aqueous flows characterize the medial zone and sheet-floods and loose of channelization occur in the distal zone

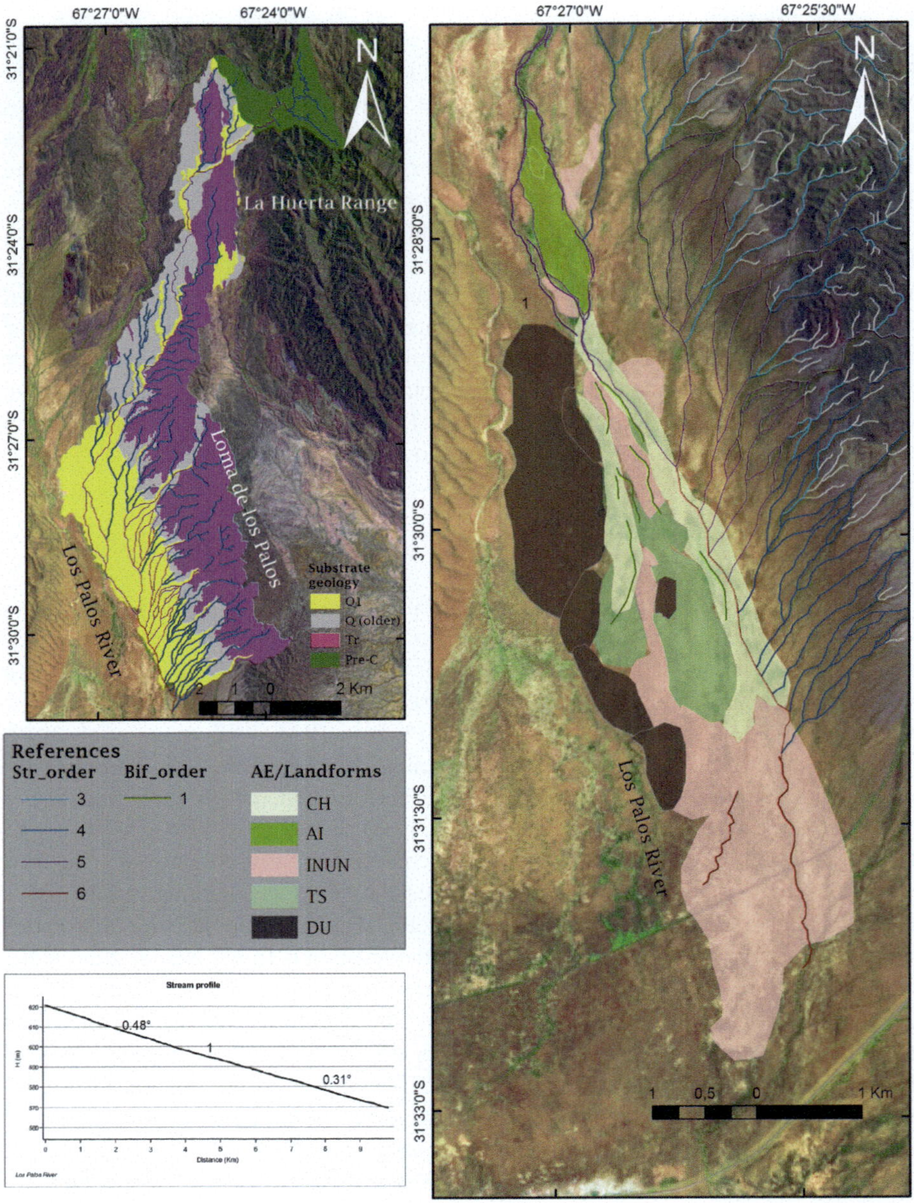


**Fig. 8.** Drainage basin rock composition, geomorphological map and stream profile of the Los Palos River. Composition of drainage basin rocks are: Pre-C: crystalline Neoproterozoic-Early Paleozoic rocks, Tr: Triassic rocks, Q (older): ancient alluvial level, Q1: recent alluvial level. Strahler order refers to stream order of tributary streams and bifurcation order refers to hierarchy order of distributary channels. Architectural elements or Landforms: CH: channels, AI: interchannel area, INUN: floodplain, SAL: saline ponds, DU: eolian dunes, TS: terminal splays.

(Colombo 2010). The DFS models also present three or four regions, depending on the authors (Nichols and Fisher 2007; Davidson et al. 2013). These regions are proximal, medial, distal and termination zones. In this case, aqueous flows and channelization are present in all the system, from proximal to distal zone, and a general organized fluvial architecture composed of well-developed and diverse channel and floodplain elements (Hartley et al. 2010). Other feature is the presence of terminal splays in the distal zone, or in the termination of the DFS, created by the disappearance of the main fluvial channel of the system (Nichols and Fisher 2007; terminal DFS, Hartley et al. 2013). Seasonal storms, high evaporation in arid climates, permeable substrates and the decreasing capacity and competence of channels in the distal zone of DFS triggered channelization losses and the formation of terminal splays (Nichols and Fisher 2007).

In the western piedmont of the La Huerta Range drainage basin morphometric studies suggest that they could favor rapid surface run-off and flood occurrence, enhanced by seasonal climate. Shape morphometric parameters display an elongated basin, which presents a northwest to southeast trend and where morphostructural units decrease in height together with piedmont slope in a N-S trend (Rothis et al. 2018). Derived parameters as the Melton ruggedness numbers (MRN) show the same decreasing trend.

Together with these north-south morphometric parameters, different alluvial environments are recognized from north to south in the western piedmont of the La Huerta Range. Starting from the Las Chacras River, where slope is 1.9°, according to a low sinuosity multi-channeled fluvial system domain, mostly in sections one and two, where longitudinal bars and gravelly lobes derived from debris flow processes are the main deposits. In section three, sheet-floods spread finer sediments over the Bermejo river floodplain forming a terminal fan-shaped deposit. The three sections described in the Las Chacras River, plus the dominance of debris flow lobes in the proximal and medial area, resemble the classical alluvial fan model, with its proximal, medial and toe zone respectively (Blair and McPherson 1994; Colombo 2010). Also, the morphometric analysis of the Las Chacras drainage basin concluded that it tends to generate fans dominated by debris flow processes. Therefore, the Las Chacras River is interpreted as an alluvial fan. The highest bifurcation order reached in the study zone corresponds to the Las Chacras River distributary network, which divergence character started in the proximal zone of the alluvial fan.

The Arenal River presents slope values of 1.4°. This river also presents three sections: section one, where drainage network gradually diverge forming a fan-shape deposit compose of debris-flow lobes; a section two, completely disrupted by a small-scale, eolian dune field crossed by many parallel distributary channels, which disappear in salt ponds; and a section three, characterized by a low sinuosity multi-channeled fluvial system, which diverges in a distributary channel network that deposits finer sediment. According to the sections described, the Arenal River presents a facial architecture similar to those proposed for the alluvial fan model, where section one, two and three correspond with proximal, medial and distal zones, respectively. Accordingly, it has a debris-flow-fan generating drainage basin. Like the Las Chacras River, the Arenal River deposits correspond to an alluvial fan with important eolian interaction.

The Puquial River has slope piedmont values of 1.15°. There are three sections identified: section one is characterized by low sinuosity multi-channeled rivers with longitudinal bars that gradually diverge and deposit gravel debris-flow lobes. In section two, eolian interaction occurs and there is a high to mid sinuosity mono-channelized fluvial system. The section three is characterized by a fan-shaped deposit, with a loose distributary channel net, compose of finer sediments and salt ponds. The section one of the Puquial River would correspond to the proximal zone of an alluvial fan, dominated by debris flows lobes and eolian dunes. As in the Arenal River, the section two of the Puquial River corresponds to the medial zone of an alluvial fan, with some eolian dune fields, but where aqueous flows and channelized systems are dominant. In section three, the Puquial River diverge in a distributary network that forms a fan-shaped deposit of fine sediments and salt ponds. Morphometric studies show that differing from the above described systems the Puquial River is a fluvial fan generating basin and comparing it to the Las Chacras or Arenal rivers, there are more fluvial architectural elements, especially in its medial zone. The three section of the Puquial River are the proximal, medial and distal zone of an alluvial fan with important eolian interaction and dominance of aqueous flows in its medial zone.

Differing from the alluvial system discussed above, the La Cañada and Los Palos river flow parallel to the axial drainage system of the basin (the Bermejo river), in a region between the transitional unit and the valley unit. This generates a low slope along the river course, below 0.5°.

In this portion of the piedmont, drainage converges in the La Cañada River and, from that point, three sections are recognized (Fig. 6). A section one characterized by low to mid sinuosity, multi-channelized fluvial system with longitudinal, alternate and lateral accreted bars. A section two presents middle to high sinuosity, mono-channelized fluvial system with scroll bars. A section three is characterized by a mid-sinuosity, multi-channelized fluvial system that diverges in a distributary net of channels developing a fan-shaped terminal deposit. The terminal fan is composed of sandy terminal splays, eolian dunes, distributary channels and ponds.

Unlike the other river systems, the La Cañada River presents significant lower slope, a wider variety of fluvial architectural elements plus the largest drainage basin area. As a result, a better fluvial architecture organization and the dominance of aqueous flows in its depositional processes are observed in the La Cañada River. Differing from the above discussed systems, section one and two present low to high sinuosity fluvial systems, decreasing capacity and competence downstream. In section three, a distributary drainage network starts from the point where the trunk channel loses confinement. Once it loses confinement, the trunk channel decreases again in capacity and competence from sandy to muddy textures. The fan shape morphology of the deposit appears in section three, where terminal splay and sheet floods occurs. Drainage basin morphometry suggests that the La Cañada River drainage basin tends to develop fans dominated by fluvial processes, like it is observed in its three sections. These features allows considering the La Cañada River as an embryonic distributive fluvial system, according to the definition of Hartley et al. (2010) and Davidson et al. (2013). Also, eolian interaction is very important, especially with floodplain elements.

The Los Palos River presents low slope and intense eolian interaction. This system disappears by infiltration and diverge forming a wide depositional area compound of fine sediments and salt ponds in a fan-shaped morphology. The presence of abundant eolian deposits could produce the infiltration of the channels, since down-flow losses are not compensated by tributary recharge. This produces the deposition of bed/suspended-load sediments, partially covering eolian dunes in wide terminal lobes that accreted downstream. Similar environments are described in the literature as flood-out or terminal flood-outs (Gore et al. 2000).

Finally, eolian interaction is a common feature observed in the piedmont environments under study and are mentioned in the literature as reworking processes in alluvial fans (Blair and McPherson 1994). The Arenal and Puquial rivers show important eolian interaction, causing disruption of fan sedimentation as in the medial zone of the Arenal River and triggering aggradation in proximal settings, as it is observed in the proximal zones of both systems (Figs. 4 and 5). In the La Cañada River, eolian dunes are observed between floodplain elements and reworking fluvial landforms as channels, levees and bars. The Los Palos River disappears by infiltration between eolian dunes, triggering rapid deposition of channels-load and forming terminal splays.

## 6 Conclusions

The Las Chacras, Arenal, Puquial, La Cañada and Los Palos rivers drain the western piedmont of the La Huerta Range to the Bermejo River valley. To classify these distributary systems many criteria were used as basin morphometry, piedmont slope, landforms and sedimentary processes. Drainage basins of these rivers show that they are from north-south debris-flow fans to aqueous-flows fans generating basins. Accordingly, from north to south, the Las Chacras river resemble as an alluvial fan, which progrades through the Bermejo River valley, where three sections that could correspond to the proximal, medial and distal fan portions of an alluvial fan could be identified. The Arenal and Puquial rivers are also interpreted as alluvial fans that present a more complex behavior, developing two cones of deposits, the first close to the mountain front composed of debris flow lobes and the second one in its distal zone. In these cases, eolian disruption and interaction triggers aggradation in the proximal zone of both systems and alter the typical fan shape morphology of the fan. The La Cañada River presents aqueous flows in its three zones, loosing capacity and competence downstream from middle to high sinuosity channels characterized by sheet floods. In the distal zone distributary channels form a large cone shape deposits, where there are terminal splays and eolian dunes. The La Cañada River deposits resembles a distributive fluvial system model, although smaller in size comparing to those defined in the literature (Nichols and Fisher 2007; Davidson et al. 2013). Finally, the Los Palos River disappears by infiltration between eolian dunes, producing high concentration of vegetation in channelized areas and depositing wide areas of fine sediments, which is considered as a flood-out system.

## References

Bastias, H., Uliarte, E., Paredes, J.D., Bastias, J.A., Perucca, P.: Neotecttónica de la provincia de San Juan. In: 11° Congreso Geológico Argentino, Actas, Buenos Aires, pp. 228–244 (1990)

Blair, T., Mc Pherson, J.: Alluvial fans and their natural distinction from rivers based on morphology, hydraulic processes, sedimentary processes, and facies assemblages. J. Sediment. Res. **A64**, 450–489 (1994)

Bossi, G.: Geología de la Cuenca de Marayes-El Carrizal (Provincia de San Juan), República Argentina. In: VI Congreso Geológico Argentino, Actas, I, Bahía Blanca, pp. 23–38 (1976)
Colombo, F.: Abanicos aluviales: secuencias y modelos de sedimentación. In: Arche, A. (ed.) Sedimentología del proceso físico a la cuenca sedimentaria, pp. 131–183. CSIC, Madrid (2010)
Costa, C.H., Audemard, M.F., Becerra, A., Lavenu, M.N., Machete, G.: An overview of the main Quaternary deformation of South America. Rev. Asoc. Geol. Argentina **4**(61), 461–479 (2006)
Damanti, J.F.: Geomorphic and structural controls on facies patterns and sediment composition in a modern foreland basin. Spec. Publ. Int. Ass. Sediment. **17**, 221–233 (1993)
Davidson, S., Hartley, A., Weissmann, G., Nichols, G., Scuderi, L.: Geomorphic elements on modern distributive fluvial systems. Geomorphology **180–181**, 82–95 (2013)
Delgadillo, A., Páez, G.: Aspectos hidrológicos, subcuencas susceptibles a crecidas, escenarios de riesgo por crecidas. Mérida, Venezuela (2008)
Doffo, N., González Bonorino, G.G.: Caracterización morfométrica de la cuenca alta del arroyo Las Lajas, Córdoba: Un análisis estadístico. Rev. Asoc. Geol. Argentina **60**(1), 16–22 (2005)
Edmonds, D., Slingerland, R.: Mechanics of river mouth bar formation: implications for the morphodynamics of delta distributary networks. J. Geophys. Res. **112**, F02034 (2007)
Fielding, C.R., Ashworth, P.J., Best, J.L., Prokocki, E.W., Smith, G.H.: Tributary, distributary and other fluvial patterns: what really represents the norm in the continental rock record? Sed. Geol. **261**, 15–32 (2012)
Gore, D., Brierley, G., Pickard, J., Jansen, J.: Anatomy of a floodout in semi-arid eastern Australia. Z. Geomorphol. **122**, 113–139 (2000)
Gravelius, H.: Flusskunde. Goschen Verlagshan dlug Berlin. In: Zavoianu, I. (ed.) Morphometry of Drainage Basins. Elsevier, Amsterdam (1914). (1985)
Hadley, R.F., Schumm, S.A.: Sediment sources and drainage basin characteristics in upper Cheyenne River basin. USGS water supply paper, US Geological Survey, 1531-B (1961)
Hartley, A.J., Weissmann, G.S., Nichols, G.J., Warwick, G.L.: Large distributive fluvial systems: characteristics, distribution, and controls on development. J. Sediment. Res. **2**, 167–183 (2010)
Hartley, A.J., Weissmann, G.S., Bhattacharayya, P., Nichols, G.J., Scuderi, L.A., Davidson, S. K., Mather, A.E.: New Frontiers in Paleopedology and Terrestrial Paleoclimatology. SEPM (Soc. Sediment. Geol.) **104**, 149–158 (2013)
Horton, R.: Drainage basin characteristics. Am. Geophys. Union Trans. **13**, 350–361 (1932)
Horton, R.: Erosional development of streams and their drainage basins. Hydrophysical approach to quantitative morphology. Geol. Soc. Am. Bull. **56**, 275–370 (1945)
Jordan, T.E., Allmendinger, R.W.: The Sierras Pampeanas of Argentina; a modern analogue of Rocky Mountain foreland deformation. Am. J. Sci. **286**(10), 737–764 (1986)
Kostaschuk, R.A., MacDonald, G.M., Putnam, P.E.: Depositional process and alluvial fan-drainage basin morphometric relationships near Banff, Alberta, Canada. Earth Surface Process. Land. **11**(5), 471–484 (1986)
Marchi, L., Dalla Fontana, G.: GIS morphometric indicators for the analysis of sediment dynamics in mountain basins. Environ. Geol. **48**(2), 218–228 (2005)
Melton, M.A.: An analysis of the relations among elements of climate, surface properties, and geomorphology. Unpublished Ph.D. Dissertation, Columbia University, New York (1957)
Miller, V.: A quantitative geomorphic study of drainage basin characteristics in the clinch mountain area. Virginia and Tennessee. Technical report 3. Office of Naval Research, Department of Geology, Columbia University, New York (1953)
Nag, S.K.: Morphometric analysis using remote sensing techniques in the Chaka Sub-basin, Purulia District, West Bengal. J. Indian Soc. Remote Sens. **26**(1–2), 69–76 (1998)

Nichols, G.J., Fisher, J.A.: Processes, facies and architecture of fluvial distributary system deposits. Sed. Geol. **195**(1), 75–90 (2007)

Olariu, C., Bhattacharya, J.: Terminal distributary channels and delta front architecture of river-dominated delta systems. J. Sediment. Res. **76**, 212–233 (2006)

Pereyra, B.: Clima de la provincia de San Juan. In: Catálogo de recursos humanos e información relacionada con la temática ambiental en la región andina de Argentina (online (2000) C-Bra Systematics, Buenos Aires ed.), GTZ, Technical Argentina-Germany cooperation (1996)

QGIS Development Team: QGIS Geographic Information System. Open Source Geospatial Foundation Project (2016). http://qgis.osgeo.org

Ramos, V.A., Cristallini, E.O., Pérez, D.J.: The Pampean flat-slab of the Central Andes. J. S. Am. Earth Sci. **15**, 59–78 (2002)

Rothis, L., Perucca, L., Santi Malnis, P., Pantano Zuñiga, A., Alcacer, J., Haro, F., Vargas, N.: Análisis morfotectónico en el piedemonte suroccidental del antepaís andino, Sierras Pampeanas Occidentales, San Juan, Argentina. Geociências **37**(1), 55–74 (2018)

Ruiz, J.: Hidrología: Evolución y vision sistémica, la morfometría de cuencas como aplicación. UNELLEZ, Barinas, Venezuela (2001)

Santi Malnis, P., Colombi, C.E., Rodríguez Posatini, N., Rothis, L.M., Limarino, C.O.: Caracterización sedimentológica de un Sistema Fluvial Distributivo de clima árido: arroyo Papagayos, en el piedemonte oriental de las Sierras La Huerta-Imanas, San Juan, Argentina. Andean Geol. **45**(2), 186–228 (2018)

Schumm, S.A.: Evolution of drainage systems and slopes in badlands at Perth Amboy, New Jersey. Geol. Soc. Am. Bull. **67**, 597–646 (1956)

Smith, K.G.: Standards for grading texture of erosional topography. Am. J. Sci. **248**(9), 655–668 (1950)

Strahler, A.: Dynamic basis of geomorphology. Geol. Soc. Am. Bull. **63**, 923–938 (1952)

Suriano, J., Limarino, C.: Sedimentación pedemontana en las nacientes del Río Jáchal y Pampa de Gualilán, Precordillera de San Juan. Rev. Asoc. Geol. Argentina **65**(3), 516–532 (2009)

Tripaldi, A., Forman, S.L.: Geomorphology and chronology of Late Quaternary dune fields of western Argentina. Palaeogeogr. Palaeoclimatol. Palaeoecol. **251**(2), 300–320 (2007)

Vujovich, G., Chernicoff, C., Tchilinguirian, P., Godeas, M., Marín, G., Pezzutti, N., Díaz, I.: Hoja Geológica 3166-III. Chepes, provincias de La Rioja, San Juan y San Luis. SEGEMAR, Buenos Aires (2007)

Weissmann, G., Hartley, A., Nichols, G., Scuderi, L., Olson, M., Buehler, H., Banteah, R.: Fluvial form in modern continental sedimentary basins: distributive fluvial systems. Geology **38**(1), 39–42 (2010). https://doi.org/10.1130/G30242.1

# Preliminary Morphometric Analysis: Río Talacasto Basin, Central Precordillera of San Juan, Argentina

Carla E. Ginesta Torcivia(✉) and Natalia N. Ríos López(✉)

CONICET - Gabinete de Neotectónica y Geomorfología, CIGEOBIO, Universidad Nacional de San Juan, Av. José I. de la Roza y Meglioli, Rivadavia, 5400, San Juan, Argentina

ginestageo@gmail.com, natalia_noemi_rios@hotmail.com

**Abstract.** Preliminary morphometric analysis: Río Talacasto basin, Central Precordillera, San Juan, Argentina.

This work analyzes various morphometric characteristics of Río Talacasto basin, situated approximately 80 km northwest of the city of San Juan, in the Central Precordillera geological province. Various morphometric parameters were analyzed in order to evaluate its flash flood hazard. For this purpose, some measurements (area, perimeter, length, width, maximum and minimum height and Strahler order, slope, aspect, curvature and convergence index, topographic position index, and roughness index) were calculated. These measurements allow a future analysis to predict approximately the behavior of the basin in the presence of a series of theoretical rainstorms that may generate unusual runoff volumes.

The importance of this evaluation is that the studied basin includes the National Route 149 and the Provincial Route 436, both being the main accesses to the northern and western portions of the province. The results obtained conclude that the basin has a significant susceptibility to the occurrence of flash floods and other landslides.

**Keywords:** Morphometry · Digital elevation models · Landslide · San Juan · Central Pre-Andean · Precordillera · Argentina

## 1 Introduction

The importance of fluvial morphometry in the genesis of flows is crucial, because it is a significant factor of the speed at which the water will circulate along the surface and therefore the basin behavior after a certain precipitation.

A flash flood is a rapid rise of water in streams, creeks and storm drains that poses a threat to life and property. Due to their almost instantaneous occurrence, together with their high capacity of transport, flash floods are one of the most significant weather-related hazards in many parts of the world (Gaume et al. 2009). Flash floods are one of the most common types of geological hazards in the region, and they are extremely dangerous because of their abrupt nature. Their characteristics are: short duration, small areal extent, high flood peaks and rapid flows, as well as causing of heavy loss to


M. M. Collantes et al. (Eds.): *Advances in Geomorphology and Quaternary Studies in Argentina*, SPRINGEREARTH, pp. 158–168, 2020.
https://doi.org/10.1007/978-3-030-22621-3_8

property. They occur because of high-intensity storms, steep slopes in the catchment, poor vegetation cover and high-velocity flows (Lin 1999).

One of the most important tools in hydrological analysis is the morphometric survey of a basin, which allows establishing evaluation parameters on the behavior of the hydrological system of the basin areas. This tool allows the establishment of parameters for the evaluation of the functioning of geomorphological processes and the hydrological system of a certain region. These systems are susceptible to statistical and/or mathematical treatment that permits a much smaller influence of the researcher subjectivity upon the conclusions derived from those parameters.

In many articles, morphometric analyses have been used for basin characterization, for instance, Miller (1953), Boulton (1968), Gregory and Walling (1973), Gardiner (1975), Costa (1987), Topaloglu (2002), Moussa (2003), Sreedevi et al. (2004), Mesa (2006), Esper Angillieri (2008), Esper Angillieri and Perucca (2014) and Wani et al. (2018).

Intensity rainfall has been registered in most years in the area, usually in January and February. These months should be considered as the most important epoch in which the flash floods processes may occur. The morphometric characteristics of the basin analyzed for this area should also be taken into consideration. The study of the mentioned parameters, among others, allows future analysis of the susceptibility to flash floods which may occur in the area.

On January 25$^{th}$ 2015, the flash floods destroyed the culvert along the national route N° 141. Therefore, the nearest towns were separated for more than two days. Another rainfall event on February 1$^{st}$, 2016 and January 17$^{th}$, 2018, also left disrupts on the route.

## 2 Study Area

The Talacasto river basin is located at Ullum, a northwestern county in the province of San Juan, Argentina. This zone is locally called the "Quebrada de La Burra". This basin covers an area of up to 300 km$^2$ and it lies between latitude 31°0′ 19, 37″ S and longitude 68°49′ 47, 26″ W. The climate is warm and extremely arid, with a mean annual temperature of 17 °C. Annual rainfall does not exceed 100 mm, occurring mostly in January and February. In winter, some snowfall may be accumulated in the higher mountains, but this is just a transitory snow cover.

The basin is located in the geological province of the Central Precordillera. The Precordillera is defined as an Andean fold and thrust belt. It is composed of Paleozoic sedimentary rocks which was impacted as a result of the thrust front which migrated toward the east. This sector was involved in subhorizontal subduction of the Nazca plate beneath the South American plate. These characteristics determined an orogenic front migration, the lack of volcanic activity, intense intraplate seismicity and remarkable neotectonic activity mainly located in this area (Ramos et al. 1986). Litologically, it is composed of Ordovician limestones, sandstones, and reddish Silurian shales. Subsequently, Devonian sandstone units occur and, finally, Neogene sandstones and conglomerates appear interbedded with shale layers (Fig. 1).

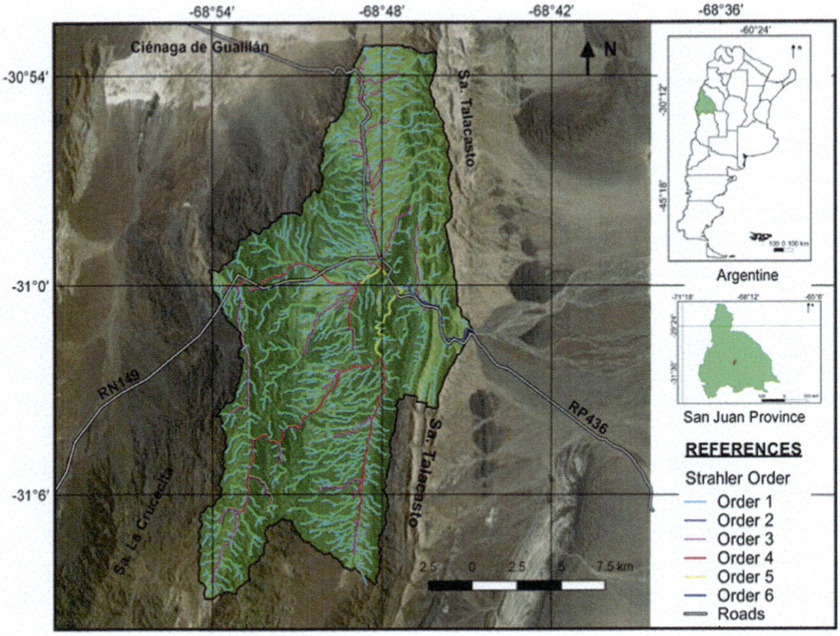


**Fig. 1.** Study area location.

## 3 Materials and Methods

A 2008 Digital Elevation Model of high resolution (12.5 m) was used, with ALOS platform and PALSAR sensor, and processed with QGis 2.18.15 and SAGA 2.2.5 software and finally the different parameters were analyzed. Considering that a river basin is a portion of the continental surface whose waters drain towards a common channel, it is possible to define a drainage basin based upon the analysis of the topography orientation. The area was defined by the number of pixels that drain their surface water towards the pixel that contains the point of interest or closure of the basin (Oñate and Bosque 2007).

The Río Talacasto Basin was characterized morphometrically by the calculation of area (A), perimeter (P), length (L), width (W), maximum and minimum elevation (H, h) and order, in the sense of Strahler (1957). The main and derivative parameters obtained were slope, aspect, curvature and convergence index, algorithms used in Table 1 (Zevenbergen and Thorne 1987), topographic position index, and roughness index. where:

$$D = \frac{\delta^2 z}{2\delta x^2} = \frac{\left[\frac{Z_4 + Z_6}{2} - z_5\right]}{L^2}, E = \frac{\delta^2 z}{\delta y^2} = \frac{\left[\frac{Z_2 + Z_8}{2} - z_5\right]}{L^2} \quad (6)$$

$$F = \frac{\delta^2 z}{\delta x \delta y} = \frac{(-z_1 + z_3 + z_7 - z_9)}{4L^2}, G = \frac{\delta_z}{\delta_x} = (-z_4 + z_6)/2L\ , H = \frac{\delta_z}{\delta_y} = (z_2 - z_8)/2L \quad (7)$$

**Table 1.** Equations used to calculate basic parameters of Talacasto river basin.

| Variable | Unit | Algorithm |
|---|---|---|
| Slope (Zevenbergen and Thorne 1987) | Degree | $\text{Tan}^{-1}\left[G^2+H^2\right]^{1/2}$ (1) |
| Aspect (Zevenbergen and Thorne 1987) | Degree | $180-\tan^{-1}\left[\left(\frac{H}{G}\right)+90\left(\frac{G}{\lvert G\rvert}\right)\right]$ (2) |
| Planar Curvature (Zevenbergen and Thorne 1987) | 1/m | $2\left(DH^2-FGH+EG^2\right)/\left(G^2+H^2\right)^{3/2}$ (3) |
| Vertical Curvature (Zevenbergen and Thorne 1987) | 1/m | $-2\left(DG^2+FGH+EH^2\right)/\left(G^2+H^2\right)\left(G^2+H^2+1\right)^{3/2}$ (4) |
| Convergence index (Olaya and Conrad 2009) | Degree | $\left(\frac{1}{8}\sum_{i=1}^{8}\theta_i\right)-90$ (5) |

where:

θ = angle between vector direction joining the center of the cell and the middle of the window.

D, E, F, G are partial derivatives of the surface Z = Z (x, y).

L = distance between the middle of two cells and the size of the cell. It has the same units of the DEM resolution.

## 4 Derivative Parameters

### 4.1 Slope

Slope is a very important driver of the largest number of processes on the land surface because it acts upon the surface and subsurface velocity flow, on the runoff rate, and on the moisture contained in the soil and its properties (Suet and Luam 2004). The slope influences flow rates of water and sediment that are controlling the rate of energy expenditure or stream power that is available to drive the flow too.

### 4.2 Aspect

The aspect defines the slope direction and therefore the direction of the flow. The knowledge of how the aspect varies throughout a basin provides the information necessary to determine what portion of the upslope land area contributes to the flow at any point in the catchment (Zevenbergen and Thorne 1987). This parameter is calculated like as it was of circular size. It begins at 0° in a northern position and it increases towards the right direction up to 360°, a value which falls again at the north. When water moves under the influence of gravity, it follows in a certain, specified direction determined by the aspect.

### 4.3 Curvature

This parameter is expressed in terms of vertical and horizontal (plane) curvature. Blaszczynski (1997) described that the vertical curvature (Cv) is like the surface curvature in the slope direction. The horizontal curvature (Ch) is like the surface curvature in the slope in a perpendicular direction. The curvature means also the change of the slope rate. Positive values express convex curvature and negative values do so for concave curvature. Values near 0 indicate an almost flat surface.

In the terrain, concavity and convexity can be associated with flow accumulation and material dispersal. Therefore, the combination of both values can estimate the flow behavior. The horizontal curvature (Ch) is the convergence or divergence of flow measurement. It is an indication of the water and solid materials distribution pattern throughout the slope direction. The surface and sub-surface flow water converges when $Ch < 0$, and it diverges when $Ch > 0$.

The vertical curvature (Cv) represents a measure of the acceleration and deceleration of the flow. The sub-surface flow soil is decelerated when $Cv < 0$, and it is accelerated when $Cv > 0$. This parameter can be rendered like erosion, runoff and thin soil development in convex areas; deeper soils deposition and development, in the concave and flat areas. The vertical curvature is important because it reflects the change in slope angle; thus, it controls the change in velocity of mass flow along the slope (Klingseisen et al. 2008).

Vertical and horizontal curvatures indicate geological alignments and can be used to determinate fault morphology (Florinsky 2008).

### 4.4 Convergence Index

This parameter is based on the aspect and it is used for discerning areas with convergent and divergent flows in DEMs. The possible values grade between 90° (convergent flow) to – 90° (divergent flow). The flatter areas without curvature do not present a defined aspect, thus, the convergence index value is 0.

This is similar to the plan curvature but the results are softer. The calculation uses aspects of surrounding cells, that is, to what degree the cells point towards the central cell. The result is given as a percentage. Negative values correspond to convergent flow conditions and positive values do so to divergent flow conditions. Less than 100 could be a crest; more than 100 would be as a beak of a cone, plus 100 represents a hole, where 0 is a slope pair.

### 4.5 Terrain Ruggedness Index

This parameter is considered as a local parameter. It indicates the land complexity and the presence of undulations. It is related to the slope. Higher values of standard deviation indicate irregular terrain, whereas lower values indicate a regular surface (Olaya 2009).

### 4.6 Topographic Position Index

This parameter is considered a compound topographic index (Moore et al. 1991). Generally, the topographical position is taken as a light variable defined to characterize the local topography relative to its spatial context. The importance of this concept lies in the classification of landforms since it is based on both morphology and the relative position of the land. Nowadays, the starting point for the analysis of landforms is related on elevation data since it implicitly contains information about shape, vertical order and magnitude of the landscape features.

There are several methods for calculating the topographic position (Zimmermann 2000; Jenness 2005; Weiss 2001). The first one has a hierarchical approach and uses $3 \times 3$ windows to calculate the difference between the average elevation of the cells surrounding the central cell and its elevation. A hierarchical integration in a single map allows appreciating the different topographical positions. The topographical position depends upon the scale; the same point on the crest of a mountain can be seen as a peak or a flat area depending upon the scale of analysis.

## 5 Results and Discussion

The basin is relatively small (Chow et al. 1979). It covers an area (A) of 269.61 $km^2$, has a perimeter (P) of 131.58 km, a total length (L) defined by the method proposed by Schumm (1956) as 31.08 km and a mean width (W) of 16.08 km. It is a $6^{th}$. order basin. The maximum elevation is 2,882.6 m a.s.l. and the minimum altitude is 1,288 m a.s.l.

The maximum slope is about 79% (38.31°) and most of it changes with significantly lower values (Fig. 2A).

In this analysis, the aspect presents the majority of values consequently accumulate near zero (Fig. 2B).

According to the histogram of curvatures, a greater data accumulation occurs in the core values that contain the 0 (Fig. 3A and B).

As regarding TRI, this parameter presents values between 0.278 and 7.06, and the standard deviation (SD) is 1.82; thus, it can be considered that the values obtained are low. The majority of them are found near standard deviation (Fig. 4A).

Concerning the Topographic Position Index (TPI) (Fig. 4B) presents values between −2.14 to 2.4. The highest percentage of obtained values are concentrated around 0.128, near cero. To a lesser extent, the values are organized into the following intervals: −2.14 to −1.01, −1.01 to 0.128, 0.128 to 1.26 and 1.26 to 2.4.

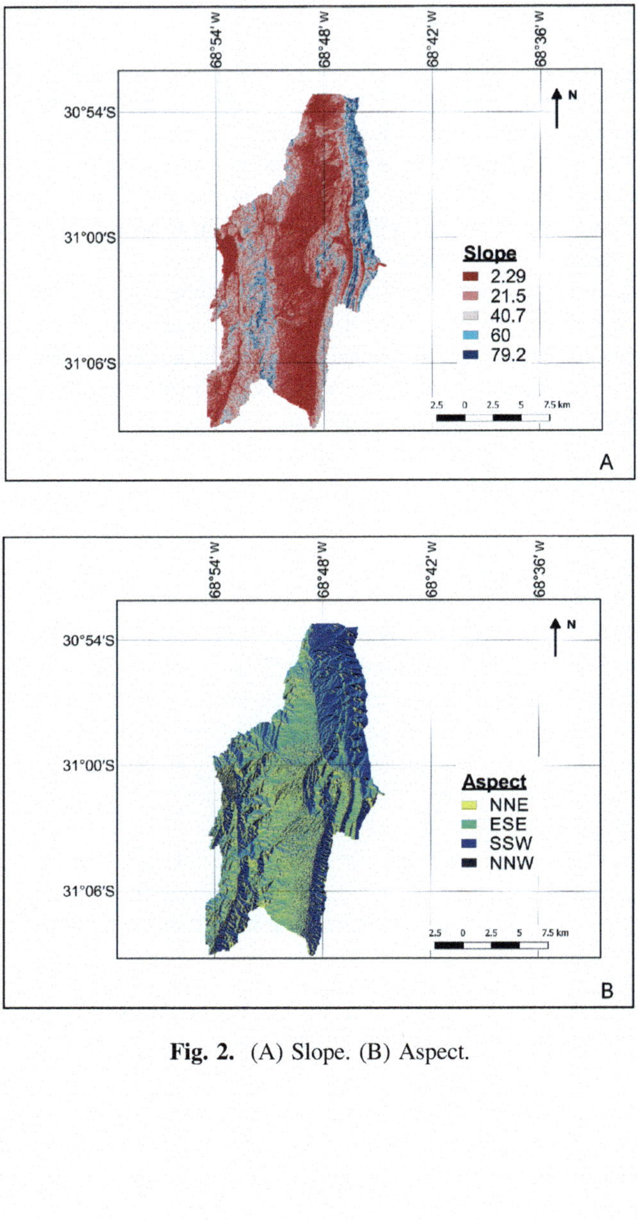


**Fig. 2.** (A) Slope. (B) Aspect.

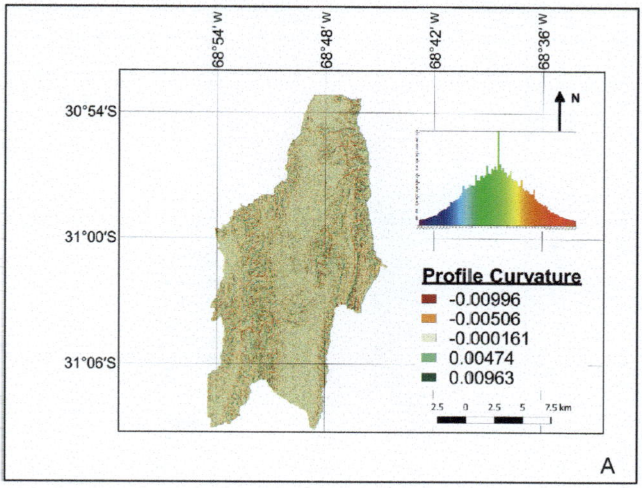


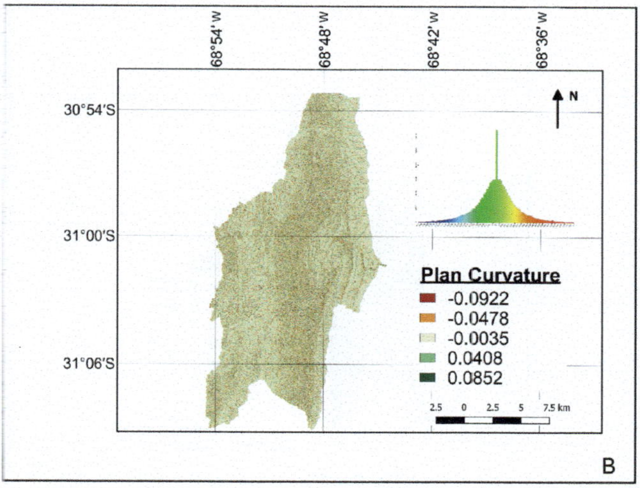


**Fig. 3.** (A) Plan curvature. (B) Vertical curvature.

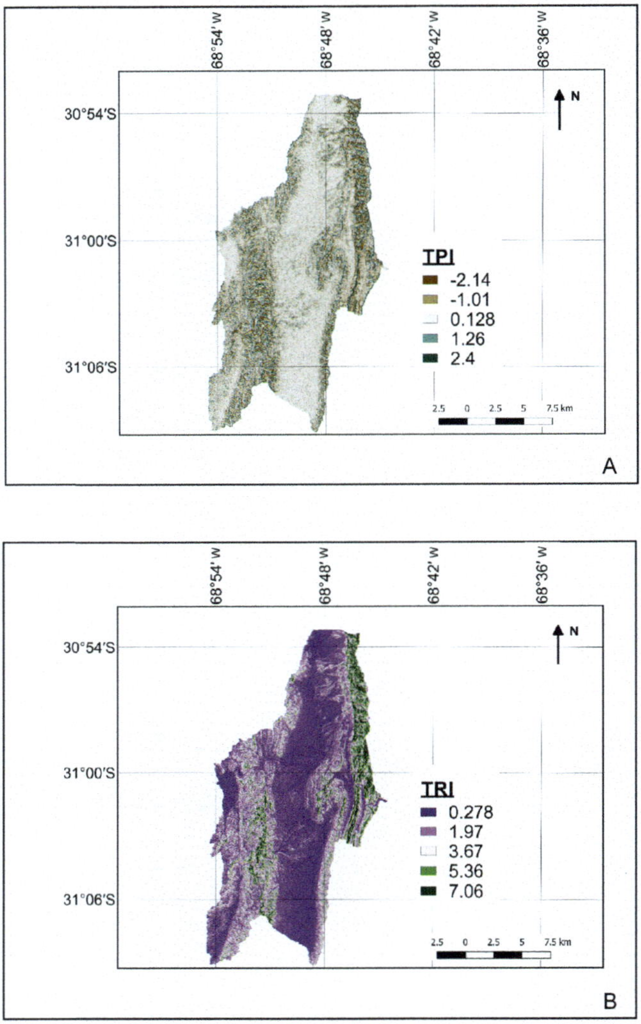


**Fig. 4.** (A) Topographic ruggedness index (TRI). (B) Topographic position index (TPI).

## 6 Conclusions

Slope suggests further erosion towards mountains. Hydrologically, the gravitational energy available produces an increase towards the center of the basin, creating an area with higher moisture, and an area where deposition may occur. It can be estimated that the highest values are concentrated in the western direction of the Talacasto mountain ranges and those of lesser values, in the eastern side. The minimum values are concentrated in the middle of the basin, that is, in the depositional piedmont environments. Thus, it can be stated that the direct relationship existing between altitudes and slopes is evident.

With respect to the aspect, it supports the aforementioned deductions. This assumption is also supported by the curvature index, which indicates the convergence towards the central zone, deceleration, and accumulation. On the other hand, there is a divergence in surrounding areas, rapid runoff and high level of erosion as well. The flow converges in the middle of the basin and diverges in the bounding mountain ranges. Therefore, in the study area, most of it is flat.

About the TRI, this means that the area is not regular enough because there are various landforms. The Talacasto mountain range is in the western bound, where the values are higher. The values are lower in the middle of the basin because it is the flattening zone, that is to say, the depositional environment.

Concerning TPI, it presents higher zones in the Talacasto mountain ranges and flat landforms in the center of the basin.

To assume the susceptibility of flash flood occurrence is higher according to a basic and general basin analysis performed.

To improve predictions, a deeper parameter study with support of high resolution images, aerial photographs, satellite images and topographic maps is strongly advised.

The study area has had recurrent torrential rains in January and February, as recorded by local newspapers. Thus, it is crucial to model rainfall analysis in order to predict the future hydrological conditions.

**Acknowledgements.** We thank Dr. Maria Yanina Esper Angillieri, Dr. Laura Patricia Perucca and Dr. Jorge Rabassa for constructive criticism and thoughtful reviews, which greatly improved our study.

## References

Blaszczynski, J.S.: Landform characterization with geographic information systems. Photogram. Eng. Remote Sens. **63**(2), 183–191 (1997)

Boulton, A.G.: Morphometric analysis of river basin characteristics. J. Hydrol. **6**(2), 227 (1968)

Chow, V.T., Eliassen, R., Linsley, R.K.: Wastewater Engineering: Treatment, Disposal. Reuse. McGraw-Hill, New York (1979)

Costa, J.E.: Hydraulics and basin morphometry of the largest flash floods in the conterminous United States. J. Hydrol. **93**, 313–338 (1987)

Esper Angillieri, M.Y.: Morphometric analysis of Colangüil river basin and flash flood hazard, San Juan, Argentina. Environ. Geol. **55**, 107–111 (2008). https://doi.org/10.1007/s00254-007-0969-2

Esper Angillieri, M.Y., Perucca, L.P.: Caracterización morfométrica de la Cuenca del Río Seco a propósito de las fuertes precipitaciones de enero de 2013, Departamento Sarmiento, San Juan, Argentina. Bol. Soc. Geol. Mex. **2014**(66), 2 (2014)

Florinsky, I.V.: Global lineaments: application of digital terrain modelling. In: Zhou, Q., Lees, B., Tang, G. (eds.) Advances in Digital Terrain Analysis. Springer, Heidelberg (2008)

Gardiner, V.: Drainage basin morphometry. British Geomorphological Group. Tech. Bull. **14**, 48 p. (1975)

Gaume, E., Bain, V., Bernardara, P., Newinger, O., Barbuc, M., Bateman, A., Daliakopoulos, I.: A compilation of data on European flash floods. J. Hydrol. **367**(1–2), 70–78 (2009)

Gregory, K.J., Walling, D.E.: Drainage Basin Form and Process: A Geomorphological Approach. Wiley, New York, 456 p. (1973)
Jenness, S.: Topographic Position Index (tpi_jen.avx) extension for ArcView 3.x. Jenness Enterprises (2005). http://www.jennessent.com/arcview/tpi.htm
Klingseisen, B., Metternicht, G., Palaus, G.: Geomorphometric landscape analysis using a semi-automated GIS-approach. Environ. Model. Softw. **23**, 109–121 (2008)
Mesa, L.M.: Morphometric analysis of a subtropical Andean basin (Tucumán, Argentina). Environ. Geol. **50**, 1235–1242 (2006)
Miller, V.C.: A quantitative geomorphic study of drainage basin characteristics in the Clinch Mountain area. Virginia and Tennessee. Technical report. 3. Office of Naval Research. Department of Geology, Columbia University, New York (1953)
Moore, I., Grayson, R., Ladson, A.: Digital terrain modelling review of hydrological, geomorphological and biological applications. Hydrol. Process. **5**, 3–30 (1991)
Moussa, R.: On morphometric properties of basin, scale effects and hydrological response. Hydrol. Process. **17**, 33–58 (2003)
Olaya, V.: Basic land-surface parameters. In: Hengl, T., Reuter, H. (eds.) Geomorphometry: Concepts, Software, Applications, pp. 141–169. Elsevier, Amsterdam-Oxford (2009)
Olaya, V., Conrad, O.: Geomorphometry in SAGA. In: Hengl, T., Reuter, H. (eds.) Geomorphometry: Concepts, Software, Applications, pp. 293–308. Elsevier, Amsterdam-Oxford (2009)
Oñate, V.F., Bosque, S.J.: Extracción de modelos digitales de elevación a partir de imágenes ASTER para la determinación de características morfométricas de cuencas hidrográficas. In: 12° Congreso de la Asociación Española de Teledetección, Mar del Plata, pp. 457–464 (2007)
Ramos, V.A., Jordan, T.E., Allmendinger, R., Mpodozis, C., Kay, S.M., Cortés, J.M., Palma, M.: Paleozoic terranes of the Central Argentine-Chilean Andes. Tectonics **5**(6), 855–880 (1986)
Schumm, S.A.: Evolution of drainage systems and slopes in badlands at Perth Amboy, New Jersey. Geol. Soc. Am. Bull. **67**, 597–646 (1956)
Sreedevi, P.D., Subrahmanyam, K., Ahmed, S.: The significance of morphometric analysis for obtaining groundwater potential zones in a structurally controlled terrain. Environ. Geol. **47**, 412–420 (2004)
Strahler, A.N.: Quantitative analysis of watershed geomorphology. Eos, Trans. Am. Geophys. Union **38**(6), 913–920 (1957)
Suet, C., Liam, Y.: Comparison of flow routing algorithms used in Geographic Informations Systems. University of Southern California, Faculty of Graduate School, 97 p. (2004)
Topaloglu, F.: Estimation of instantaneous peak flows in Seyhan river basin using regional regression procedures. Turk. J. Agric. For. **26**(1), 47–55 (2002)
Wani, M.B., Ali, S.A., Ali, U.: Flood assessment of lolab valley from watershed characterization using remote sensing and GIS techniques. In: Hydrologic Modeling, pp. 367–390. Springer, Singapore (2018)
Weiss, A.: Topographic position and landforms analysis. Poster presentation, ESRI User Conference, San Diego, California (2001)
Lin, X.: Flash floods in arid and semi-arid zones. International hydrological programme, Technical documents in hydrology No. 23. UNESCO, Paris (1999)
Zevenbergen, L.W., Thorne, C.R.: Quantitative analysis of land surface topography. Earth Surf. Proc. Land. **12**(1), 47–56 (1987)
Zimmermann, N.E.: toposcale.aml (2000). http://www.wsl.ch/staff/niklaus.zimmermann/biophys/#Topo.html

# Risk Analysis Due to the Variation of Chemical Parameters Induced by Lithology. Fiambalá, Chaschuil River, Catamarca, Argentina

Erlinda V. Ortiz[1,2(✉)], Adriana E. Niz[1], Marcelo E. Savio[1], and Cinthia A. Lamas[1]

[1] Instituto de Monitoreo y Control de La Degradación Geoambiental (IMCoDeG), Facultad de Tecnología y Ciencias Aplicadas, Universidad Nacional de Catamarca, Maximio Victoria 55, CP4700 Catamarca, Argentina
valleortiz@tecno.unca.edu.ar
[2] CONICET, Buenos Aires, Argentina

**Abstract.** The water bodies represent a fundamental resource for the growth and development of a region, which must be well studied and managed so as not to limit its availability. In predominantly semiarid environments, such as the town of Fiambalá, the water resource directly impacts on the socio-economic development of the region. The demand of the resource increases its vulnerability and it can produce a deterioration in quality and quantity due to different environmental and anthropic factors. Therefore, it is necessary to carry out studies that improve the knowledge of water resources and help a more efficient planning in the management of these resources for the development and the general welfare of the population. The Abaucán River is mainly supplied by the Fiambalá and Chaschuil rivers, the latter being the most important tributary. Lithology affects the location and exploitation of aquifers; therefore, it is necessary to analyze the fluvial contributions of the area, the Quaternary deposits, and the physical-chemical quality of the water that flows over them. For its study, samples were obtained for physical-chemical analysis considering the different lithology that the channel crosses, in order to determine the incidence of these deposits in the chemical conditions of the water. To understand the geochemical processes, piper plot has been used for the study. From the analysis carried out, it is concluded that the geochemical alteration due to the lithological influence is almost negligible, which is why it presents strong performance in the agricultural development of the region.

**Keywords:** Water quality · Lithology · Chaschuil · Catamarca · Andes mountains · Argentina

## 1 Introduction

The determination of the extent of the effects of geochemistry on aquifers imply to define physical, chemical, biological and anthropic factors of the environment. Lithology is one of the factors that controls the concentration of anions and cations, as


M. M. Collantes et al. (Eds.): *Advances in Geomorphology and Quaternary Studies in Argentina*, SPRINGEREARTH, pp. 169–181, 2020.
https://doi.org/10.1007/978-3-030-22621-3_9

well as the variations that this can have as it circulates downstream. This is why it is necessary to have control points and analyze the fluvial contributions of each sub-basin or where lithological changes are observed (such as volcanic soils, Tertiary sediments and Quaternary deposits). These factors influence the water quality, which may be carriers of salinity or other elements not suitable for human and agrarian consumption. This is important as a first approach to the knowledge of the physicochemical conditions of the water resources in a certain region. The investigation directed to the quality of the water and its use is very important to establish its aptitude of use, for example, the destiny for human or agricultural consumption, among others. If the study is oriented towards the use for irrigation, the water quality will depend on the type and amount of dissolved salts that are present in relatively small but significant amounts (FAO 2002). In general, these salts originate from the slow dissolution or erosion of rocks and soil, including the dissolution of lime, gypsum and other minerals. The most important factors to take into account to determine the quality of the water used for agricultural purposes are pH, salinity, sodium (sodium adsorption ratio, SAR) carbonates and bicarbonate in relation to the content of Ca and Mg.

Therefore, the study of the lithological incidence in water quality allows the determination of the aptitude of the resource for the analysis of production and development in agriculture, the maintenance of soil productivity in a sustainable manner and the future analysis of the conservation of resources and the protection of the environment. Knowing the quality of surface waters from the follow-up of easy-to-determine physical-chemical indicators, means a great contribution to the socioeconomic development of the region.

In the present work, it is intended to determine the lithological incidence in water quality in the area with the greatest possibility of agricultural development, fixed in terms of the spatial and temporal location of the physicochemical parameters that indicate to a large extent the quality of the water and its use for irrigation or human consumption.

## 2 Study Area

The study area, the valley of the Chaschuil River, is located west of the town of Fiambalá, 65 km from the homonymous city, in the Tinogasta Department of the Province of Catamarca, Northwestern Argentina. It is located in the Andean Precordillera, at 4,500 m a.s.l. (meters above sea level). The name of this valley is given by the Chaschuil River, also called Guanchín river (Fig. 1; Turner 1972).

The studied sector is located within the following geographical coordinates: NW boundary: 26°42’59.84” S, 68°37’45.91” W; NE boundary: 26°42’08.36” S, 67° 51’46.56” W; SE boundary 27°55’07.26” S, 67°51’31.28” W - SW boundary 27° 55’18.74” S, 68°35’08.24” W. This sector covers an approximate area of 1,100 $km^2$. It is accessed from the city of San Fernando del Valle de Catamarca to the Chaschuil Valley by National Route 38, towards the town of Chumbicha that intersects with National Route No. 60 to the international boundary of Argentina and Chile.

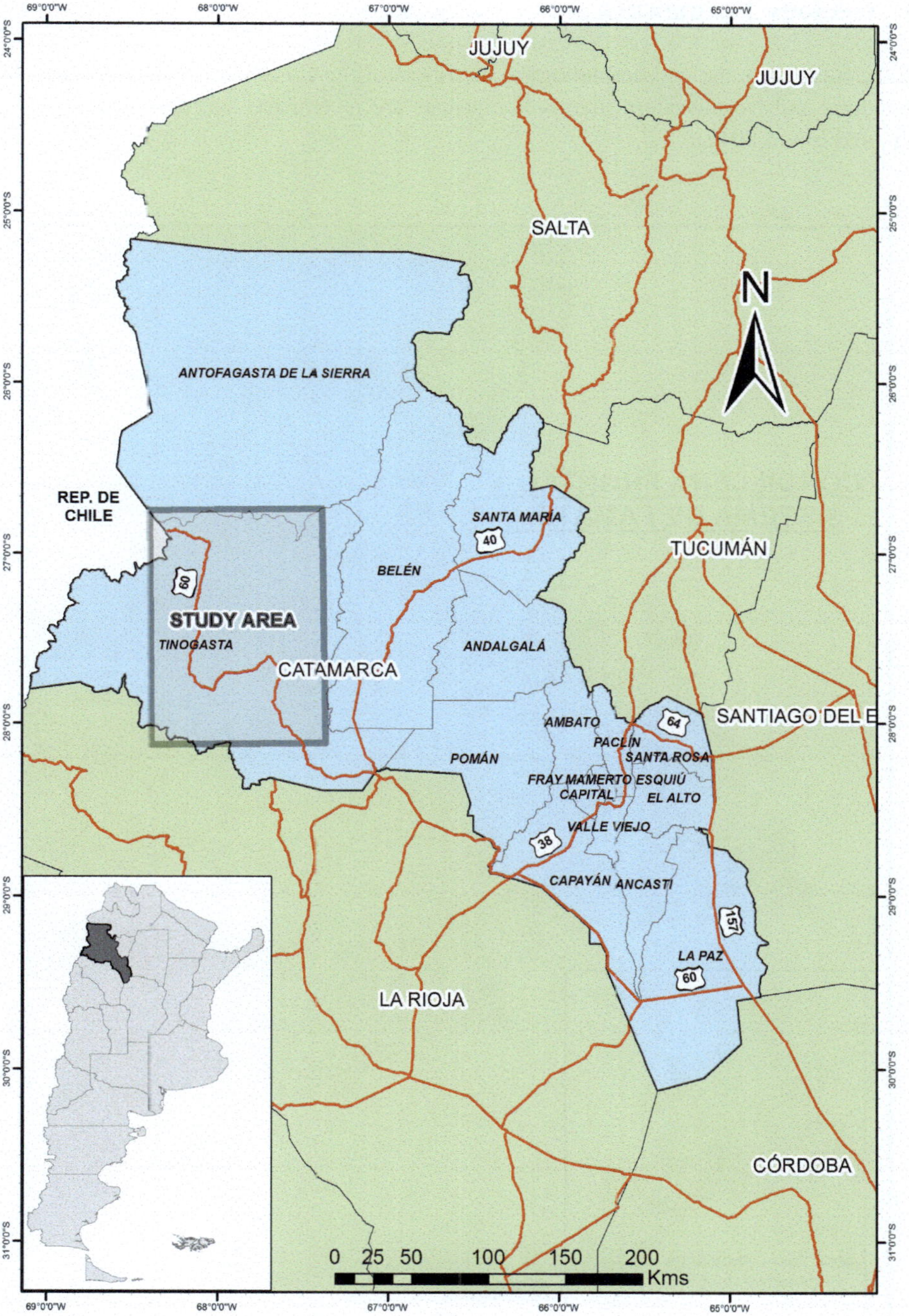


**Fig. 1.** Map of the sampling area in the Valley of Chaschuil, Catamarca, Argentina.

# 3 Geology Framework

In the study area, the geomorphological provinces of the Cordillera Frontal, Sistema de Famatina and Northwestern Sierras Pampeanas are recognized and described (Fig. 2; Niz and Savio 2017).

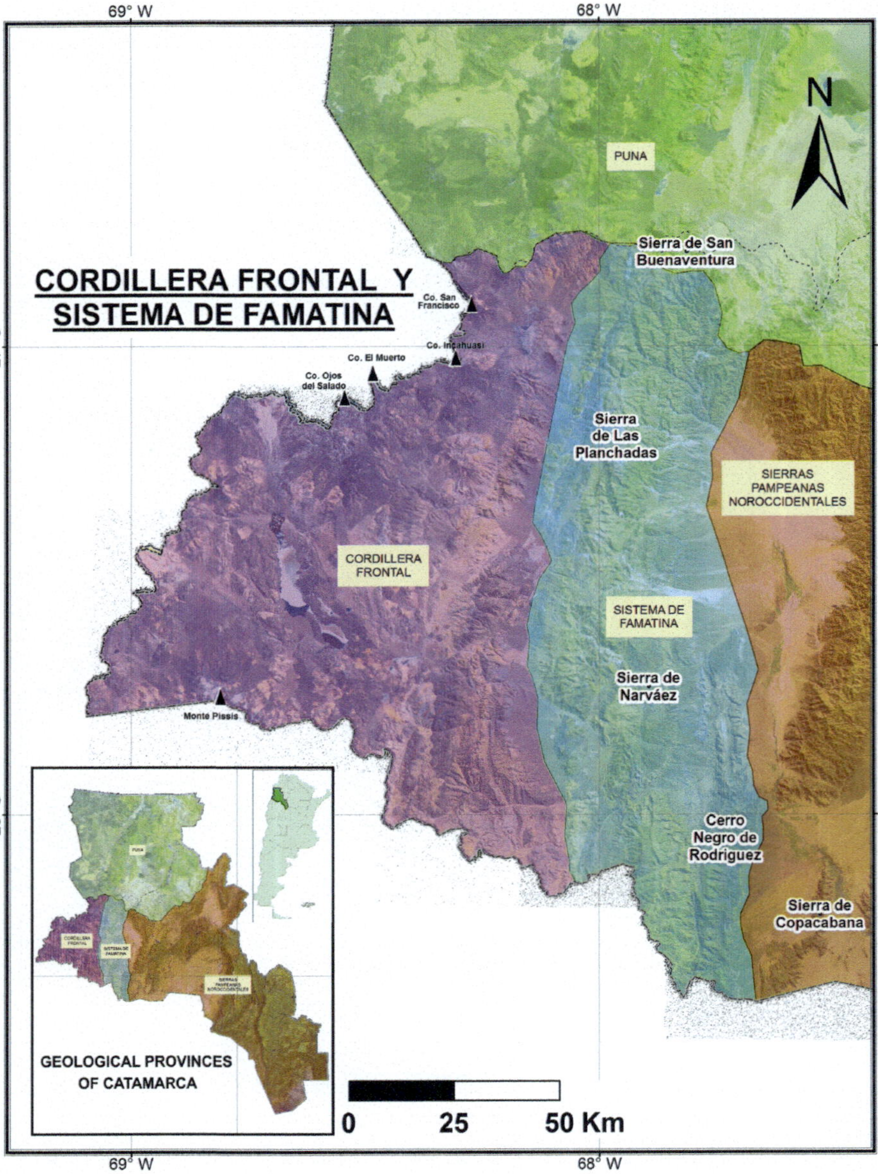


**Fig. 2.** Geomorphological provinces of the study area

The geomorphological province of the Cordillera Frontal is a structural stripe oriented in a north-south direction, between 50 to 70 km wide, born in the southern limit of the Puna, that extends to the Diamante River in the province of Mendoza. It differs from the Puna by the geological characteristics of the rocky units that compose it, that is, folded and fractured Paleozoic sedimentary rocks. The Cordillera Frontal and the Precordillera geological units are rather undefined in this northern area as a single morphostructural unit, but gradually they become individualized towards the south, outside the territory of the province of Catamarca. It is composed of independent tectonic blocks of high altitude, without a uniform general direction. These are impressive volcanic vents, of the stratovolcano type, such as the Monte Pissis, Cerro Incahuasi, Cerro El Muerto, Cerro Ojos del Salado, and Cerro San Francisco, among many others. Smaller orographic elevations are also common and they give origin to deep fluvial valleys. Because the arid climate prevails, in general, the limit of permanent snow is established above 5,000 m a.s.l., where processes and landforms typical of glacial and periglacial environments are frequent.

In the Geomorphological Province of the Sistema de Famatina its principal landforms are oriented NNW-SSE. From a geological perspective, it is characterized by its low-grade metasedimentary basement rocks and an important sedimentary and volcano-sedimentary Ordovician sequence, and the Tertiary rocks present in the Costa de Reyes Formation, located at the latitude of the homonymous locality. This is an outcrop of eruptive rocks, composed of conglomerates of porphyry and andesitic pumice. These rocks could be related to an eruptive center originated as a result of the fracture La Aguadita - Santa Cruz and the El Potrerillo Formation, which is a sedimentary package partially covered by piedmont fanglomerates. These gravels consist mainly of fine-grained, reddish brown, light grey and yellowish sandstones, with siltstones and sandy clays. The Paleozoic (Ordovician) sedimentary rocks are also present, which link it with the Precordilleran environment, and distinguish it from that of the Sierras Pampeanas. It is represented by the Sierra de las Planchadas (which reaches 4,300 m a.s.l.) and the Sierra de Narváez (as high as 3,700 m a.s.l.) which to the south of the town of Tinogasta is called Cerro Negro de Rodríguez. These ranges constitute the boundary between the valleys of Tinogasta and Fiambalá with the Chaschuil Valley. The Sistema de Famatina is characterized by its low-grade metasedimentary basement and a powerful sedimentary sequence of Ordovician volcano-sedimentary rocks, which differentiate it from both the Precordillera and Sierras Pampeanas (Niz 2003).

The geomorphological province of the Northwestern Sierras Pampeanas, as they were named by Stelzner (1873), encompasses a series of mountain ranges formed by metamorphic or crystalline schists that protruded from the adjacent "pampas" or flat lowlands. Later studies perceived differences within this province, so that at present it is subdivided into two, the Eastern Sierras Pampeanas and the Western Sierras Pampeanas. The latter includes the Sierras Pampeanas of Tucumán, La Rioja, San Juan, and the western sector of San Luis and Catamarca (Ramos 1988). The current morphology is the result of the elevation along high-angle reverse faults, generated by the Andean orogeny during the Neogene, which defines the particular feature that most of these mountain ranges have a morphological asymmetry, in which the western slopes show steep slopes that are much steeper than the eastern slopes. The internal structure of the rocks is complex, the product of several old cycles of deformation reinforced by the placement of granitic bodies of various categories and by the regional and local metamorphic recrystallization of the matching rocks.

As it has been mentioned, the study area belongs to the geological province Sistema de Famatina and the Cordillera Frontal transition zone between them. In the stratigraphy, its geological composition includes phyllites and schists intruded by granitic rocks from the Precambrian, with its entourage of porphyry dykes (Table 1).

Paleozoic sedimentary rocks are unconformably arranged upon these formations. These rocks are partly of marine origin, like those of the Suri Formation with intercalations of tuffs and continental sedimentary rocks of the Agua Colorada and de la Cuesta formations. Between the two entities, and even after the second, there have been intrusions and eruptions, which are exposed in the form of dykes.

During the Mesozoic era there was a long period of erosion and intrusions in the shape of layers, in the sandstones corresponding to the la Cuesta Formation.

The sediments pertaining to the Cenozoic era are all of continental origin, corresponding the oldest to the Neogene, followed by materials from the Tambería and Guanchín formations. Between the Late Neogene and the Early Quaternary, eruptions of basalts and andesites occurred, and finally the Quaternary was differentiated into three different entities.

**Table 1.** Stratigraphic chart of the study area (Turner 1967)

| Age | | | Stratigraphic | Lithology |
|---|---|---|---|---|
| Cenozoic | Quaternary | Recent and Current | | Sandy loess |
| | | Late | | Agglomerates, sandstones and sandstones clays and conglomerates poorly consolidated; gravel limestone |
| | | | Unconformity | |
| | | Early | Tunas Formation | Andesite |
| | | | Filo Negro Formation | Basalt |
| | | | Unconformity | |
| | Neogene | Pliocene | Gaunchín Formation | Sandstones, conglomerates and tuffs |
| | | Miocene | Tambería Formation | Dykes<br>Sandstones and conglomerates select |
| | | | unconformity | |
| | | Early | Aparejos Formation | Sandstones |
| Mesozoic | | | Unconformity | |
| | | | | Andesite |
| | | | | Basalt |
| | | | | porphyry dykes |

(*continued*)

**Table 1.** (*continued*)

| Age | | | Stratigraphic | Lithology |
|---|---|---|---|---|
| Paleozoic | Permian | | Cuestas Formation | Sandstones, conglomerates, tuffs and volcanic agglomerate |
| | | | Unconformity | Dykes |
| | Carboniferous | | Agua Colorada Formation | Sandstones and Conglomerates |
| | | | Unconformity | |
| | Devonian - Ordovician | | Las Planchadas Formation | Extrusive volcanic rocks: dacite, porphyry quartz. Diabase |
| | | | Unconformity | |
| | Ordovician | | Suri Formation | Shales and tuffs |
| | | | Unconformity | |
| Precambrian | | | Narváez Formation | Porphyry dykes |
| | | | Jumes Formation | Granite rock |
| | | | Laguna Amarga Formation | Phyllite |
| | | | | Schist |

The landscape of the region has been mainly modeled by the endogenous processes which were partially modified by the fluvial processes. The region presents a configuration of a folded and running belt that is superimposed to arc volcanism in recent times. Consequently, there remains a series of blocks composed of different lithologies: Precambrian basement, granitoids, sedimentary rocks and volcanic rocks, which are associated rocks each of them having differentiated response to the action of denudation processes, in particular, weathering and fluvial action.

In general the volcanism is related to an arc magmatism, that towards the upper part acquires certain characteristics of interpolate volcanic episodes. With interruptions, volcanism was active in the region since the Miocene, recognizing lavas and volcanic bodies of the Early and Late Miocene, Pliocene and Quaternary age.

The volcanism in the area of the international boundary develops a series of large stratovolcanoes or composite volcanoes, formed by superposition of andesitic-dacitic flows, formation of domes and accumulation of tephras. The arrangement of domes and volcanic cones suggests the existence of boilers of various sizes in the area.

Finally, the particularities of the climate have determined the generalized presence of sandy accumulations and/or pebbles from the differential deflation of materials. It is also important to mention that in the orogeny that includes the Chaschuil valley, evidence of glacial processes and action is quite clear. In terms of hydrology, this basin is located in the western region of the hanging valley and with an elevation of more than 3,500 m a.s.l., it receives permanent water supply from the Las Peladas, Las Lozas and Cazadero rivers (Tezón 1963). Along the axis of the valley there are areas of natural springs (locally known as "vegas") that are due to the morphology of the mountainous ranges, where compact or impervious rocks limit the flow of groundwater. The head of the basin is located in those "vegas" or wetlands located at the foothills of the San Francisco volcano that converges with the Chaschuil -Guanchín river, downstream with the Fiambalá river as a result of the end of the Abaucán river basin (Fig. 1). The Chaschuil River is a permanent stream with abundant flow. These characteristics makes it extraordinary considering the type of relief that the region presents, given that it is fed by the thawing that occurs in its upper basin (in the mountain range), in addition to the average annual rainfall of around 100 mm allows the settlement of a stable population and excellent possibilities for sustainable agriculture.

## 4 Materials and Methods

The field sampling points were geo-referenced by GPS (Table 2). These points were selected based upon the predominant lithological types or changes of orientation that the channel has adopted at the moment. Ten representative points were selected for taking samples from the Chaschuil-Guanchín River; the sampling was carried out in November 2015. They represent a homogeneous spatial distribution throughout the basin (Fig. 3). The water samples were collected in clean polyethylene bottles according to the standard methods for the analysis of various water quality parameters of the American Public Health Association (1989). During sampling, the bottles were rinsed three times with the same river water before the final intake. Then they were stored at 4 °C during transport to the soil laboratory of the Faculty of Agricultural Sciences of the National University of Catamarca for the analysis of electrical conductivity, sodium adsorption ratio (SAR), pH, calcium, magnesium, sodium, potassium, carbonates, bicarbonate, sulfate, chloride, total dissolved solids (TDS) and total hardness.

Geochemical plots were determined using the Diagrammes software that identified the chemical processes and water type of the area and are studied by plotting the analyzed results of major cations (such as Ca, Mg, Na and K) and major anions (Cl, $SO_4$, $CO_3$ and $HCO_3$).

**Table 2.** General description of the sampling sites of the Valley of Chaschuil.

| Sampling code | Location | Latitude | Longitude | Elevation (m) |
|---|---|---|---|---|
| 1 | Medanitos | −27°31'36.8019 | −67°3459.1643 | 1616.000 |
| 2 | Medanitos | −27°31'34.9759 | −67°35'1.7392 | 1617.000 |
| 3 | Quebrada La Angostura | −27°42'8.6842 | −67°55'25.1894 | 2542.793 |
| 4 | Quebrada La Angostura | −27°42'37.2178 | −67°58'1.5198 | 2748.995 |
| 5 | Chaschuil | −27°47'8.7181 | −68°4'42.7479 | 3073.437 |
| 6 | Chaschuil | −27°47'9.1189 | −68°4'42.9151 | 3075.360 |
| 7 | Cortaderas | −27°33'18.8485 | −68°8'41.4877 | 3371.444 |
| 8 | Las Lozas | −27°12'51.9577 | −68°6'30.1076 | 3731.214 |
| 9 | Las Grutas | −26°54'40.7458 | −68°7'46.6063 | 4039.075 |
| 10 | Las Grutas | −26°54'49.0620 | −68°7'45.1289 | 4041.958 |

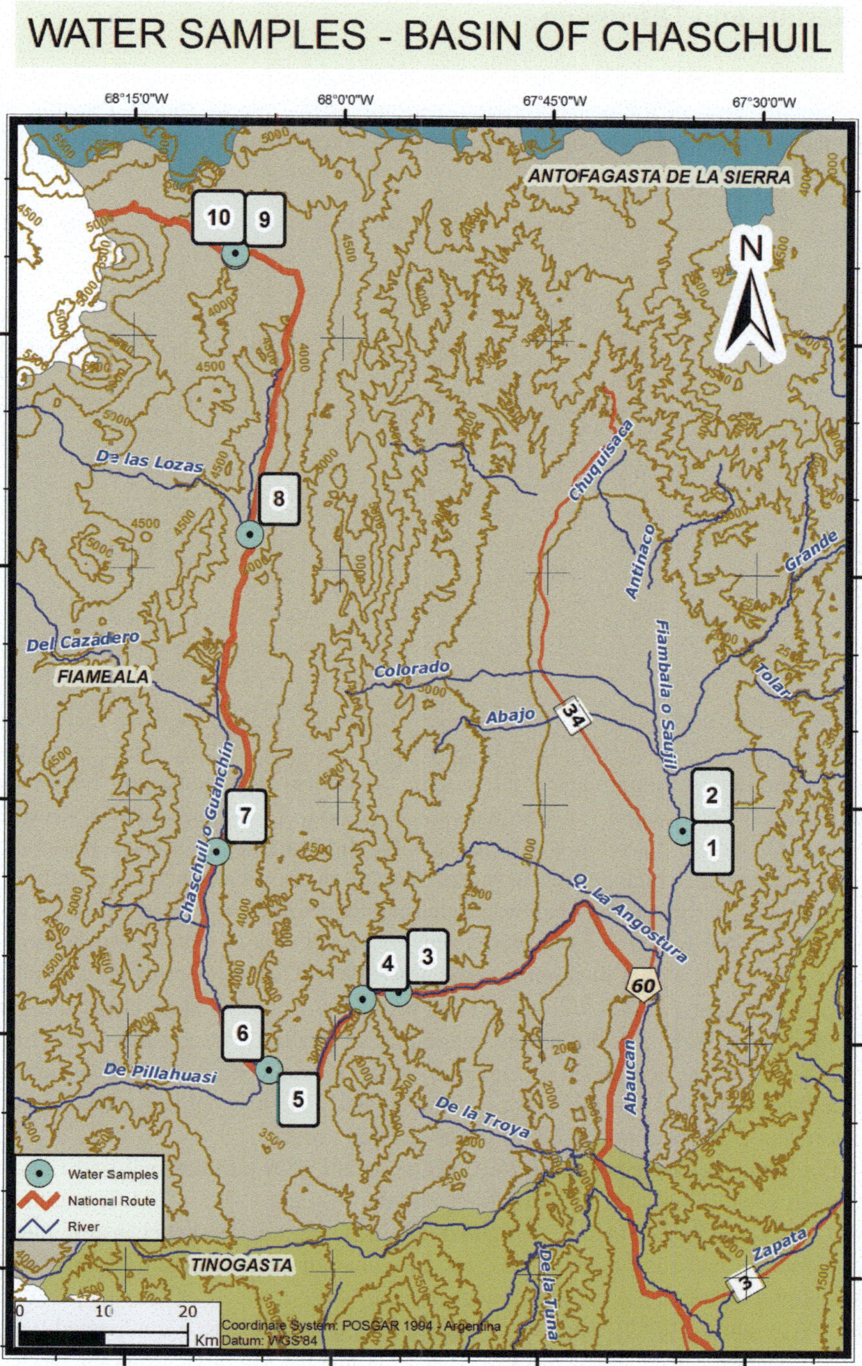


**Fig. 3.** Map of the locations of sampling sites. Square tags mark the main sampling sites.

# 5 Results and Discussion

The results of the physical and chemical analyzes of the water were compared with the water quality criteria for consumption according to the World Health Organization (WHO 1993) (Table 3).

**Table 3.** Concentration of the physico-chemical parameters analyzed at the sampling points.

| Sampling code | C.E. (µS/cm) | SAR | PH | $Ca^{++}$ (mg/l) | $Mg^{++}$ (mg/l) | $Na^{+}$ (mg/l) | $K^{+}$ (mg/l) | $CO_3^{--}$ (mg/l) | $HCO_3^{-}$ (mg/l) | $SO_4^{--}$ (mg/l) | $Cl^{-}$ (mg/l) | TDS | $Co_3Ca$ |
|---|---|---|---|---|---|---|---|---|---|---|---|---|---|
| 1 | 825 | 4.9 | 8 | 40.4 | 8.2 | 130.7 | 0.4 | 12 | 207.5 | 61.5 | 78 | 528 | 134.6 |
| 2 | 849 | 4.3 | 7.7 | 40.4 | 16.3 | 129.3 | 0.4 | 24 | 470.9 | 78.3 | 63.8 | 543.4 | 168.5 |
| 3 | 1344 | 3.9 | 8.3 | 44.9 | 59.9 | 171.4 | 14.7 | 48 | 231.9 | 134.4 | 145.3 | 860.2 | 359.9 |
| 4 | 1403 | 4 | 8.1 | 49.4 | 54.5 | 171.9 | 15 | 36 | 280.7 | 136.8 | 163.1 | 897.92 | 348.6 |
| 5 | 1389 | 4.3 | 8.2 | 58.4 | 40.9 | 175.8 | – | 36 | 146.4 | 183.4 | 148.9 | 889 | 314.7 |
| 6 | 1405 | 4.1 | 8.2 | 40.4 | 62.6 | 177.2 | 17.3 | 36 | 268.5 | 137.3 | 159.5 | 899.2 | 359.9 |
| 7 | 1334 | 3.8 | 8.2 | 44.9 | 62.6 | 166.8 | 16.7 | 48 | 244.1 | 124.4 | 159.5 | 853.8 | 371.1 |
| 8 | 1268 | 3.6 | 8.2 | 62.8 | 43.6 | 151.6 | 5 | 36 | 231.9 | 121.5 | 184.3 | 811.5 | 337.2 |
| 9 | 549 | 1.6 | 8.3 | 35.9 | 30 | 54.2 | 0 | 36 | 158.6 | 40.9 | 7.1 | 351.4 | 213.6 |
| 10 | 605 | 1.6 | 8.2 | 44.9 | 30 | 57.8 | – | 40 | 162.7 | 53.9 | – | 387.2 | 236 |
| WHO* | 400* | | 6.5 –8.5* | | 50* | | 12* | | | | | 100.0* | |

The electrical conductivity of the surface continental waters oscillates between 15 and 3300 µS/cm. The WHO (1993, 2004) considers that drinking water should not have a conductivity higher than 400 µS/cm. Water with a conductivity higher than 2000 µS/cm is considered not adequate for irrigation. The values of this study are between 549 and 1405 µS/cm, which indicate that the water analyzed has a low content of dissolved salts, which makes it acceptable for many uses. The concentration of dissolved salts in the water is below the detrimental limits.

The pH value in the water varied from 7.7 to 8.3. These waters are moderately alkaline, which are within the quality standards of WHO and who consider the optimal parameter between 7.0–8.5 (WHO 2004). For this study, water throughout the basin has a high pH value due to lithofacies integrating a pyroclastic lava section (Filo Negro Formation), Tertiary sediments - Quaternary (the Gaunchin Formation) and the alteration of calcoalkaline plutonic bodies, differentiated into two entities with unconformities in floor and ceiling according to the deposit order.

The hardness of the water, that is the $CaCO_3$ content in the samples analyzed, obtained values between 134.6 and 371.1 mg/liter, and the average value is 284.41 mg/liter of $CaCO_3$. A value higher than 100 mg/liter of $CaCO_3$ is adequate, because if there is acidification of the water due to some type of contamination, the alkalinity that it possesses would stabilize the pH. On the other hand, there is also a relationship between conductivity and alkalinity since lower values of conductivity are accompanied by low alkalinity values. This analysis is repeated at all points the sampling along the river.

The observed values for chlorides range between 7.1–184.3 mg/liter. It can be assumed that this increase in chloride content in the samples is due to the fact that in areas near the catchment the upwelling of sodium sulphate and a predominance of sodium carbonate or soda that the locals call "coipa". Likewise, the high pH (7.87) in comparison with the other points evaluated has suggested the present authors to presume to estimate that there is the presence of anions that could increase this pH. The chloride values found are within the limits established by the standard (max 600 ppm or mg/liter).

With regard to anion toxicity, no alterations are observed, since they are within the "acceptable" values. Requiring a greater study in the presence of bicarbonates (HCO3) that from 1.5 meq/liter, it may impact the proper absorption and translocation of this nutrient in the case of the use for irrigation.

Piper diagram (Piper 1944) is the most convenient method for plotting the results of multiple analyses on the same graph, which may disclose grouping of certain samples and show different hydrochemical phases or water origin.

Figure 4 shows cases of water types, their classification and the processes or rocks related to them. The Piper diagram shows the composition of the water in the basin. There are three groups:

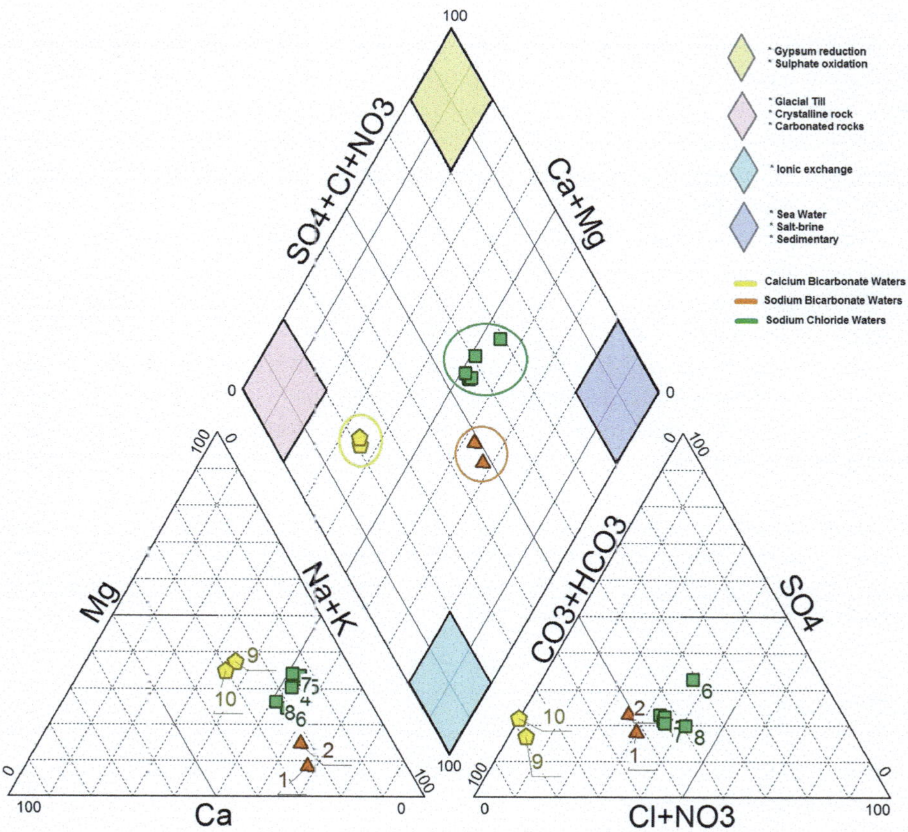


**Fig. 4.** Distribution of hydrochemical data Piper diagram Surface water.

1. Calcium bicarbonate waters: superficial cold water, with a great variety of anions and cations influenced by meteoric recharge, the crystalline and carbonated rocks.
2. Sodium bicarbonate waters: ionic exchange along the basin.
3. Sodium chloride waters: presence of geothermal fluids at depth, related to sedimentary environments.

It is observed that most of the samples are within the classification of sodium chlorides. There are samples that are within the range of calcium bicarbonates, which are cold surface water products from the meteoric recharge of the glacier area.

The points corresponding to groups 2 and 3 are located very close to each other, preserving their composition with the contribution of surface water from their tributaries, unlike group 1, which varies due to the influence of the surroundings in the volcanic mountain high zone.

## 6 Conclusions

From the study carried out in this paper, the water quality of the Chaschuil Valley could be analyzed based on the geological analysis and the results of the samples. It is concluded that:

According to the results of the analysis of each sampling point, the importance of lithology as a contributing factor to water geochemistry is low and it has little influence. Thus, in general, the use of water is acceptable for development agricultural of the region.

Piper diagram provided an efficient way for analyzing the water origin and hydrochemical processes that affected water chemistry. The main water types are Ca-Mg-$HCO_3$, Na-Cl and Na-$HCO_3$.

They have an acceptable quality for irrigation with moderate restriction of use taking into account the carbonate/bicarbonate concentration.

The present geochemical - geological behavior is related to the lithology that corresponds to an area with arid - semiarid characteristics.

It is considered that the Chaschuil Valley is an excellent water resource that has not yet been studied completely, the geochemical alteration due to the lithological influence is almost negligible, which is why it presents strong performance in the agricultural development of the area.

**Acknowledgements.** The authors wish to acknowledge the assistance of the Consejo Nacional de Investigaciones Científicas y Técnicas (CONICET) and the Instituto de Monitoreo y Control de la Degradación Geoambiental of the Universidad Nacional de Catamarca, both of which support facilities that were used in this investigation.

## References

American Public Health Association (APHA-AWWA-WPCF): Standard methods for the examination of water and wastewater (1989)

Food and Agriculture Organization of the United Nations (FAO): Agua y cultivos: logrando el uso óptimo del agua en la agricultura. Roma, Italia (2002)

Niz, A.: Geomorfología del sector meridional del Departamento Tinogasta, Provincia de Catamarca, Argentina. Editorial Científica Universitaria, UNCA 214. Catamarca, Argentina (2003)

Niz, A., Savic, M.: Geomorfología general de la provincia de Catamarca. In: Muruaga, C.M., Grosse, P. (eds.) Ciencias de la Tierra y Recursos Naturales del NOA. Relatorio del XX Congreso Geológico Argentino, San Miguel de Tucumán, pp. 804–828 (2017)

Piper, A.M.: A graphic procedure in the geochemical interpretation of water analyses. Am. Geophys. Union Trans. **25**, 914–923 (1944)

Ramos, V.: The tectonics of the Central Andes; 30° to 33°S Latitude. Geol. Soc. Am. Spec. Pap. **218**, 31–54 (1988)

Stelzner, A.: Comunicaciones sobre la geología y mineralogía de la República Argentina. Anales de Agricultura, vol. 123, pp. 133–142, Buenos Aires (1873)

Tezón, R.: Texto Explicativo del Mapa Hidrogeológico de la República Argentina. Dirección Nacional de Geología y Minería, Buenos Aires (1963)

Turner, J.C.: Descripción geológica de la Hoja 13b, Chaschuil, provincia de Catamarca. SEGEMAR, Boletín N°106. Buenos Aires (1967)

Turner, J.C.: Puna. In: Leanza, A.F. (ed.) Academia Nacional de Ciencias. Primer Simposio de Geología Regional Argentina, pp. 91–116, Córdoba (1972)

World Health Organization (WHO): Guías para la calidad del agua potable, Recomendaciones. vol. 1, Geneva, Switzerland (1993)

World Health Organization (WHO): Guidelines for drinking water quality: training pack. Sample Heading (Third Level), WHO, Geneva, Switzerland (2004)

# Identification of Knickpoints in Littoral Basins of Argentine Patagonia: Geomorphic Markers in a Passive Margin

Gabriella M. Boretto[1,2(✉)], Marcela Cioccale[1], José Tello[1], Eduardo García Aráoz[1,3], and Sandra Gordillo[1,2]

[1] Facultad de Ciencias Exactas, Físicas y Naturales, Universidad Nacional de Córdoba, Córdoba, Argentina
gmboretto@yahoo.com.ar

[2] Consejo Nacional de Investigaciones Científicas y Tecnológicas (CONICET), Centro de Investigaciones en Ciencias de la Tierra, (CICTERRA), Av. Vélez Sársfield 1611, Edificio CICTERRA, 2° Piso of. 18, X5016CGA, Ciudad Universitaria, Córdoba, Argentina

[3] Consejo Nacional de Investigaciones Científicas y Técnicas (CONICET), Departamento de Geología Aplicada. Facultad de Ciencias Exactas, Físicas y Naturales, Universidad Nacional de Córdoba, Córdoba, Argentina

**Abstract.** Knickpoints are an expression of disequilibrium in a fluvial system and they occur in both alluvial and bedrock rivers. A significant amount of research in the past decade around the globe demonstrated the usefulness of the knickpoints as geomorphic markers, specially in active margins, which due to its formation is related to tectonic, climatic and geomorphic processes. However, less attention is paid to the fluvial systems in well-developed along the passive coastal margin as the Atlantic Ocean coast in Patagonia. For the first time in this region, knickpoint analysis together with hypsometric curves and longitudinal stream profiles studies were conducted to understand the landform evolution in coastal basins along the passive margin in Bustamante and Solano Bays (Chubut province, Patagonia, Argentina) during the recent past. The methodology employed digital elevation models (DEMs) and satellite images which have been processed by ArcGIS software using such specific tools as (1) ArcHydro and knickpointfinder, (2) CalHypso, and (3) TopoToolbox (which contains a set of MATLAB functions) to complete the proposed studies. A total of 590 knickpoints were identified along longitudinal fluvial profiles in 18 basins. The results suggest that knickpoints frequency and density vary according to the lithological-structural and tectonic conditions, and it is possible to separate two sectors in the study area, in both the northern and the southern portion, being the southern zone more dynamic linked to active alignments in the depocenter of the San Jorge Gulf. This work provides a method for using knickpoints as geomorphic markers in coastal basins associated to the Atlantic Ocean margin.

**Keywords:** Knickpoints · Hypsometry · Patagonia · Argentina · Longitudinal profile


M. M. Collantes et al. (Eds.): *Advances in Geomorphology and Quaternary Studies in Argentina*, SPRINGEREARTH, pp. 182–208, 2020.
https://doi.org/10.1007/978-3-030-22621-3_10

## 1 Introduction

Geodynamic and tectonic processes can be quantified by measuring rock and surface deformation using geomorphic and stratigraphic markers (Willet et al. 2006; Pavano et al. 2016). Stream long profiles and drainage networks are particularly sensitive to base-level fall (Whipple 2004; Perron et al. 2012; Willett et al. 2014), creating geomorphic markers such as terraces (Pazzaglia 2013) and knickpoints (Berlin and Anderson 2007) by unsteady and transient river incision. In this sense, the unsteady base-level fall at the mouths of rivers, such as those that flow directly into the sea, generates knickpoints that migrate upstream through the drainage network (Pavano et al. 2016). These knickpoints will vertically climb all stream channels at the same rate if their migration follows a detachment-limited stream power law, where the power dependency of the slope term is $\sim 1$ and the drainage area is steady (Howard and Kerby 1983; Seidl and Dietrich 1992; Howard 1994; Seidl et al. 1994; Sklar and Dietrich 1998; Stock and Montgomery 1999; Whipple and Tucker 1999; Kirby and Whipple 2001; Snyder et al. 2000; Dietrich et al. 2003; Zaprowski et al. 2001; Pavano et al. 2016; Perucca et al. 2018). Besides knickpoints, other geomorphological index are rewarding to control changes and evolution in the landscape along time such as hypsometric curves (Pérez-Peña et al. 2009; Pavano et al. 2016; Geach et al. 2017) and the longitudinal stream profile (Whipple and Tucker 1999; Snyder et al. 2000; Kirby and Whipple 2001; Castillo and Lugo-Hubp 2011; Pavano et al. 2016; Castillo et al. 2017; Perucca et al. 2018). The hypsometric curve for a drainage basin represents the relative proportion of the watershed area below (or above) a given elevation (Strahler 1952; Schumm 1956). This surface-elevation curve is a useful tool for characterising the topographic relief within a drainage basin, and, hence, establishing comparisons between different basins (Pérez-Peña et al. 2009). These analyses have been widely used in geomorphology, hidrology and engineering practice (Ciccacci et al. 1992; Willgoose and Hancock 1998; Chen et al. 2003; Pérez-Peña et al. 2009; Geach et al. 2017). Steady-state streams show concave-upward power-function long profiles (i.e. equilibrium profile; Mackin 1948; Hack 1957; Pavano et al. 2016), traditionally attributed to the well-known inverse discharge-slope relationship necessary to generate the shear stresses that drag and move sediments (Gilbert 1877; Baker and Ritter 1975; Leopold and Bull 1979; Bull 1991; Pavano et al. 2016). Deviation from this condition is commonly indicative of a transient phase of landscape change, represented by knickpoints in the river profile. For settings where rock uplift and rock type throughout the watershed are uniform, the channel reach upstream of the knickpoint remains graded to the higher, paleo–base-level condition (Pavano et al. 2016; Yunus et al. 2016). In this case, every knickpoint can be considered as a time-dependent geomorphic marker that migrates through the landscape, lowering the river profile to the new base level (Pavano et al. 2016; Yunus et al. 2016). The specific erosion processes, commonly modeled as a detachment-limited stream power dependent relationship (Crosby and Whipple 2006; Berlin and Anderson 2007), dictate the rate at which the knickpoint migrates upstream through the channel. The analyses of the longitudinal profiles are necessary to understand the evolution of the fluvial environment system together with the other proxies.

In this context, the Patagonian coast of Argentina preserves impressive traces of sea-level oscillations represented by marine deposits and erosional forms of high scientific value for understanding climatic changes and tectonic uplift history in this area (Feruglio 1950; Codignotto et al. 1992; Rutter et al. 1990; Rostami et al. 2000; Schellmann and Radtke 2000, 2003, 2010; Pedoja et al. 2011a, 2011b; Ribolini et al. 2011; Isola et al. 2011; Zanchetta et al. 2012; Boretto et al. 2014). Particularly, the Bustamante Bay and surrounding areas such as Solano Bay are located in the northern and central sectors of the San Jorge Gulf (Chubut Province), where an outstanding succession of Quaternary raised beach deposits is exposed for a total area of more than 1000 km$^2$, extending up to 10 km landward and reaching 100 m of elevation (m a.s.l.) (Isola et al. 2011). These sandy-gravelly beach ridges may occur in singular form or as sets characterised by crests located at the same elevation. Their curvature varies from almost straight (in the higher and older forms) to very curved, tracking the variation of the shoreline geometry in time and they are dissected by fluvial erosion whose base level fall in the channels generates knickpoints, that is to say sudden breaks of the slope of the river bed along its longitudinal profile (Brush and Wolman 1960; Gardner 1983; Bishop et al. 2005; Quezada et al. 2010; Castillo and Lugo-Hubp 2011; Pavano et al. 2016, among others). The knickpoints are dynamic relief elements that reflect climatic, lithological, morphological and tectonic processes. Their formation and subsequent recession denote the presence of a new regime to which the preexisting erosion system must adjust (Castillo and Lugo-Hubp 2011). The purpose of this work is to use geomorphological markers as knickpoints besides hypsometric curves and longitudinal stream profiles to understand the landform evolution on the passive margin in the Bustamante and Solano Bays during the recent past.

## 2 Setting

### 2.1 Regional Setting

The study area is located between 44°20′–45°54′ S and 68°00′–65°14′ W (Fig. 1), covering the locality of Bustamante Bay and surroundings, central and northern area of the San Jorge Gulf. Due to its latitudinal location and to the presence of the Andean Cordillera, the Patagonian climate is characterized by two main meteorological features such as very strong winds and low precipitation (Coronato et al. 2008). Constant dry winds blow strongly from the west, particularly in the summer months, and eolian erosion shapes rocky outcrops and deflates loose sediments (Isola et al. 2011). The low amount of precipitation (<300 mm/yr) and moderate thermal amplitude allow the growth of a sparse cover of grass and shrubby vegetation, consistent with the semiarid climate of Patagonia (Isola et al. 2011). Wave action is the most important morphoclimatic agent responsible for the past and present landscape. This area is characterized by a high-energy system with intense storms and a macrotidal regime. A minimum tidal range of >4 m can be reasonably assumed for most of the Patagonian coast (Isola et al. 2011).

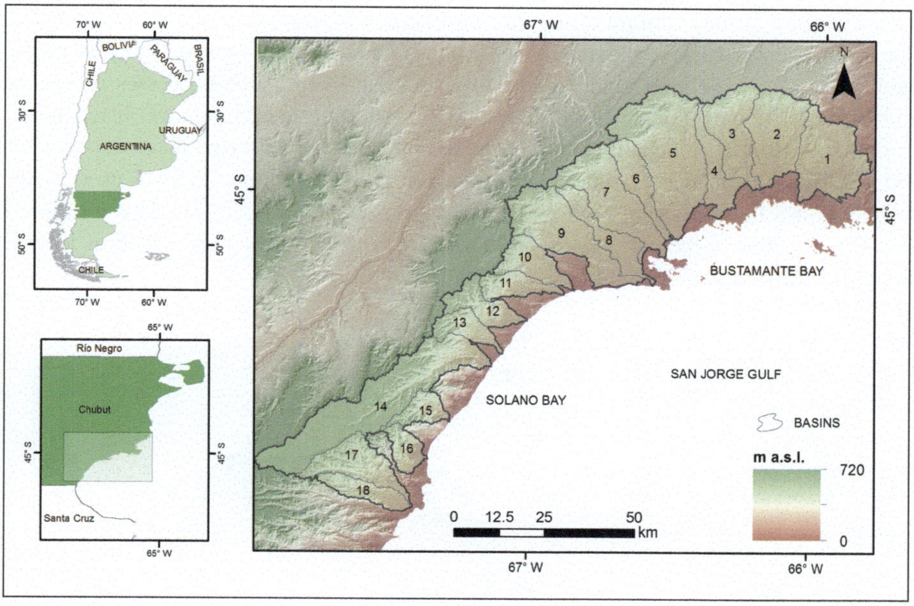


**Fig. 1.** Location of the study area.

## 2.2 Geological Setting

The oldest rock unit, the Marifil Complex (Malvicini and Llambías 1974), consists of Jurassic rhyolites, ignimbrites and volcanoclastic conglomerates (Franchi et al. 1975; Lesta et al. 1980; Sciutto et al. 2000; Lema et al. 2001). Subordinate outcrops expose the Paleocene marine deposits of the Salamanca Formation and the Paleocene continental deposits of the Río Chico Formation (Andreis et al. 1975; Sciutto et al. 2000; Lema et al. 2001). Above it, the continental Sarmiento Group extends in conformably contact and it is followed in erosion unconformity by the marine sedimentary rocks of the Patagonia Formation (Riggi 1979) or the Chenque Formation, from Oligocene to Early Miocene age (Caviglia 1979; Riccardi and Rolleri 1980), which fill the deeply embayed basin that reached the base of the Andean Cordillera. These rocks are covered by the Santa Cruz Formation, deposited in continental environments. Above a clear erosion unconformity the oldest level of the Montemayor Formation (Panza et al. 2002), also called the "Rodados Patagónicos" (i.e., Patagonian gravels) was deposited. They are composed of gravels of rounded to sub-rounded clasts, essentially acid volcanic rocks, with abundant coarse to fine sandy matrix, of massive aspect or with faintly marked stratification, with ages ranging from the Late Pliocene to the Early Pleistocene (Fidalgo and Riggi 1970; Franchi et al. 1975; Haller 1979; Cortés 1981; Cionchi 1988). A second level of "Rodados Patagónicos", mostly re-worked sediments, covers the pediments which develop between the well-defined plateau and the coastal zone along the Pleistocene (Sciutto et al. 2000). Finally, eolian and coastal marine deposits associated with regression episodes, named in the Bustamante Bay as the Caleta Malaspina and Zanjón El Pinter formations, pertaining to the Late Pleistocene and the Holocene, respectively

(Cionchi 1985, 1988), and mass-wasting sediments of Quaternary age were deposited. The most recent detrital sediments make up current beaches, composed of gravel, fine to medium sandy gravel and sand, genetically associated with slope processes combined with fluvial events (Cionchi 1988). The general tectonic conditions are distensive in general, constituting blocks that sink deeper to the southern part (the basin depocenter) and superficial expression faults in generally east-west direction which are used by the drainage lines that outlet into the Atlantic Ocean (Sciutto et al. 2000). The geological map (Fig. 2) shows the formations and structures of the study area, which is located along the northern-central sector of the San Jorge Gulf.

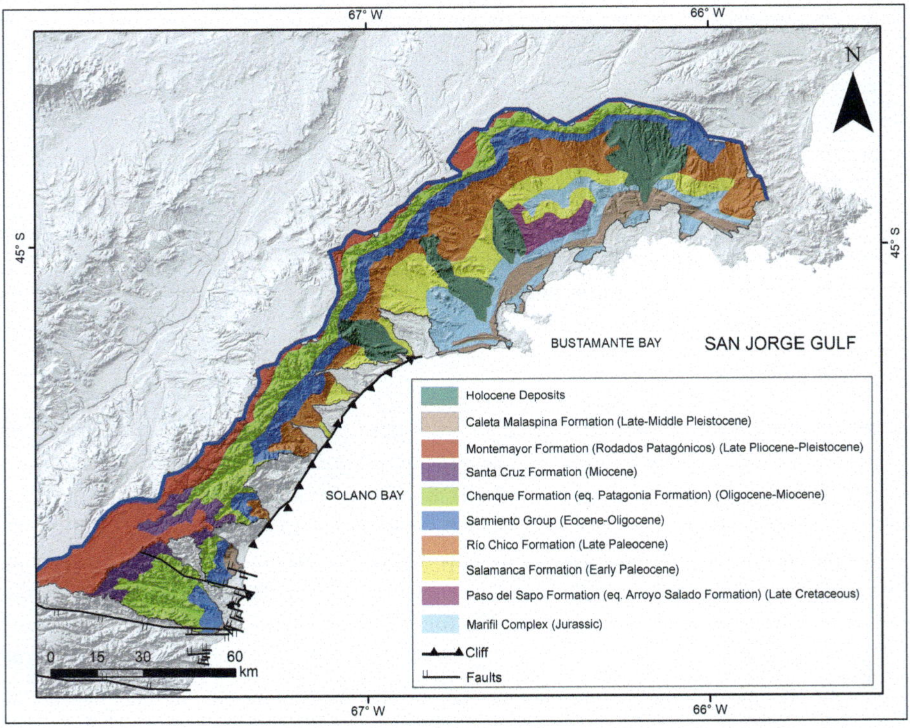


**Fig. 2.** Geological map of the study area, modified from Sciutto et al. (2000) and considering the database from the "Servicio Geológico Minero Argentino" (http://sigam.segemar.gov.ar).

### 2.2.1 Geological History of the San Jorge Gulf Basin

The San Jorge Gulf, which is the most important hydrocarbon-producing sedimentary basin of Argentina, has a complex geological history (Fitzgerald et al. 1990; Sylwan 2001). The sedimentary fill of the basin is closely related to its tectonic history. In this section the most relevant episodes that led to the formation of this basin are highlighted.

(1) Its basement was structured by a large compressive strain associated with a Neopaleozoic collision, followed by important extensional periods linked to the Jurassic opening of the Weddell Sea in Antarctica (Ramos 2015).

(2) Northwest-southeast trending grabens formed and filled during a Triassic and Early Jurassic rift phase, climaxing with a pervasive Middle Jurassic volcanic episode, continued growth and filling of the basin occurred during a Late Jurassic-earliest Cretaceous late rift phase and Early and Late Cretaceous sag phases (Fitzgerald et al. 1990).
(3) The extensional fracturing is superimposed over the region in an almost orthogonal orientation which is associated with the opening of the South Atlantic Ocean (Ramos 2015). This extensional regime prevailed during the Early Cretaceous, this time produced by subduction with extension governed by the absolute motion of the South American plate (Ramos 2015). The onset of compression in the Late Cretaceous controlled the tectonic load subsidence of the continental deposits of the Chubut Group in a foreland basin, therefore, along this sector with a highly segmented basement and no confined margins also favored the existence of extensional faults with subparallel orientation to the main compressive stress during the Late Cretaceous (Ramos 2015).
(4) The extensional system prevailed during the Early Paleogene, with a conspicuous stress rotation, culminating in the Oligocene with an incipient primary extension associated with an intraplate alkali basaltic volcanism. This was associated with a decrease in the absolute displacement of the South American plate (Ramos 2015). Late Cretaceous-Early Tertiary extension set up many of the present-day structural traps along the normal faults (Fitzgerald et al. 1990; Sylwan 2001; Giacosa et al. 2004).
(5) Middle Tertiary Andean compression produced the narrow, north-south San Bernardo structural belt, which exhibits reversed movements on older, normal, graben-bounding faults on local, low-angle thrust faults. Marked Early to Middle Tertiary erosion produced a significant unconformity within Cretaceous beds around basin margins (Fitzgerald et al. 1990; Sylwan 2001; Giacosa et al. 2004). The structure of the basin obtained the present geometry in the Neogene with the final development of the Bernárdides fold and thrust belt in the west-center portion of the San Jorge basin (Giacosa et al. 2004; Ramos 2015).
(6) More recently and considering just the Atlantic Ocean margin of the San Jorge Gulf, the coast feature sequences of Neogene and Quaternary age, which are shaped by a combination of sea level oscillations and uplifting tectonics (Pedoja et al. 2011a, 2011b, 2014). The vertical ground motion associated with the longlasting development of paleoshore sequences as the imprint of glacial cycles on globally uplifted passive margins (Rostami et al. 2000; Schellmann and Radtke 2000, 2010; Pedoja et al. 2011a, 2011b, 2014).

## 2.3 Geomorphological and Tectonic Setting

The study area presents the typical "meseta" like morphology of the Patagonian region adjacent to the marine coast (Cionchi 1988). It is characterized by the development of relatively flat and sloping plateaus (locally named as "pampas"), descending on gentle slopes until reaching sea level (Cionchi 1988). Cionchi (1985, 1988) and Sciutto et al. (2000) distinguished three major features in the landscape of the study area:

(1) High plateaus ("mesetas altas"): in the northwestern edge of the region, they have an elevation of 450 meters above sea level and are named as Pampa Malaspina, Pampa Las Violetas and Pampa de Salamanca;
(2) Low plateaus ("mesetas bajas"): they correspond to smooth hillocks, which greater relative height, being good examples the followings hills: Tetas de Pineda, Cóndor, Cerro del Gualicho, among others; these landforms constitute discordant elements of the local landscape;
(3) Littoral zone: it shows presence of Quaternary marine littoral beach ridges which are parallel or subparallel to the coastline and they have different elevation (those closest to the coast reach 10 m a.s.l., where those further away can reach more than 100 m a.s.l.), among which there are depressions associated with relicts of older coastal lagoons.

The elevated littoral deposits contain pebbly gravel and sand with highly fossiliferous marine strata (Feruglio 1950; Cionchi 1988; Codignotto 1990; Codignotto et al. 1988, 1990; Rutter et al. 1989, 1990; Schellmann and Radtke 2000, 2010; Aguirre et al. 2005). Schellmann and Radtke (2000, 2010) discriminated the spatial distribution of the Bustamante deposits, classified as beach ridge systems and littoral terraces and valley-mouth terraces based on features (geomorpho-pedo-stratigraphic differentiation) and their relative ages. Their elevation is probably a consequence of glacioeustatic sea-level fluctuations (Schellmann and Radtke 2000) or of tectonic movements (Rostami et al. 2000; Pedoja et al. 2011a, 2011b, 2014; Bini et al. 2017). Schellmann and Radtke (2000) dated these Quaternary deposits using different methods: electron spin resonance (ESR) and α-spectrometric $^{230}$Thorium/$^{234}$Uranium for the Pleistocene beach ridges and radiocarbon dating ($^{14}$C) for the Holocene deposits. These authors suggested that these beach ridge systems have been formed starting at around 400 ka BP, and representing episodes equivalent to the maximum sea-level highstands of the last interglacials. On the other hand, throughout the central and southern sector of the study area, Solano Bay and surroundings, the coastal marine deposit reached 15 m a.s.l. and they belong to the Holocene (Codignotto et al. 1988).

## 3 Materials and Methods

Multi-Error-Removed Improved-Terrain digital elevation model (MERIT-DEM) at 90 m spatial resolution (Iwahashi et al. 2017; Yamazaki et al. 2017), Google Earth imagery and ESRI imagery (1 m spatial resolution) were used as primary data sources to obtain a geomorphological map of the study area, considering also the previous geomorphological cartography made by different authors (Cionchi 1985, 1988; Codignotto et al. 1988, 1990; Codignotto 1990; Sciutto et al. 2000; Ardolino et al. 2008; Isola et al. 2011). Each set of map was converted to the Universal Transverse Mercator (UTM) projection (WGS 84 Datum, 20S zone). Geomorphological features were digitally drawn in a geographical information system (GIS) by analysing satellite images and shaded images derived from DEMs throughout ARC GIS and SAGA softwares.

Concerning the hypsometric curves (Strahler 1952; Schumm 1956), they are calculated from DEM using the CalHypso program which is an extension tool for ArcMap (ArcGIS desktop) to access spatial analyst algoritms (Pérez-Peña et al. 2009). This surface-elevation curve is a useful tool for characterizing the topographic relief within a drainage basin. Its shape is related with the stage of geomorphic development of the basin: (1) convex hypsometric curves are typical of a youthful stage, (2) sigmoidal curves are linked to maturity stage (equilibrium), and (3) concave curves are indicative of a peneplain stage (Strahler 1952; Schumm 1956).

Stream profiles carry information on tectonic and/or climatic perturbations in the form of knickpoints migrating in upstream direction (Pavano et al. 2016; Castillo et al. 2017). For bedrock channels, considering steady, uniform flow and that discharge scales linearly with drainage basin area (A, $m^2$) (Seidl and Dietrich 1992), the general form of the stream power erosion (E, m/yr) model is:

$$E = KA^m S^n \tag{1}$$

where S (dimensionless) is the stream slope and the m and n exponents are positive constants depending on the erosion processes, hydrological features of the stream, and channel geometry (Howard and Kerby 1983; Howard 1994; Whipple and Tucker 1999; Crosby and Whipple 2006), and K [m(1–2 m)/yr] is the detachment-limited erosion coefficient linked to both climatic and lithological factors (Snyder et al. 2000; Pavano et al. 2016). The time rate of change in the elevation of the river channel (dz/dt) is simply the difference between the rate of rock uplift (U) and the rate of incision,

$$dz/dt = U - E \tag{2}$$

By substitution into Eq. 1, with slope (S) expressed as dz/dx,

$$dz/dt = U - KA^m (dz/dx)^n \tag{3}$$

A general expression that can be rearranged for the equilibrium case where dz/dt = 0, and, solved for slope, reveals

$$S = (U/K)^{1/n} \cdot A^{-m/n} \tag{4}$$

Equation 4 has the same form as Hack's law (Hack 1957; Flint 1974; Snyder et al. 2000; Kirby and Whipple 2001; Pavano et al. 2016),

$$S = k_s \cdot A^{-\theta} \tag{5}$$

and

$$k_S = (U/K)^{1/n} \tag{6}$$

where $k_s(m^{2\theta})$ is the steepness index that describes the stream gradient and θ is the dimensionless concavity index (m/n), a morphometric parameter that defines the rate of change of slope in the channel gradient as a function of the drainage area growing downstream. Comparing Eqs. 4 and 5 implies that $k_s$ scales with rock uplift and/or rock type (K) (Whipple and Tucker 1999; Snyder et al. 2000). Given a constant K, several studies have suggested a linear, or near-linear, relation between $k_s$ and U, whereas they have confirmed that θ remains relatively constant for a wide range of tectonic and climatic settings, commonly assuming values ranging between ~0.3 and 0.6 with a mean value of 0.45 (Snyder et al. 2000; Kirby and Whipple 2001; Kirby et al. 2003). To calculate the concavity index (θ) and steepness index ($k_s$) the methodological proposal used by Schwanghart and Kuhn (2010) is used throughout the TopoToolbox in which a set of MATLAB functions that provide utilities for relief analysis for GIS is contained. The tools have been developed to support the work flow in combined spatial and non-spatial numerical analysis, they offer flexible software for hydrological and geomorphological research that involves digital elevation model analysis and focuses on material fluxes and spatial variability of water, sediment, chemicals and nutrients (Schwanghart and Kuhn 2010).

Due to geomorphological investigations a series of processes have been identified that explain the presence of knickpoints in the fluvial bed (Whipple and Tucker 1999; Snyder et al. 2002; Castillo and Lugo-Hubs 2011; Kirby and Whipple 2012, among others). Castillo and Lugo-Hubs (2011) proposed a type of knickpoint classification which provide information links to the morphogenesis and morphodynamics (Fig. 3). Kirby and Whipple (2012) also explored the analyses considering the slope-area along the longitudinal profiles which allow separate different kinds of knickpoint, i.e. lithological, tectonic or eustatic control.

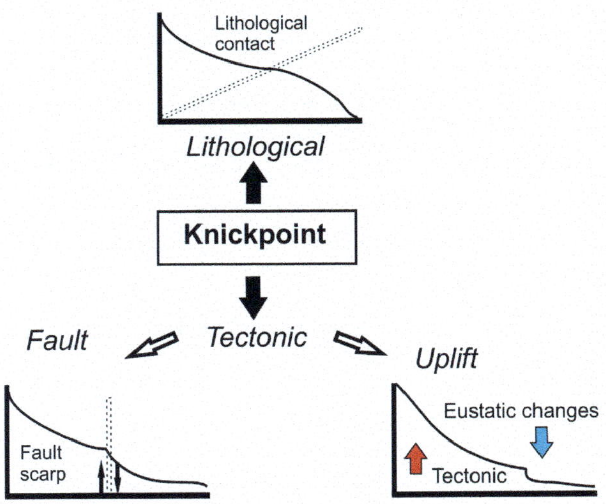


| Type | Subtype | Process |
|---|---|---|
| Lithological knickpoint | (1) Differential erosion<br>(2) Structural | Lithological contacts.<br>Structural and topography control. |
| Tectonic knickpoint | (1) Fault<br>(2) Uplift | Normal and reverse fault displacement.<br>Increase of the uplifting rate, base level fall, eustatic changes. |

**Fig. 3.** Type of knickpoints. This figure was modified from Castillo and Lugo-Hubs (2011) and Kirby and Whipple (2012).

Castillo and Lugo-Hubs (2011) suggested the indirect detection of knickpoints by processing a DEM, previously having selected the fluvial channels that will be analyzed by the tracing or extraction of a topographic profile and subsequent calculation of their slope throughout segments. The slopes values should be plotted on the ordinate axis and on the abscissa axis the values of distance or drainage area should be so in a logarithmic scale to obtain a longitudinal profile. The processing of the data was carried out by the ArcGIS software to delimit drainage network and watershed, where the specific tools ArcHydro and knickpointfinder (Queiroz et al. 2015) were used to recognize the knickpoints. Queiroz et al. (2015) indicated that the theoretical approach to the development of the software was based on a geomorphometric parameter proposed by Hack (1973), the stream length-gradient (SL) index. This parameter pertains to the longitudinal profiles of rivers or drainage lines and it is calculated by multiplying the slope gradient of the stretch of river by the distance between this stretch and the river sources, which then determine the knickpoints of interest in morphotectonic, morphostructural and neotectonic studies (Queiroz et al. 2015). Etchebehere et al. (2004) proposed a derivative of the SL index, the relative slope-extension (RDE) index, which gives an indication of the current energy in a particular drainage stretch and varies with the slope of its surface and the discharge of water at the end of the stretch. According to these authors, the RDE index for a stretch (RDEs) may be calculated using the following relationship:

$$\mathrm{RDEs} = (\Delta \mathrm{H}/\Delta \mathrm{L}).\ \mathrm{L} \tag{7}$$

where ΔH is the elevation difference between the two extremities of a stretch along a river; ΔL is the length of the stretch being studied; and L is the distance between the lower end of this stretch and the source of the river.

The total RDE index (RDEt), which in turn refers to the total length of a river, takes into account the total slope between the source and mouth and the natural logarithm of its entire length (Seeber and Gornitz 1983; Etchebehere et al. 2006; Queiroz et al. 2015). This index is calculated using the relationship:

$$\mathrm{RDEt} = (\Delta \mathrm{H}/\Delta \mathrm{L}).\ln(\mathrm{L}) \tag{8}$$

After measuring the RDES indexes of various stretches and the RDET indexes of their respective drainages, the aim is to compare them to determine which stretches have anomalous slopes (Queiroz et al. 2015). Consequently, when the value of the ratio RDEs/RDEt $\geq 2$, the drainage stretch under analysis can be considered anomalous. A value of RDEs/RDEt between 2 and 10 represents an anomaly of the 2nd order, whereas a value of RDEs/RDEt greater than 10 is an anomaly of the 1st order (Queiroz et al. 2015).

## 4 Results

Figure 4 shows the geomorphological map obtained in which 3 large sectors have been distinguished: (1) High Zone, (2) Middle Zone, and (3) Marginal Zone. This classification could be comparable with that exposed in Cionchi (1985, 1988).

1. High Zone: it is characterized by the development of plateaus (structural surfaces) with elevations around ~450 m a.s.l covered by the "Rodados Patagónicos", and from this, a scarp of 15–40° is generated. In this sector the headwaters of the study area channels begin (Figs. 4 and 5).
2. Middle Zone: this is the largest sector; it is about ~23 km wide, and the general terrain slope is 2–10°. The drainage network of the study area is developed in this area, which acquires a typical dentritic design from sedimentary lithologies, the channels are linked to valleys and in the field it is possible to identify fluvial terraces that are usually covered by modern sediments. The pediments stand out and they form a structural and erosion surface similar to terraced morphology and they are located at the flanks of the fluvial valleys. Besides, accumulations and exposure of alluvial and other aeolian sediments distributed in the whole area are observed (Figs. 4 and 5).
3. Marginal Zone: this sector is related to the coastal marine environment, the dominant landform are the Quaternary coastal beach ridges, which are linked with changes in the coastline over time, whose heights vary between ~100 m a.s.l. The more internal and older crests, up to ~2 m a.s.l. which are located parallel to the current coastline, are the most recent. Between these deposits, inter-beach ridge

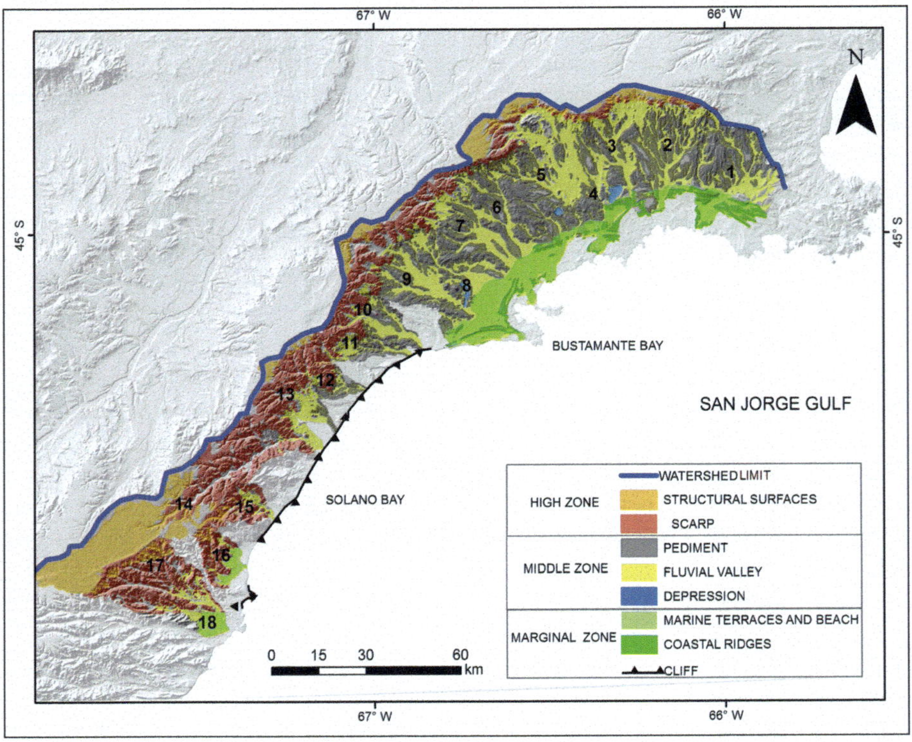


**Fig. 4.** Geomorphological map obtained for the study area through DEM and satellite image processed.

**Fig. 5.** Geomorphological zones of the study area. A, B. High zone: high plateaus where the headwaters begin and a panoramic view of the Middle zone is available. C. Sediment covered section of the Middle Zone. D. Marginal zone: view of the plateaus, beach and volcanic basament outcrops. E. View of the plateaus, Middle zone and Marginal zone characterized by Holocene beach ridges.

depressions develop, and these are associated with old coastal lagoons and tidal plains. The drainage network acquires a subparallel design and accommodates the morphology of the coastal landforms, extending through the mentioned depressions. Marine, alluvial and eolian sediments (such as dune accumulations) appear in the area. On the coastal zone the more highlighted geomorphological features are: sandy-gravel beaches, promontories composed by basement rocks on which cliffs are worked, abrasion platform formed on volcanic rocks which extend from the cliff, and it is feasible to find *in situ* a marshy environment restricted to small zones which are composed by fine sediments, abundant organic matter and covered by halophytes (Figs. 4 and 5).

In this work, 18 basins with similar conditions were delimited going through the areas defined above (geomorphological map) (Figs. 4 and 6). Their peculiar feature is the development of the hydrological network, their ephemeral character is in phase with the arid characteristics of the region, in which the channels cut the beach ridges supporting an important continental contribution towards the coastal marine sector. The undulating relief leads to the channels to assume a serpentine course, subdendritic design, which becomes rectilinear and sub-parallel to the coast in the terminal sector, reaching the sea through the most recent crests. The mean height of the headlands is around ~355 m a.s.l. Continental and marine sediments are often mixed, forming complex systems, covered in some cases by recent eolian materials.

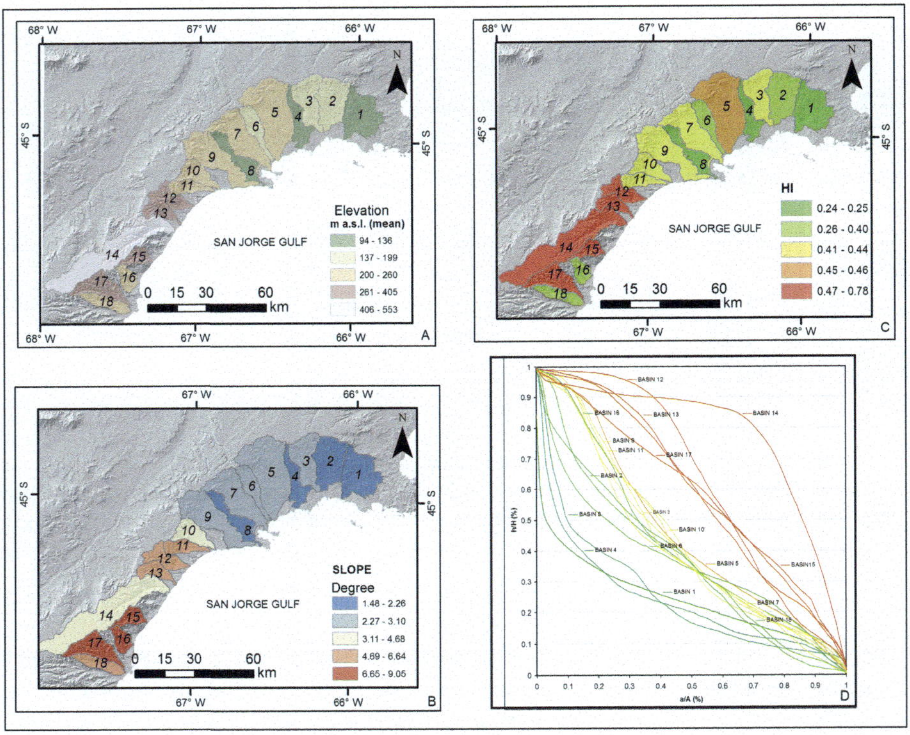


**Fig. 6.** A. The 18 basins differentiated in the study area. B. Slope map. C. Hypsometric integral (HI) value. D. Hypsometric curves.

Through the CalHypso tool the hypsometric curves were extracted for the whole basins (Fig. 6). This analysis allows the identification of different geomorphic development associated with the maturity stage of erosion cycle. At first sight, the curves from the northern basins numbered as 1, 4 and 8 present convex shapes, which are linked to the old stage (Fig. 6), whereas the curves from the southern area, named as the 12, 13, 14 and 15 basins, have a convex shape, and they are linked to a youthful phase. Between these extremes, the curves from the other basins have asymmetric sigmoidal shapes that are above the hillslope fluvial equilibrium curve. The hypsometric integral (HI) ranges from 0.78 to 0.24 with a mean of 0.51. As a whole, the trend of the hypsometric curves is consistent with catchment slope and topography. As suggested by the slope map (Fig. 6), it is possible to distinguish three areas, the lowest slope values are recorded in the northern sector with an average of ~2.44°, along the central area the average slope reaches an average value of ~5.22°, and in the southern zone the highest values occur, with a mean of ~7.25°.

The longitudinal profiles obtained show the capacity of erosion and load from its source to the base level, appreciating different skills that the flow can acquire depending on the topography of the bedchannel. Clearly two opposite sectors are distinguished, the northern and southern portion of the study area, the concave shape and the regularity of the profiles of basins 1 to 10 as opposed to the irregularities that are acquiring the profiles in the highest basins towards the southern sector (Fig. 7). A loss of balance caused by unevenness in the topography of the river valley was observed, with a break in the scaling of the bed slopes being visible. From the drainage basin-scale observations the stream channels are characterized, their long profiles, concave shape, steepness, and associated knickpoints. Long profiles have concave shape values θ ranging around 0.30 (Table 1). Taking into account the theoretical values (see Sect. 3, Materials and Methods), in which θ range around 0.3 to 0.6, our lower values could be linked with the lithology characteristics of the local rocks, whereas the rivers flow over sedimentary terrains and have correspondingly lower concavities. Similarly, the channel steepness index ($k_s$) has relatively low values between 0.17 to 12.06, and a mean of 3.04 (Table 1). Nevertheless, it is possible to discriminate two sectors, the northern one (1 to 10 basins), with a mean of θ value is 0.31 and a $k_s$ index of 3.57, whereas for the southern area (11 to 18 basins), the index concavity average is 0.25 and the $k_s$ mean value is 2.39. In this sense, Wobus et al. (2006) and Kirby and Whipple (2012) indicated that lower values in the concavity index (θ) is relatively insensitive to differences in rock uplift rate, climate or substrate lithology at steady-state (provided such differences are uniform along the length of the channel), they considered that the steepness index ($k_s$) varies with these factors makes this latter index a useful metric for tectonic geomorphic studies. To have a better understanding about the disequilibrium along the stream channels we carried out the identification of knickpoints along the longitudinal fluvial profiles.

The fall of the base level in the stream channels generates knickpoints (Fig. 8), distinguishing these abrupt ruptures of the slope in the channel along its longitudinal profile. These are dynamic relief elements that reflect tectonic, climatic and geomorphological processes, thus discreet changes in the steepness of the river profile whose form and behavior depend on both the nature of the perturbation and the mechanics of river incision (Kirby and Whipple 2012). In the study area 590 knickpoints were identified (Fig. 8), which have been filtered considering 5 categories, with values between 2–4, 5–6, 7–9 being called of 2nd order, and those between 10–17 and 18–33 values are named as of 1st order (Queiroz et al. 2015), considering knickpoints with values ≥ 9 relevant to understand changes in the fluvial profile. It is observed in

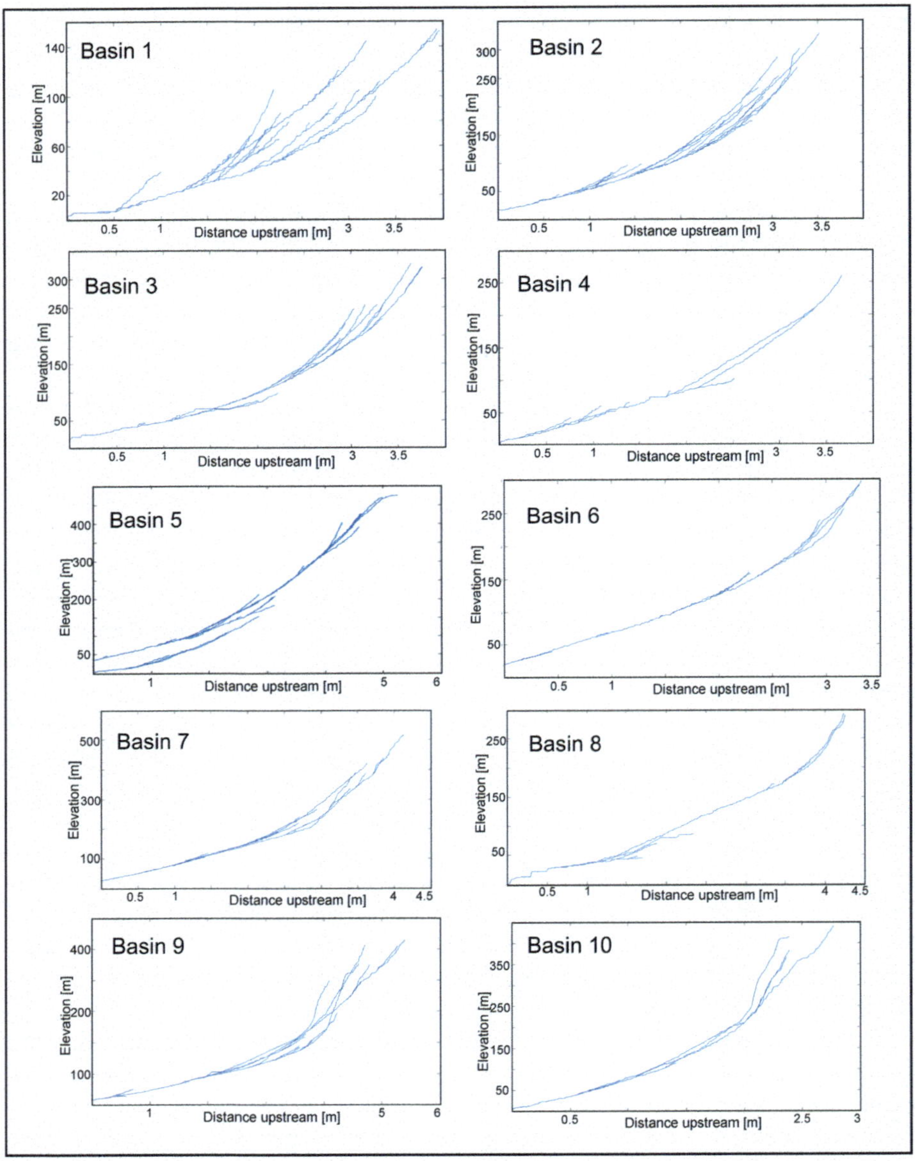


**Fig. 7.** Long stream profile for basins 1–10 (northern sector).

Fig. 9 how the knickpoints are grouped according to lithological-structural conditions and the general slope of the terrain: (1) considering the whole study area, the different kinds of knickpoints are grouped in the High Zone, which is linked with the topography and elevation of the terrain, the points of average values are arranged towards the

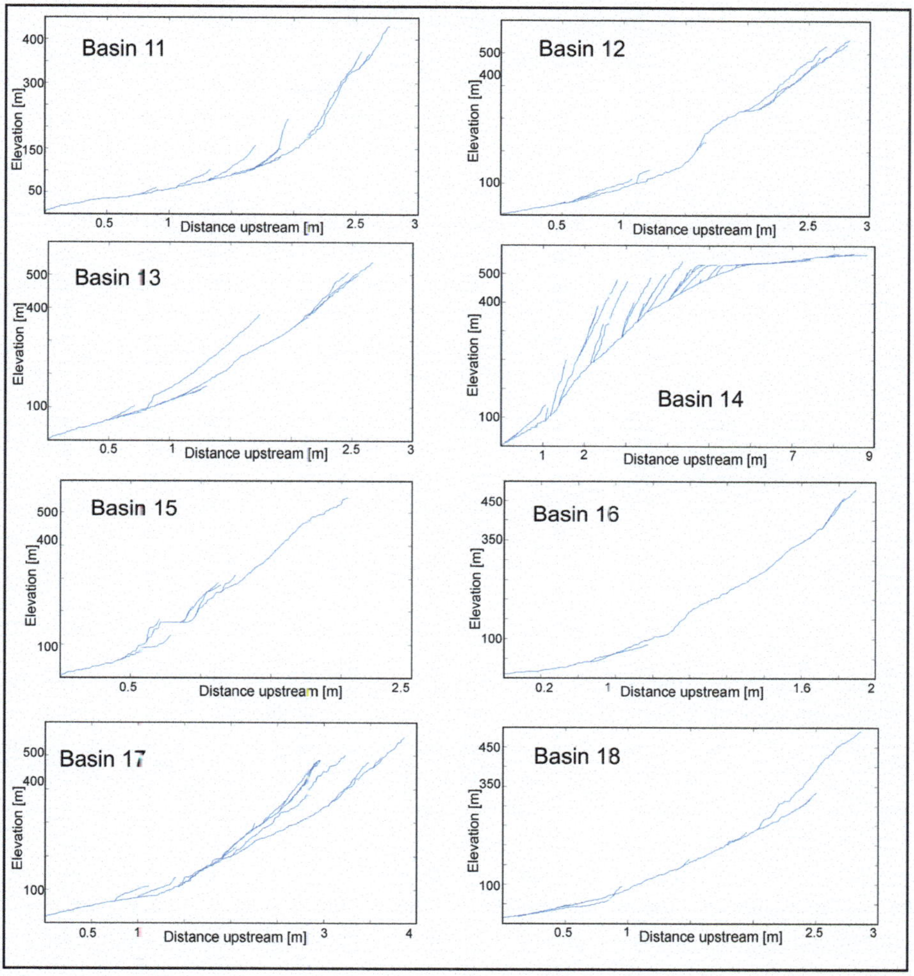


**Fig. 8.** Long stream profile for basins 11–18 (southern sector).

Middle Zone, decreasing in the terminal sector (Marginal Zone), and (2) the knickpoints of higher values are concentrated in the southern sector of the area of study, particularly in basins 12, 14, 15 and 16. This peculiarity is also visible with the development of the hypsometric curves belonging to the juvenile stage and in the values of the concavity and steepness indexes (Table 1).

## 5 Discussion

The results obtained record constant dynamics in these fluvial basins associated to a passive margin being able to recognize different stages of maturity or erosion cycles. Geomorphologically, the whole study area is characterized by the same pattern,

**Table 1.** Morphometric indexes of the 18 studied basins.

| Basin | HI | Area | Perimeter | Stream | Slope | Height | Θ | Ks | Knickpoint | RDEs | RDEs/RDEt | RDEt |
|---|---|---|---|---|---|---|---|---|---|---|---|---|
| | % | $km^2$ | km | km | grade | m a.s.l. | | | amount | | | |
| 1 | 0,24 | 362,76 | 139,72 | 2,5 | 2,15 | 94,03 | 0,33 | 1,36 | 10 | 64,31 | 5,42 | 12,67 |
| 2 | 0,40 | 329,13 | 118,16 | 3,2 | 2,26 | 171,88 | 0,27 | 0,98 | 33 | 98,68 | 4,48 | 21,54 |
| 3 | 0,43 | 268,17 | 128,55 | 2,4 | 2,70 | 199,25 | 0,31 | 1,95 | 33 | 125,77 | 4,47 | 23 |
| 4 | 0,24 | 163,40 | 128,02 | 1,6 | 1,48 | 107,51 | 0,45 | 12,06 | 9 | 109,33 | 4,57 | 25,51 |
| 5 | 0,46 | 628,36 | 180,60 | 4,9 | 2,47 | 227,67 | 0,16 | 0,17 | 54 | 105,28 | 5,89 | 18,51 |
| 6 | 0,37 | 155,06 | 110,01 | 2,3 | 2,64 | 184,30 | 0,27 | 1,03 | 14 | 89,86 | 4,6 | 20,32 |
| 7 | 0,41 | 382,03 | 175,60 | 3,6 | 3,10 | 221,19 | 0,23 | 0,60 | 44 | 129,2 | 5,69 | 22,82 |
| 8 | 0,25 | 177,51 | 142,48 | 3,1 | 2,16 | 135,52 | 0,39 | 4,38 | 13 | 101,03 | 4,24 | 26,11 |
| 9 | 0,43 | 412,09 | 178,76 | 4,9 | 3,06 | 248,70 | 0,29 | 1,75 | 44 | 154,87 | 5,37 | 30,22 |
| 10 | 0,44 | 161,15 | 93,85 | 1,4 | 4,68 | 259,68 | 0,37 | 11,41 | 32 | 185,9 | 5,75 | 33,04 |
| 11 | 0,44 | 117,10 | 85,30 | 1,3 | 5,94 | 256,15 | 0,32 | 4,58 | 31 | 143,71 | 4,76 | 29,89 |
| 12 | 0,63 | 146,66 | 106,60 | 1,9 | 5,76 | 369,85 | 0,13 | 0,20 | 29 | 173,83 | 5,62 | 29,39 |
| 13 | 0,60 | 113,02 | 81,36 | 1,7 | 5,89 | 362,61 | 0,17 | 0,45 | 33 | 159,2 | 4,72 | 33,04 |
| 14 | 0,78 | 698,36 | 273,65 | 7,0 | 4,04 | 553,30 | 0,20 | 0,39 | 92 | 230,05 | 6,78 | 28,19 |
| 15 | 0,61 | 96,66 | 78,99 | 8,8 | 9,05 | 389,05 | 0,25 | 1,97 | 27 | 208,82 | 5,45 | 35,91 |
| 16 | 0,40 | 92,98 | 65,46 | 1,3 | 7,76 | 257,01 | 0,29 | 3,61 | 21 | 175,91 | 6,15 | 30,94 |
| 17 | 0,58 | 223,01 | 131,18 | 2,7 | 8,77 | 405,17 | 0,29 | 2,68 | 48 | 160,75 | 4,87 | 34,12 |
| 18 | 0,37 | 122,88 | 84,25 | 1,6 | 6,64 | 258,24 | 0,34 | 5,24 | 23 | 127,82 | 4,69 | 27,3 |

differentiating three zones (High, Middle and Marginal Zones), and lithologically the sedimentary rocks (marine and/or continental origin) dominate the terrain, although in certain sectors some volcanic rocks appear, printing a trait peculiar to the region, the coverage of loose sediments that are distributed linked to fluvial, alluvial, eolian and coastal processes. Through the hypsometric analysis it is determined that fluvial basins in a state of equilibrium extend throughout the study area; however, it is clear that hypsometric curves linked to old stages of maturity are located in the northern area and the youth cycles of erosion are evident in the southern sector. In the same way, longitudinal profiles analysis distinguish this differentiated behavior, standing out abrupt concavities in basins 1 to 10, with average values of $\theta \sim 0.31$ and $k_s \sim 3.57$. On the other hand, in the upper basins located in the southern sector, the breaks in the slopes are well marked and reflect changes in the equilibrium profile, with average values of $\theta \sim 0.25$ and $k_s \sim 2.39$. In this sense, the analysis of knickpoints from the 18 basins indicate that they occur anywhere along the profiles, but the highest values (knickpoints of second order) are recorded in the headland areas, linked to the elevation and gradient, classified as lithological-structural knickpoints (Castillo and Lugo-Hubs 2011; Kirby and Whipple 2012), and in steep sections of the longitudinal profile in the southern basins, they are categorized as tectonic knickpoints linked to mechanisms of fault and crust uplifting (Castillo and Lugo-Hubs 2011; Kirby and Whipple 2012). Likewise, the slope map of the terrain reflects this range of zones, probably linked to a differential tectonic conjunction and if geomorphology, lithology and climate are considered as constants, high slope values are recorded in the southern sector and the lowest ones occur in the northern sector.

In general, the study region is characterized by a relative tectonic stability associated with a passive margin (Sciutto et al. 2000, Schellman and Radtke 2000, 2010, Pedoja

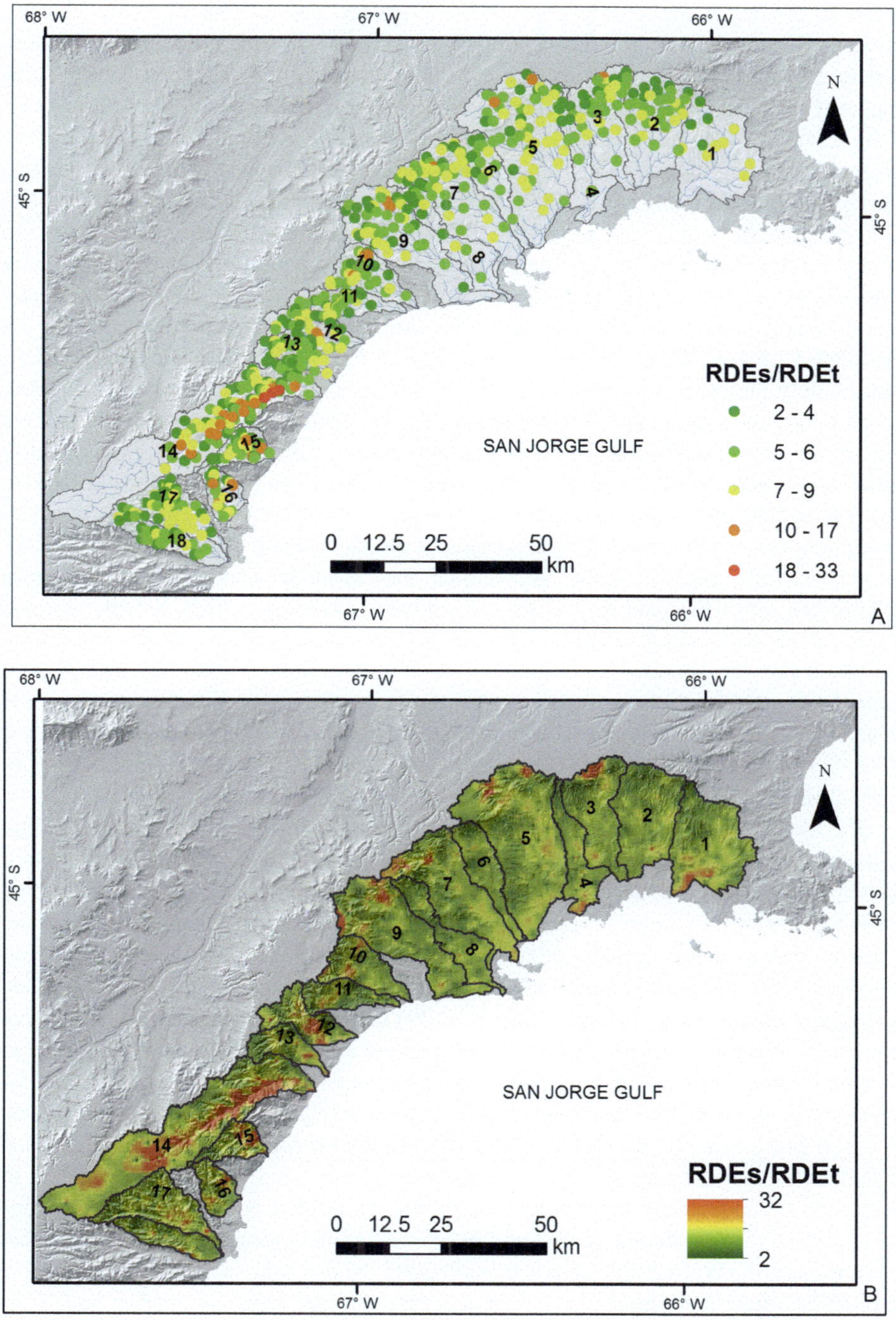


Fig. 9. A. Knickpoint map. B. Interpolation map RDEs/RDEt knickpoint density.

et al. 2011a, 2011b, 2014; Bini et al. 2017); however, two tectonic factors are distinguished: (1) studies carried out by geophysical methods (seismic reflection, refraction, gravimetry and magnetometry) recognized that the basement is staggered and this has impacted on the overlying sedimentary column (Fitzgerald et al. 1990; Sciutto et al. 2000; Giacosa et al. 2004; Ramos 2015), (2) evidence of tectonic uplifting during the Quaternary throughout coastal feature sequences as beach ridges and marine terraces (Feruglio 1950; Rostami et al. 2000; Schellmann and Ratdke 2000, 2010; Zanchetta et al. 2012; Pedoja et al. 2011a, 2011b, 2014; Folguera et al. 2016; Bini et al. 2017). On the surface it can be seen a slight regional dip to the east and a well marked one to the southern sector, proving the progressive disappearing of the basement and its coetaneous coverage towards the depocenter of the San Jorge Gulf basin, produced along the tensional Mesozoic faults of west-east orientation that have suffered different reactivations throughout its geological history (Fitzgerald et al. 1990; Sciutto et al. 2000; Sylwan 2001; Giacosa et al. 2004; Ramos 2015). The shorelines from the entire passive margin of eastern South American coast can be used as a tectonic benchmark to quantify uplift rates (Pedoja et al. 2011, 2014; Folguera et al. 2016). Pedoja et al. (2011a, 2011b, 2014) indicated by several investigations that the eastern Patagonia uplift is constant throughout Quaternary times but the process of the mechanism is a topic of discussion, associated with glacio-eustatic, geotectonic and isostatic compensation processes (Clapperton 1993; Rostami et al. 2000; Schellmann and Radtke 2000, Schellmann and Radtke 2010; Pedoja et al. 2011a, 2011b, 2014; Folguera et al. 2016; Bini et al. 2017). This work provides another perspective on the dynamics of the region, since it has been studied by many authors considering the climate and sea level changes during the Quaternary with special interest in coastal marine deposits (Feruglio 1950; Cionchi 1988; Schellmann and Radtke 2000; Aguirre et al. 2005; Codignotto 2005; Fucks et al. 2009, 2011; Ribolini et al. 2011; Zanchetta et al. 2012; Boretto et al. 2014).

The identification of the knickpoints along the fluvial profile and the understanding of their characteristics have been widely analyzed during the last decades in the landscape of active margins, where the erosion dynamics is high, such studies show how these break points are linked to the fluctuation of the base level, climate change and the tectonic control (Burbank and Anderson 2001; Zaprowski et al. 2001; Hayakawa and Matsukura 2003; Hayakawa and Oguchi 2006, 2009; Hayakawa et al. 2008; Quezada et al. 2010; Pavano et al. 2016; Castillo et al. 2017; Limber and Barnard 2017; Perucca et al. 2018, among others). However, the knickpoint analyses along the passive margin is being carried out recently and interesting results have been obtained as expected for active environments (Catalano and De Guidi 2003; Frankel et al. 2007; Yunus et al. 2016; De Sordi et al. 2018). The lithological-structural knickpoints have been described in both active and passive margins, these relevant forms tend to be more abundant in steep sections upstream of the longitudinal profile (Hayakawa and Oguchi 2006; Yunus et al. 2016). A strong hydraulic system linked to the climate generates these breaks due to intense runoff (Wohl 2000; Chin and Wohl 2005; Yunus et al. 2016). In the study area semi-arid climate predominates; it presents scant shrubby cover vegetation and high-intensity and short-duration runoffs are the befalling processes (Fig. 5), such as those described in the Arabian basins by Yunus et al. (2016). Therefore, in this type of climate, the knickpoints respond to the fluvial erosive processes, which in turn vary according to the resistance, disposition and degree of fracture presented by the rocks (Castillo and

Lugo-Hubs 2011). The main mechanism that triggers the formation of this kind of knickpoint is the fluvial incision into a lithological contact zone, that is, associated with differential erosion (Castillo and Lugo-Hubs 2011). Likewise, the erodability of the materials has a direct relationship with the slope bed adjustment; thus, abrupt changes of the gradient close to the lithological contacts can be seen along the fluvial profile (Hack 1973; Goldrick and Bishop 1995, 2007; Larue 2008; Castillo and Lugo-Hubs 2011).

In this context, the development of knickpoints in the study area and their differentiation are highlighted considering two sectors: the northern and southern areas. The longitudinal profiles analyzed go through different stratigraphic formations, generating noted knickpoints in all the basins which are associated with the topographic configuration as well. The basins named as 1–10 (northern area) record more density and intermediate values of knickpoints in the geomorphological High Zone, thus a link with the headwaters, being the dominant lithology the sedimentary rocks of the Montemayor Formation (the "Patagonian Gravels"), the Chenque Formation (marine sedimentary rocks) and the Sarmiento Group (continental sedimentary rocks). Towards the Middle and Marginal Zones the density and degree of knickpoints decrease notably, the channels cross lithological contacts between the occurring sedimentary rocks (the Río Chico and Salamanca formations) and the ensuing igneous-volcanic-sedimentary basement that emerges there (the Marifil Formation), crowning along the Quaternary coastal beach ridges (the Caleta Malaspina and Zanjón El Pinter formations). These break points have been classified as lithological-structural since the abrupt changes occur along the longitudinal profile associated with the contact between formations and the structural features topography. Nevertheless, the knickpoints recorded in the southern sector are classified as of tectonic origin; this type has been described by Castillo and Lugo-Hubs (2011) through two mechanisms: (1) faulting and (2) crust uplifting. The faulting, of the normal or of the thrust type, impacts on the fluvial profile producing a sudden change in the base level which causes an unevenness expression (Burbank and Anderson 2001; Castillo and Lugo-Hubs 2011), being this a direct process that generates imbalance in stream beds (Burbank and Anderson 2001), or in surrounding areas such as those studied by Molin et al. (2004) and Whittaker et al. (2007) in Italy, Larue (2008) in the French central massif that drains into the Mediterranean Sea, Quezada et al. (2010) in the Chilean coastal region, just to name a few examples. The crust uplifting is linked to the rate of increase of the tectonic uplift and its relationship with the spreading of the disproportion and imbalance, being this the best documented case for the formation of knickpoints (Castillo and Lugo-Hubs 2011), the analysis of the landscape indicates that in the fluvial system there is a communication between the erosion processes and the fluvial base level changes the tectonics (Whipple and Tucker 1999; Loget and Van den Driessche 2009; Castillo and Lugo-Hubs 2011). In this work, the tectonic knickpoints indicate that locally the zone presents distinctive dynamics which could be associated with strain faults (Fig. 2) (Sciutto et al. 2000) and the Patagonia regional uplift (Pedoja et al. 2011a, 2011b, 2014; Folguera et al. 2016) that rejuvenated the landscape causing an imbalance of these basins. The longitudinal profiles from basins 11 to 18 cross sedimentary rocks and the development of a cliff dominates the Marginal zone. It is remarkable the different behavior of these breaks points in the southern sector with respect to the northern side.

The identification of knickpoints using MERIT DEM from the automated approach allows a study of major knickpoints zones in the San Jorge Gulf Basin and its usefulness to understand the continental-coastal dynamics in the region. This work provides new techniques for geomorphological mapping in areas of difficult access and with limited databases. However, the recognition of knickpoint formation in smaller areas and the analysis of tributary basins requires a higher resolution DEM. In this sense, Foster and Kelsey (2012) and Yunus et al. (2016) discussed the effectiveness of a high-quality DEM in the knickzones research by comparing a 1 m DEM LIDAR and a 10 m DEM; such more detailed studies for the Patagonian Atlantic region are expected to be carried forward in future work.

## 6 Final Remarks

1. This work demonstrates the usefulness of the knickpoints, longitudinal profiles and hypsometry analysis generated from a DEM to detect areas of imbalance and contribute to the study area characterization to understand the lithological, structural-topographic, climatic and tectonic influences of the landscape evolution from a coastal passive margin.
2. The knickpoints are an expression of the disequilibrium on a fluvial system and they are present in all the basins, differing their concentration according to lithological-structural characteristics in the whole area and depending on the tectonic conditioning.
3. The lithology with the relief control seems to have oriented the development of the knickpoints at least in three ways: (i) presence of sedimentary rocks in the head-waters, (ii) occurrence of lithological contact (both hard and soft rocks) in the Middle and Marginal zones, and (iii) uneven erosion within the sedimentary rocks.
4. The study area is linked to the crust uplifting of the passive margin that implies the entire South American Atlantic margin. However, the southern basins (11–18) differ from those of the northern sector because this process together with the presence of extension faults involving the basement (which bounds the San Jorge Basin) would produce the imbalance in the river basins originating the tectonic knickpoints; therefore these circumstances would generate the landscape rejuvenation recorded by the hypsometric curves. In the northern basins the knickpoints are considered as of litohological-structural origin.
5. The analyses of the knickpoints have indicated that frequencies and densities differ according to the tectonic control more than the lithology-structural arrangement does along the basins.
6. This work is intended to focus in the near future the study of knickpoints in other coastal fluvial basins of the Patagonian passive margin to determine knickzones, their relationship with landscape morphology, environmental changes and tectonic control at a larger scale, both temporal and spatial.

**Acknowledgments.** The authors wish to thank Prof. Dr. D. Yamazaki for MERIT DEM images and to Prof. Dr. J.V. Pérez-Peña for the CalHipso tool software. Their contributions have been fundamental to carry out this work. Also, we thank Dr. Enrique Fucks for the photographies and Dr. Ernesto Brunetto for his comments and suggestions which undoubtedly improved this manuscript.

## References


Aguirre, M.L., Negro Sirch, Y., Richiano, S.: Late Quaternary molluscan assemblages from Bahía Bustamante coastal area (Patagonia, Argentina). J. S. Am. Earth Sci. **20**, 13–32 (2005)

Andreis, R., Mazzoni, M., Spalletti, L.: Estudio paleoestratigráfico y ambiental de las sedimentitas terciarias entre Pico Salamanca y Bahía Bustamante, Provincia de Chubut, Argentina. Rev. Asoc. Geol. Argentina **31**, 85–103 (1975)

Ardolino, A., Busteros, A., Fauqué, L., Franchi, M., Lema, H.: Depósitos cuaternarios de las bahías Bustamante y Camarones. In: Ardolino, A., Lema, H. (eds.) Sitios de Interés Geológico de la República Argentina. Comisión de Sitios de Interés Geológico de la República Argentina, Servicio Geológico Minero Argentino. Anales, vol. 46, pp. 705–718 (2008)

Baker, V.R., Ritter, D.F.: Competence of rivers to transport coarse bedload material. Geol. Soc. Am. Bull. **86**, 975–978 (1975)

Berlin, M.M., Anderson, R.S.: Modeling of knickpoint retreat on the Roan Plateau, western Colorado. J. Geophys. Res. **112**, F03S06 (2007). https://doi.org/10.1029/2006JF000553

Bini, M., Isola, I., Zanchetta, G., Pappalardo, M., Ribolini, A., Ragaini, L., Baroni, C., Boretto, G., Fucks, E., Morigi, C., Salvatore, M.C.: Mid-Holocene relative sea-level changes along Atlantic Patagonia: new data from Camarones, Chubut, Argentina. Holocene 0959683617714596 (2017)

Bishop, P., Hoey, T.B., Jansen, J.D., Artza, I.L.: Knickpoint recession rate and catchment area: the case of uplifted rivers in Eastern Scotland. Earth Surf. Process. Land. **30**, 767–778 (2005)

Boretto, G.M., Baranzelli, M.C., Gordillo, S., Consoloni, I., Zanchetta, G., Morán, G.: Shell morphometric variations in a Patagonian Argentina clam (*Ameghinomya antiqua*) from the Mid-Pleistocene (MIS 7) to the present. Quatern. Int. **352**, 48–58 (2014)

Brush, L., Wolman, G.: Knickpoint behaviour in non-cohesive material: a laboratory study. Geol. Soc. Am. Bull. **71**, 59–74 (1960)

Bull, W.B.: Geomorphic Responses to Climatic Change, 326 p. Oxford University Press, New York (1991)

Burbank, D., Anderson, R.: Tectonic Geomorphology, 274 p. Blackwell Science, Malden (2001)

Castillo, M., Lugo-Hubp, J.: Estado actual del conocimiento, clasificación y propuesta de inclusión del término knickpoint en el léxico geológico-geomorfológico del español. Bol. Soc. Geol. Mex. **63**, 353–364 (2011)

Castillo, M., Ferrari, L., Muñoz-Salinas, E.: Knickpoint retreat and landscape evolution of the Amatlán de Cañas half-graben (northern sector of Jalisco Block, western Mexico). J. S. Am. Earth Sci. **77**, 108–122 (2017)

Catalano, S., De Guidi, G.: Late Quaternary uplift of northeastern Sicily: relation with the active normal faulting deformation. J. Geodyn. **36**, 445–467 (2003)

Caviglia, S.E.: Discusión del denominado "Piso Patagónico" sobre la base de la edad de presencia de cetáceos. In: VII Congreso Geológico Argentino (Buenos Aires). Actas, II, pp. 385–392 (1979)

Chen, Y.C., Sung, Q., Cheng, K.Y.: Along-strike variations of morphotectonics features in the Western Foothills of Taiwan: tectonic implications based on stream-gradient and hypsometric analysis. Geomorphology **56**(1–2), 109–137 (2003)

Chin, A., Wohl. E.: Toward a theory for step pools in stream channels. Prog. Phys. Geogr. **29**, 275–296 (2005)

Ciccacci, S., D'Alessandro, L., Fredi, P., Palmieri, E.L.: Relations between morphometric characteristics and denudational processes in some drainage basins of Italy. Z. Geomorphol. **36**, 53–67 (1992)

Cionchi, J.L.: Geomorfología y estratigrafía del Cuaternario de Bahía Bustamante y zonas adyacentes, provincia de Chubut. Unpublished Doctoral thesis, Facultad de Ciencias Naturales y Museo. Universidad Nacional de La Plata, La Plata, 179 p. (1985)

Cionchi, J.L.: Geomorfología de Bahía Bustamante y zonas adyacentes, Chubut. Rev. Asoc. Geol. Argentina **43**, 51–62 (1988)

Clapperton, C.M.: Quaternary Geology and Geomorphology of South America, 768 p. Elsevier, Amsterdam (1993)

Codignotto, J.O.: Evolución del Cuaternario alto del sector de costa y plataforma submarina entre Río Coig, Santa Cruz y Punta María, Tierra del Fuego. Rev. Asoc. Geol. Argentina **45**, 9–16 (1990)

Codignotto, J.O.: Vulnerabilidad al ascenso del nivel del mar y cambio climático en la Costa Argentina. Rev. Asoc. Argentina Geol. Apl. Ing. **21**, 25–28 (2005)

Codignotto, J.O., Marcomini, S., Santillana, S.: Edad entre las terrazas marinas entre Puerto Deseado y Bahía Bustamante. Rev. Asoc. Geol. Argentina **43**, 43–50 (1988)

Codignotto, J.O., Césari, O., Beros, C.A.: Morfocronología secuencial evolutiva holocena en Bahía Solano, Chubut. Rev. Asoc. Geol. Argentina **45**, 205–212 (1990)

Codignotto, J.O., Kokot, R.R., Marcomini, S.C.: Neotectonism and Sea-Level Changes in the Coastal Zone of Argentina. J. Coastal Res. **8**, 125–133 (1992)

Coronato, A., Coronato, F., Mazzoni, E., Vázquez, M.: The physical geography of Patagonia and Tierra del Fuego. In: Rabassa, J. (ed.) The Late Cenozoic of Patagonia and Tierra del Fuego. pp. 13–55. Elsevier (2008)

Cortés, J.M.: Estratigrafía cenozoica y estructura al Oeste de la Península de Valdes, Chubut. Consideraciones tectónicas y paleogeográficas. Rev. Asoc. Geol. Argentina **36**, 424–445 (1981)

Crosby, B.T., Whipple, K.X.: Knickpoint initiation and distribution within fluvial networks: 236 waterfalls in the Waipaoa River, North Island, New Zealand. Geomorphology **82**, 16–38 (2006)

De Sordi, M.V., Salgado, A.A.R., Siame, L., Bourlès, D., Paisani, J.C., Léanni, L., Braucher, R., do Couto, E.V., ASTER Team: Implications of drainage rearrangement for passive margin escarpment evolution in southern Brazil. Geomorphology **306**, 155–169 (2018)

Dietrich, W.E., Bellugi, D., Sklar, L.S., Stock, J., Heimsath, A., Roering, J.J.: Geomorphic transport laws for predicting landscape form and dynamics. In: Wilcock, R.P., Iverson, R.M. (eds.) Prediction in geomorphology. American Geophysical Union Geophysical Monograph, vol. 135, pp. 103–132 (2003)

Etchebehere, M.L.C., Saad, A.R., Fulfaro, V.J., Perinotto, J.A.J.: Aplicação do Índice "Relação Declividade-Extensão – RDE" na Bacia do Rio do Peixe (SP) para Detecção de Deformações Neotectônicas. Revista do Instituto de Geociências – Universidades de São Paulo, Série Científica, São Paulo, vol. 4, pp. 43–56 (2004)

Etchebehere, M.L.C., Saad, A.R., Santoni, G., Casado, F.C., Fulfaro, V.J.: Detecção de prováveis deformações neotectônicas no vale do Rio do Peixe, região ocidental paulista, mediante aplicação de índices RDE (Relação DeclividadeExtensão) em segmentos de drenagem. Rev. Geociênc. **25**, 271–289 (2006)

Feruglio, E.: Descripción Geológica de La Patagonia, vol. 3. Buenos Aires, Dirección General de Y.P.F. 431 p. (1950)

Fidalgo, F., Riggi, J.C.: Consideraciones geomórficas y sedimentológicas sobre los Rodados Patagónicos. Rev. Asoc. Geol. Argentina **4**, 430–443 (1970)

Fitzgerald, M.G.R., Mitchum, M., Uliana, M.A., Biddle, K.T.: Evolution of the San Jorge Basin. Argentina. Am. Assoc. Petroleum Geol. Bull. **74**(6), 879–920 (1990)

Flint, J.J.: Stream gradient as a function of order, magnitude, and discharge. Water Resour. Res. **10**, 969–973 (1974)

Folguera, A., Gianni, G., Sagripanti, L., Vera, E.R., Colavitto, B., Orts, D., Ramos, V.A.: Active deformation, uplift and subsidence in southern South America throughout the quaternary: a general review about their development and mechanisms. In: Gasparini, G., et al. (eds.) Marine Isotope Stage 3 in Southern South America, 60 KA BP-30 KA BP. Earth System Science Series, pp. 107–127. Springer, Cham (2016)

Franchi, M.R., Haller, M.J.F., Lapido, O.R., Page, R.F.N., Pesce, A.H.: Geología de la región nororiental de la provincia de Chubut, Argentina. Segundo Congreso IberoAmericano de Geología Económica (Buenos Aires). Actas, vol. 4, pp. 125–139 (1975)

Frankel, K.L., Pazzaglia, F.J., Vaughn, J.D.: Knickpoint evolution in a vertically bedded substrate, upstream-dipping terraces, and Atlantic slope bedrock channels. Geol. Soc. Am. Bull. **119**, 476–486 (2007)

Foster, M.A., Kelsey, H.M.: Knickpoint and knickzone formation and propagation, South Fork Eel River, northern California. Geosphere **8**, 403–416 (2012)

Fucks, E., Aguirre, M., Richiano, S., Boretto, G., Zanchetta, G., Consoloni, I., Isola, I., Pappalardo, M., Ribolini, A.: Depósitos litorales en el sector continental del norte del Golfo San Jorge. In: IV Congreso Argentino de Cuaternario y Geomorfología, XII Congresso da Associação Brasileira de Estudos do Quaternário and II Reunión sobre el Cuaternario de América del Sur, Actas, 114 p. (2009)

Fucks, E., Charó, M., Aguirre, M., Farinati, E.: Aspectos estratigráficos y geomorfológicos del litoral patagónico bonaerense. In: XVIII Congreso Geológico Argentino, Buenos Aires, pp. 275–276 (2011)

Gardner, T.W.: Experimental study of knickpoint and longitudinal profile evolution in cohesive, homogeneous material. Geol. Soc. Am. Bull. **94**, 664–672 (1983)

Geach, M.R., Stokes, M., Hart, A.: The application of geomorphic indices in terrain analysis for ground engineering practice. Eng. Geol. **217**, 122–140 (2017)

Giacosa, R.E., Paredes, J.M., Nillni, A., Ledesma, M., Colombo, F.: Fallas normales de alto ángulo en el Neógeno del margen Atlántico de la Cuenca del Golfo San Jorge (46 S–67 30 O, Patagonia Argentina). Bol. Geol. Min. **115**(3), 537–550 (2004)

Gilbert, G.: Report on the geology of the Henry Mountains. U.S. Government Printing Office Monograph. Washington, D.C. 160 p. (1877). http://pubs.er.usgs.gov/publication/70039916

Goldrick, J., Bishop, P.: Differentiating the roles of lithology and uplift in the steepening of bedrock river long profiles: an example from southeastern Australia. J. Geol. **103**, 227–231 (1995)

Goldrick, J., Bishop, P.: Regional analysis of bedrock stream long profiles: evaluation of Hack's SL form, and formulation and assessment of an alternative (the DS form). Earth Surf. Proc. Land. **32**, 649–671 (2007)

Hack, J.T.: Studies of longitudinal stream profiles in Virginia and Maryland. U.S. Geological Survey Professional Paper, vol. 294, pp. 45–97 (1957)

Hack, J.T.: Stream-profile analysis and stream-gradient index. U.S. Geol. Surv. J. Res. **1**, 421–429 (1973)

Haller, M.J.: Estratigrafía de la región al poniente de Puerto Madryn, provincia del Chubut, República Argentina. In: 7° Congreso Geológico Argentino (Buenos Aires). Actas, vol. 1, pp. 285–297 (1979)

Hayakawa, Y., Matsukura, Y.: Recession rates of waterfalls in Boso Peninsula, Japan, and a predictive equation. Earth Surf. Process. Land. **28**, 675–684 (2003)

Hayakawa, Y.S., Oguchi, T.: DEM-based identification of fluvial knickzones and its application to Japanese mountain rivers. Geomorphology **78**, 90–106 (2006)

Hayakawa, Y.S., Oguchi, T.: GIS analysis of fluvial knickzone distribution in Japanese mountain watersheds. Geomorphology **111**, 27–37 (2009)

Hayakawa, Y.S., Yokoyama, S., Matsukura, Y.: Erosion rates of waterfalls in post-volcanic fluvial systems around Aso volcano, Southwestern Japan. Earth Surf. Process. Land. **33**, 801–812 (2008)

Howard, A.D.: A detachment-limited model of drainage basin evolution. Water Resour. Res. **30**, 2261–2285 (1994)

Howard, A.D., Kerby, G.: Channel changes in badlands. Geol. Soc. Am. Bull. **94**, 739–752 (1983)

Isola, I., Bini, M., Ribolini, A., Pappalardo, M., Consoloni, I., Fucks, E., Boretto, G., Ragaini, L., Zanchetta, G.: Geomorphologic map of northeastern sector of San Jorge Gulf (Chubut, Argentina). J. Maps **7**, 476–485 (2011)

Iwahashi, J., Yamazaki, D., Matsuoka, M., Thamarux, P., Herrick, J., Yong, A., Mital, U.: Global terrain classification using Multiple-Error-Removed Improved-Terrain (MERIT) to address susceptibility of landslides and other geohazards. AGU Fall Meeting Abstracts (2017)

Kirby, E., Whipple, K.X.: Quantifying differential rock-uplift rates via stream profile analysis. Geology **29**, 415–418 (2001)

Kirby, E., Whipple, K.X.: Expression of active tectonics in erosional landscapes. J. Struct. Geol. **44**, 54–75 (2012)

Kirby, E., Whipple, K.X., Tang, W., Chen, Z.: Distribution of active rock uplift along the eastern margin of the Tibetan Plateau: inferences from bedrock river profiles. J. Geophys. Res. **108**, 2217 (2003)

Larue, J.P.: Effects of tectonics and lithology on long profiles of 16 rivers of the southern Central Massif border between the Aude and the Orb (France). Geomorphology **93**, 343–367 (2008)

Lema, H., Busteros, A., Franchi, M.: Hoja Geológica 4566-II y IV, Camarones (1:250.000), Instituto de Geología y Recursos minerales. Bol. Servicio Geol. Min. Argentino **261**, 1–44 (2001)

Leopold, L.B., Bull, W.B.: Base level aggradation and grade. In: American Philosophical Society Proceedings, vol. 123, pp. 168–202 (1979)

Lesta, P., Ferello, R., Chebli, G.: Chubut extraandino. In: XI Simposio de Geología Regional Argentina (Córdoba), pp. 1307–1387 (1980)

Limber, P.W., Barnard, P.L.: Coastal knickpoints and the competition between fluvial and wave-driven erosion on rocky coastlines. Geomorphology **306**, 1–12 (2017)

Loget, N., Van Den Driessche, J.: Wave train model for knickpoint migration. Geomorphology **106**, 376–382 (2009)

Mackin, J.H.: Concept of the graded river. Geol. Soc. Am. Bull. **59**, 463–512 (1948)

Malvicini, L., Llambías, E.J.: Geología y génesis de los yacimientos de manganeso Arroyo Verde, provincia de Chubut, República Argentina. V Congreso Geológico Argentino (Buenos Aires). Actas, vol. 2, pp. 185–202 (1974)

Molin, P., Pazzaglia, F.J., Dramis, F.: Geomorphic expression of active tectonics in a rapidly-deforming forearc, Sila massif, Calabria, Southern Italy. Am. J. Sci. **304**, 559–589 (2004)

Panza, J.L.A., Sacomani, L.E., Rodríguez, M.F., Aragón, E., Parisi, C.: Campamento Villegas. Servicio Geológico Minero Argentino. Instituto de Geología y Recursos Minerales (2002)

Pavano, F., Pazzaglia, F.J., Catalano, S.: Knickpoints as geomorphic markers of active tectonics: a case study from northeastern Sicily (southern Italy). Lithosphere **8**, 633–648 (2016)

Pazzaglia, F.J.: Fluvial terraces. In: Wohl, E. (ed.) Fluvial Geomorphology. Treatise on Geomorphology, vol. 9, pp. 379–412: Academic Press, San Diego (2013)

Pedoja, K., Regard, V., Husson, L., Martinod, J., Guillaume, B., Fucks, E., Iglesias, M., Weill, P.: Uplift of quaternary shorelines in eastern Patagonia: darwin revisited. Geomorphology **127**, 1–142 (2011a)

Pedoja, K., Husson, L., Regard, V., Cobbold, P.R., Ostanciaux, E., Johnson, M.E., Delcaillau, B.: Relative sea-level fall since the last interglacial stage: are coasts uplifting worldwide? Earth Sci. Rev. **108**, 1–15 (2011b)

Pedoja, K., Husson, L., Johnson, M.E., Melnick, D., Witt, C., Pochat, S., Nexer, M., Delcaillau, B., Pineginai, T., Poprawski, Y., Authemayou, C., Elliot, M., Regard, V., Garestier, F.: Coastal staircase sequences reflecting sea-level oscillations and tectonic uplift during the Quaternary and Neogene. Earth Sci. Rev. **132**, 13–38 (2014)

Pérez-Peña, J.V., Azañón, J.M., Azor, A.: CalHypso: An ArcGIS extension to calculate hypsometric curves and their statistical moments. Applications to drainage basin analysis in SE Spain. Comput. Geosci. **35**, 1214–1223 (2009)

Perron, J.T., Richardson, P.W., Ferrier, K.L., Lapôtre, M.: The root of branching river networks. Nature **492**, 100–103 (2012)

Perucca, L.P., Espejo, K., Angillieri, M.Y.E., Rothis, M., Tejada, F., Vargas, M.: Neotectonic controls and stream piracy on the evolution of a river catchment: a case study in the Agua de la Peña River basin, Western Pampean Ranges. Argentina. J. Iberian Geol. **44**, 1–18 (2018)

Queiroz, G.L., Salamuni, E., Nascimento, E.R.: Knickpoint finder: a software tool that improves neotectonic analysis. Comput. Geosci. **76**, 80–87 (2015)

Quezada, J., Cerda, J.L., Jensen, A.: Efectos de la tectónica y el clima en la configuración morfológica del relieve costero del norte de Chile. Andean Geol. **37**, 78–109 (2010)

Ramos, V.A.: Evolución de la cuenca golfo San Jorge: su estructuración y régimen tectónico. Rev. Asoc. Geol. Argentina **72**(1), 12–20 (2015)

Ribolini, A., Aguirre, M., Baneschi, I., Consoloni, I., Fucks, E., Isola, I., Mazzarini, F., Pappalardo, M., Zanchetta, G., Bini, M.: Holocene beach ridges and coastal evolution in the Cabo Raso Bay (Atlantic Patagonian Coast, Argentina). J. Coastal Res. **27**, 973–983 (2011)

Riccardi, A., Rolleri, E.: Cordillera Patagónica Austral. In: XI Simposio Geológico Argentino (Córdoba). Actas, vol. 2, pp. 1173–1196 (1980)

Riggi, J.C.: Nomenclatura, categoría litoestratigráfica y correlación de la Formación Patagonia en la costa atlántica. Rev. Asoc. Geol. Argentina **34**, 243–248 (1979)

Rostami, K., Peltier, W.R., Mangini, A.: Quaternary marine terraces, sea-level changes and uplift history of Patagonia, Argentina: comparisons with predictions of the ICE-4G (VM2) model of the global process of glacial isostatic adjustment. Quatern. Sci. Rev. **19**, 1495–1525 (2000)

Rutter, N., Schnack, E., Del Río, L., Fasano, J., Isla, F., Radtke, U.: Correlation and dating of Quaternary littoral zones along the Patagonian coast, Argentina. Quatern. Sci. Rev. **8**, 213–234 (1989)

Rutter, N., Radtke, U., Schnack, J.: Comparison of ESR and Amino Acid Data in correlating and dating Quaternary shorelines along the Patagonian Coast, Argentina. J. Coastal Res. **6**, 391–411 (1990)

Schellmann, G., Radtke, U.: ESR dating of stratigraphically well-constrained marine terraces along the Patagonian Atlantic coast (Argentina). Quatern. Int. **68**, 261–273 (2000)

Schellmann, G., Radtke, U.: Coastal terraces and Holocene sea-level changes along the Patagonian Atlantic coast. J. Coastal Res. **19**, 963–996 (2003)

Schellmann, G., Radtke, U.: Timing and magnitude of Holocene sea-level changes along the middle and south Patagonian Atlantic coast derived from beach ridge systems, littoral terraces and valley-mouth terraces. Earth Sci. Rev. **103**, 1–30 (2010)

Sciutto, J.C., Césari, O., Escribano, V., Pezzuchi, H.: Hoja Geológica 4566-III. Comodoro Rivadavia, Provincia de Chubut. Inst. Geol. Recursos Min. Bol. Servicio Geol. Min. Argentino **244**, 1–53 (2000)

Schumm, S.A.: Evolution of drainage systems and slopes in badlands at Perth Amboy, New Jersey. Bull. Geol. Soc. Am. **67**, 597–646 (1956)

Schwanghart, W., Kuhn, N.J.: TopoToolbox: a set of Matlab functions for topographic analysis. Environ. Model. Softw. **25**, 770–781 (2010)

Seeber, L., Gornitz, V.: River profiles along the Himalayan arc as indicators of active tectonics. Tectonophysics **92**, 335–367 (1983)

Seidl, M.A., Dietrich, W.E.: The problem of channel erosion into bedrock. CATENA **23**, 101–124 (1992)

Seidl, M.A., Dietrich, W.E., Kirchner, J.W.: Longitudinal profile development into bedrock: an analysis of Hawaiian channels. J. Geol. **102**, 457–474 (1994)

Sklar, L., Dietrich, W.E.: River longitudinal profiles and bed-rock incision models: stream power and the influence of sediment supply. In: Tinkler, K.J., Wohl, E.E. (eds.) Rivers Over Rock: Fluvial Processes in Bedrock Channels. American Geophysical Union Geophysical Monograph, vol. 107, pp. 237–260 (1998)

Snyder, N.P., Whipple, K.X., Tucker, G.E., Merritts, D.J.: Landscape response to tectonic forcing: digital elevation model analysis of stream profiles in the Mendocino triple junction region, northern California. Geol. Soc. Am. Bull. **112**, 1250–1263 (2000)

Snyder, N.P., Whipple, K.X., Tucker, G.E., Merritts, D.M.: Interactions between onshore bedrock-channel incision and nearshore wave-base erosion forced by eustacy and tectonics. Basin Res. **14**, 105–127 (2002)

Stock, J.D., Montgomery, D.R.: Geologic constraints on bedrock river incision using the stream power law. J. Geophys. Res. **104**, 4983–4993 (1999)

Strahler, A.N.: Hypsometric (area-altitude) analysis of erosional topology. Geol. Soc. Am. Bull. **63**, 1117–1142 (1952)

Sylwan, C.A.: Geology of the Golfo San Jorge Basin, Argentina. J. Iberian Geol. **27**, 123–157 (2001)

Whipple, K.X.: Bedrock rivers and the geomorphology of active orogens. Annu. Rev. Earth Planet. Sci. **32**, 151–185 (2004)

Whipple, K.X., Tucker, G.E.: Dynamics of the stream power river incision model: implications for height limits of mountain ranges, landscape response timescales and research needs. J. Geophys. Res. **104**, 17.661–17.674 (1999)

Whittaker, A.C., Cowiel, P.A., Attall, M., Tucker, G.E., Roberts, G.P.: Bedrock channel adjustment to tectonic forcing: implications for predicting river incision rates. Geology **35**, 103–106 (2007)

Willet, S.D., Hovius, N., Brandon, M.T., Fisher, D. (eds.) Tectonics, climate, and landscape evolution. Geological Society of America Special Paper, vol. 398, p. 449 (2006)

Willett, S.D., McCoy, S.W., Perron, J.T., Goren, L., Chen, C.Y.: Dynamic reorganization of river basins. Science **343**, 1248765 (2014)

Willgoose, G., Hancock, G.: Revisiting the hypsometric curve as an indicator of form and process in transport-limited catchment. Earth Surf. Proc. Land. **23**, 611–623 (1998)

Wobus, C.W., Crosby, B.T., Whipple, K.X.: Hanging valleys in fluvial systems: controls on occurrence and implications for landscape evolution. J. Geophys. Res. **111**, F02017 (2006). https://doi.org/10.1029/2005JF000406

Wohl, E.E.: Mountain Rivers. American Geophysical Union, Washington, D.C. (2000)

Yamazaki, D., Ikeshima, D., Tawatari, R., Yamaguchi, T., O'Loughlin, F., Neal, J.C., Sampson, C., Kanae, S., Bates, P.D.: A high accuracy map of global terrain elevations. Geophys. Res. Lett. **44**(11), 5844–5853 (2017)

Yunus, A.P., Oguchi, T., Hayakawa, Y.S.: Remote identification of fluvial knickzones and their imprints on landscape morphology in the passive margins of Western Arabia. J. Arid Environ. **130**, 14–29 (2016)

Zanchetta, G., Consoloni, I., Isola, I., Pappalardo, M., Ribolini, A., Aguirre, M., Fucks, E., Baneschi, I., Bini, M., Ragaini, L., Terrasi, F., Boretto, G.: New insights on the Holocene marine transgression in the Bahía Camarones (Chubut, Argentina). Ital. J. Geosci. **131**, 19–31 (2012)

Zaprowski, B., Evenson, E.B., Pazzaglia, F.J., Epstein, J.B.: Knickzone propagation in the Black Hills and northern High Plains: a different perspective on the late Cenozoic exhumation of the Laramide Rocky Mountains. Geology **29**, 547–550 (2001)

# Granitic Caves from the Achala Batholith in the Province of Córdoba, Argentina: A Study on Three Particular Stories with a Common Past

Eduardo García Aráoz[1(✉)], Nicolás Madelón[2], Micaela Pleitavino[3], Gabriella Margherita Boretto[4], and Marcela Cioccale[1]

[1] CONICET, Department of Applied Geology, Facultad de Ciencias Exactas, Físicas y Naturales, Universidad Nacional de Córdoba, Av. Vélez Sarsfield 1611, X5016 Córdoba, Córdoba, Argentina
garciaaraoze@hotmail.com

[2] Benito Soria 345, X5000 Córdoba, Córdoba, Argentina

[3] Bv. San Juan 798, X5000 Córdoba, Córdoba, Argentina

[4] CICTERRA-CONICET, Av. Vélez Sarsfield 1611, X5016 Córdoba, Córdoba, Argentina

**Abstract.** The igneous or magmatic rocks, especially those with granitic composition, constitute a special type of geomorphological substrate. As from its exposure on the Earth´s surface and the modelling that eventually takes place on them, different landscapes with specific characteristics emerge where inherited and present shapes coexist making possible to reconstruct the landscape evolution of the massif. Traditionally, in the studies of granitic landscapes, only the forms observed on the surface are described, whereas those that develop underground are not often taken into consideration. In this paper three caves are analysed in order to specify their particular origins and classify it according to its morphological characteristics and the genetic processes involved. These caves are located on the same scarp of the rock massif and, therefore, share a geological and geomorphological common history, which is shown in their hypothesis of evolution. In addition, based on the type of cavity and certain observations of their environments it is possible to establish a relative chronological order in their ages of formation. In this context, it is proven that these underground environments (as the other geographical accidents and granitic landforms) are not chaotic but - on the contrary- they are structured in geological and geomorphological processes over time and they respond to a common evolutionary framework to the entire rock massif. More detailed studies on these environments can develop in new methodological tools to understand the geological history of the granitic landscape they are located in.

**Keywords:** Geomorphology · Granite landscapes and landforms · Sierras de Córdoba · Argentina · Caves · Caverns · Achala batholith


M. M. Collantes et al. (Eds.): *Advances in Geomorphology and Quaternary Studies in Argentina*, SPRINGEREARTH, pp. 209–232, 2020.
https://doi.org/10.1007/978-3-030-22621-3_11

# 1 Introduction

The igneous or magmatic rocks, especially those with granitic composition, constitute a special type of geomorphological substrate. As from its exposure at the Earth´s surface and the modelling that eventually takes place on them, different landscapes with specific characteristics emerge where inherited and present shapes and landforms coexist making possible to reconstruct the landscape evolution of the massif (Campbell 1997; Campbell and Twidale 1995; Twidale and Vidal Romaní 2005; Migón 2006; Twidale 2012). The study of the granitic forms was approached from a systematic perspective where they can be differentiated as from their size in major and minor landforms (Twidale 1982; 1989; Twidale and Vidal Romaní 2005; Migón 2006; Twidale 2012). The major landforms (Twidale 1989; Vidal Romaní 1989) can be integrated within the macro-events that determined the exogenous evolution of the rock massif, which can be correlated with geodynamic events within a regional scope and, therefore, of simple detection in the geomorphological history of the area where they are located (Vidal Romaní and Yepes 2004). The minor forms are, on the other hand, details of the major forms and, thus, their genesis depends on the primary characteristics of the rock and its display of the geomorphological history of the area, where its integration to the evolutionary scheme is more complex (Vidal Romaní and Yepes 2004).

Traditionally, in the studies of granitic landscapes, only the landforms observed at the surface are described, whereas those that develop underground are not often taken into consideration. This could be due mostly to the difficult access that these underground environments present. Nevertheless, there are remarkable references about caves and caverns in granitic massifs in different parts of the world, among them Sjöberg (1969), Finlayson (1981); Ollier and Smith (1994); Willems et al. (2002); Lazarte (2002); Vaqueiro (2005); Vaqueiro et al. (2006); Gaal and Bella (2008); Lazaridis and Pipera (2008); Roqué et al. (2011); Song et al. (2011; 2015); Scheuerer et al. (2013); and García Aráoz et al. (2018).

In these contributions it is shown that granitic cavities are absolutely dependent on the structure of the field rocks. Bearing this criterion in mind, most authors agree that there are three basic types of cavities or granitic caverns (Vidal Romaní and Twidale 1998; Twidale and Vidal Romaní 2005; Vidal Romaní and Vaqueiro 2007; Song et al. 2011; 2015): (i) fissural cavities, (ii) block cavities and (iii) tafoni. The fissural cavities are those associated with fracture planes where water runs, washing down the weathered and/or altered material and thus increasing the underground space. A particular case is that of those fissural cavities filled with blocks, locally known as underground rivers for their strong association with water courses. These develop over a fracture of the massif, where the river carves its bed and it is eventually buried due to mass removal from the roof of the cavity (Twidale and Bourne 2008). The block caves are cavities related to residual blocks that are formed as the weathered material is removed and passages developed in the holes and empty spaces. In this case, the slide of the altered rock leaves behind tubular voids delimited by residual nuclei initially included in the mass of altered rock. Such erosion is carried out by subsurface water currents that resurface downstream. The tafoni are cavities with rounded morphology and openings mostly tubular, which develop in favor of a complex system of discontinuities. Tafoni are often found within connected networks that have carved into cliffsides, hills, and rocky areas. Several theories have been presented to explain tafoni formation, including freezing-thawing, wetting-drying, biological activity, load concentration, wind erosion, and salt weathering. However, most authors do not consider that there is only one single agent that produces the development of tafoni (Mayor Rodríguez 2011).

In this work, three caves are analyzed. They are located in the Los Gigantes granitic massif of the Achala batholith, in the province of Córdoba, Argentina. The main target of this study is to specify the particular origin of each cavity and classify it according to its morphological characteristics and the genetic processes involved. The following aspects have been taken into account in the analysis of each cave: (1) the landforms that integrate them, (ii) the geomorphological context in which they are found, and (iii) the prevailing processes that originated them.

The singularity of this input is that the three caves are located on the same scarp of the rock massif and, therefore, they share a geological and geomorphological common history.

According to the definition established by Bates and Jackson (1987) the word “cavern” is used as a reference to an immense cave –or system of interconnected caves-, considering “cave” a natural space opened underground, generally with an opening towards the surface.

## 2 Geological Settings

The research zone is located on the eastern slope of the Los Gigantes massif, belonging to the Achala batholith, on the central section of the Sierras Grandes de Córdoba in the province of Cordoba, Argentina (Fig. 1).

The sites of interest are:

(i) The grotta called La Virgen (LV): located on the Achala plains, western of Mogote Grande. Geographical coordinates are 31°25.914’ S/64°48.794’ W (WGS84).
(ii) The cavern of Valle de Los Lisos (LL): located in the valley of such name, south of Cerro La Cruz. Geographical coordinates are 31°26.170’ S/64°47.488’ W (WGS84).
(iii) The underground stream named La Bestia (LB): located 200 m downstream of the confluence of the Batán and El Cajón rivers, at the base of Los Gigantes scarp. Geographical coordinates are 31°26.468’ S/64°44.783’ W (WGS84).

As it can be seen in Fig. 1C and D, these caves are located on the same scarp of the rock massif, but they are so in different landscape positions.

The Sierras Grandes de Córdoba are formed by a series of basement blocks, limited by reverse faults that form the mountain range with N–S orientation (Carignano et al. 1999). These blocks show a mild slope to the east and they are cut short by four erosional flat surfaces (Fig. 1B), limited by compound scarps (erosive and structural) (Carignano et al. 2014).

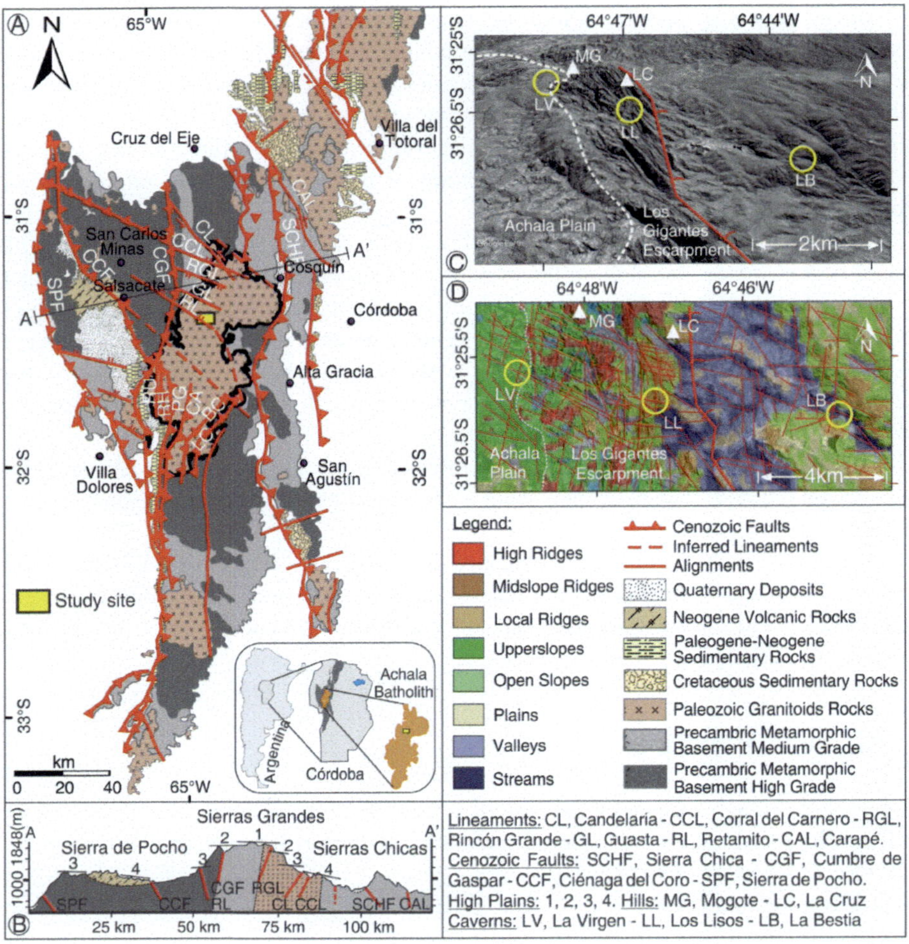


**Fig. 1.** (A) Geological map of the Sierras de Córdoba and location of study site (modified from Martino et al. 2012a). (B) Structural profile of the Sierras de Córdoba, showing the main high plains (modified from Carignano et al. 1999). (C) Caverns location in relation to the Los Gigantes scarpment (Satellite Image provided by Google Earth). (D) Geomorphological and structural map of the study site. The geomorphological classification was made from the MED-Ar image of 30 m resolution (provided by the IGN) with the TPI based landform classification module of the SAGA GIS program.

Due to its terraced disposition – mainly over 1,000 m a.s.l. - these surfaces are locally named “high plains”. Their morphology is mildly wavy, which has enabled the post-Pliocene sediment accumulation that has only been characterized in specific sites (Montes 1956; Manzur 1995; Tauber and Goya 2006; Carignano et al. 2014; Krapovickas and Tauber 2016).

According to Carignano et al. (1999; 2014), the general geomorphological evolution of these hills can be summarized in three main events:

1. A prolonged stable period during the Early and Middle Mesozoic generating the etchplain under very warm and wet climate.
2. Erosion and development of pediments and scarps associated with a change of base level after distensive processes (rifting) together with a progressively warmer climate, during the Cretaceous.
3. General upraise of the mountain range, originated by the Andean tectonics (during the Cenozoic) and then, the establishment of the present climatic conditions.

Within this general context, the Achala batholith (Fig. 1A) can be seen as a granitic massif of a definite elliptic geometry, covering an approximate area of 2,500 $km^2$, situated with a 20° NE orientation in the mid-high degree metamorphic rocks of the crystaline basement of the Sierras Grandes of Córdoba (Gordillo and Lencinas 1979). The age of this crystallization is Late Devonian (379 $\pm$ 4 Ma, U-Pb zircon, SHRIMP, Rapela et al. 2008).

Martino et al. (2017) described two main geomorphological dominances with distinctive structural features within the Achala batholith: the Northern Achala Dominance (DAN), which is caracterized by a softer granitic shape, and the Southern Achala Dominance (DAS) that presents a sharper granitic shape.

Both dominances are separated by a dextral transtensional fault with northwest orientation and almost vertical position. This fault (Fig. 1A) is named as the Retamito Alignment (RA) (Martino et al. 2012b; 2017).

This lineament raises the granitic massif of Los Gigantes (2,400 m a.s.l.) producing the tilting to the northern side of the DAN block and to the eastern section of the DAS block.

Within the DAN environment, a series of alignments (Fig. 1A), that form deep V-shaped valleys and bound elongated blocks with a northwest orientation, can be distinguished. These alignments are called –from north to south-: the Candelaria (CA), Corral del Carnero (CCA), Rincón Grande (RGA), Guasta (GA) and Retamito (RA) alignments (Martino et al. 2012b; 2017).

On the other hand, the DAS is subdivided by a series of parallel faults with northern orientation (Fig. 1A), called –from west to east-: the Rugapampa Fault (RU), Río Hondo (RH), Potrero del Gero (PG), Cumbre de Achala (CA), Boca del Río (BC) and La Cumbrecita (LC). This produces a progressive staggering towards the centre of the unit (Martino et al. 2012b; 2017).

Regarding the climatic conditions of the study area, the annual average rainfall ranges between 900 and 1,000 mm and it is concentrated between the months of November and March, whereas the annual average temperature is 10°C, with maximum temperatures ranging from 25 °C to 30 °C and minimum temperatures ranging from −10°C to −25°C (Capitanelli 1979).

## 3 Evolution of the Granitic Landforms of the Achala Batholith

The main events that Carignano et al. (1999; 2014) considered decisive in the general evolution of the Sierras de Córdoba have also been acknowledged in different specific studies about the granitic landscapes of the Achala batholith.

Particularly, the contributions of Cerutti (2002); García Aráoz and Cioccale (2015); Crespi (2015) and García Aráoz et al. (2018) point out that it is possible to establish a chronological sequence in the observed geoforms, thus allowing the reconstruction of a geomorphological history of the massif, which should be coherent with these cited events.

According to these authors, the geological history of these landforms would begin with the intrusion of the Paleozoic batholith approximately 367 Ma (Dorais et al. 1997). In these first stages, different structures such as flows, magmatic foliations and polygonal cracking would be generated, which would complete its development and would be exposed long after (Vidal Romaní and Yepes 2004).

At least at 120 Ma ago (Lobens et al. 2011; Bense et al. 2013) the Achala plain would have been a positive element, made of granitic rocks affected by orthogonal joints. These joints would have been the open spaces through which the processes of chemical weathering (under a subtropical to tropical climate) would have acted during a very long period of tectonic stability in order to produce the Jurassic etchplain (Carignano et al. 1999; 2014). The weathering front would spread in a differential form, controlled by the joint density and the differential characteristics of the granites.

Approximately 80 Ma ago (Lobens et al. 2011; Bense et al. 2013), the deep weathering valleys generated would have been eroded to the current weathering front, due to a change in the environmental conditions towards a climate progressively changing to drier conditions during the Cretaceous. Portions of more resistant rocks would remain as massif relicts (Carignano et al. 1999; 2014). Thus, the nubbins, castles koppies and bornhardts would persist in the general landscape of the plains, remaining like residual inherited forms from a previous stage (Rabassa et al. 1996; Rabassa and Ollier 2014; Crespi, 2015).

Following with the evolutionary sequence, a second generation of discontinuities would form the dormant structures after a concentration of weight which, when being exposed, would transform in primary endogenous blocks, tafoni and gnammas (Vidal Romaní and Yepes 2004; Crespi 2015).

As a consequence of the general upraising of the Sierras de Córdoba, originated by the Andean tectonic during the Late Cenozoic (18 to 11 Ma) (Yañez et al. 2001; Ramos et al. 2002) and the installation of the present climatic conditions, the erosional forms that respond to the establishment of runoff systems –river pylons, trails, or grooves- would be formed. The last generation would be composed of recent or present geoforms, where the influence of the mechanic alteration would produce broken blocks, and the chemical alteration would produce speleothems and the present weathering profile.

# 4 Materials and Methods

The three studied caverns were analyzed following the methodology proposed by García Aráoz et al. (2018). In the first instance, a geomorphological-structural characterization of their respective environments was carried out based on the photointerpretation of the satellite images provided by Google Earth. The main alignments of the sector and the general characteristics of the landscape were recognized. From the MED-Ar image, with 30 m of resolution, the geomorphological classification of the terrain was carried out by means of the TPI based landform classification module of the SAGA GIS program, and the average, maximum and minimum slopes were determined, as well as the orientations of the main slopes. All the observations were confirmed and validated in the field.

Afterwards, the interior of each cavern was surveyed by tracing open traverses in the main galleries. At each station, the corresponding orientation and distances readings to nearby stations were taken, and the distances to the walls were measured at 360° radius in horizontal and vertical direction with a frequency of 30°. These tasks were carried out with a geological compass and a digital telemeter (distance meter).

In each instance, a detailed survey of discontinuities was made, both of the surroundings and of the interior of the cavities, and the photographic record was carried out.

Finally, all the information was analyzed and interpreted in the lab in order to arrive at the corresponding hypotheses for the origin of each described cavities.

# 5 La Virgen Grotto

## 5.1 Description of Its Surroundings

The La Virgen (LV) grotto is located in the Los Caracoles gully, on the Achala plain, western of Mogote Grande (Fig. 2).

The Achala plain is seen as an elongated upland with a NW to SE direction, formed by flat hills with convex and wide summits, which have been leveled by erosion processes and sediment accumulation. The western, northern and southern slopes of these hills have a very soft slope with averages under 15°, whereas towards the eastern slope they are connected with the scarp of the Sierras de Los Gigantes, showing a straight and abrupt side, with slopes of up to 40°.

Several gullies with an NW-SE orientation (Fig. 2B) divide this soft landscape forming deep gashes of sub-vertical walls (Fig. 2C). These valleys show a flat floor and are filled with colluvial and alluvial sediments (transported from the slopes) formed by weathered material from the granite (Fig. 2D). In the proximity of the cave, the Los Caracoles gully presents an asymmetric section, with an abrupt southern side (with slopes over 50°) and the smooth northern side (with slopes under 30°) and unevenness (local relief) of 130 m between the summit and the bottom of the valley.

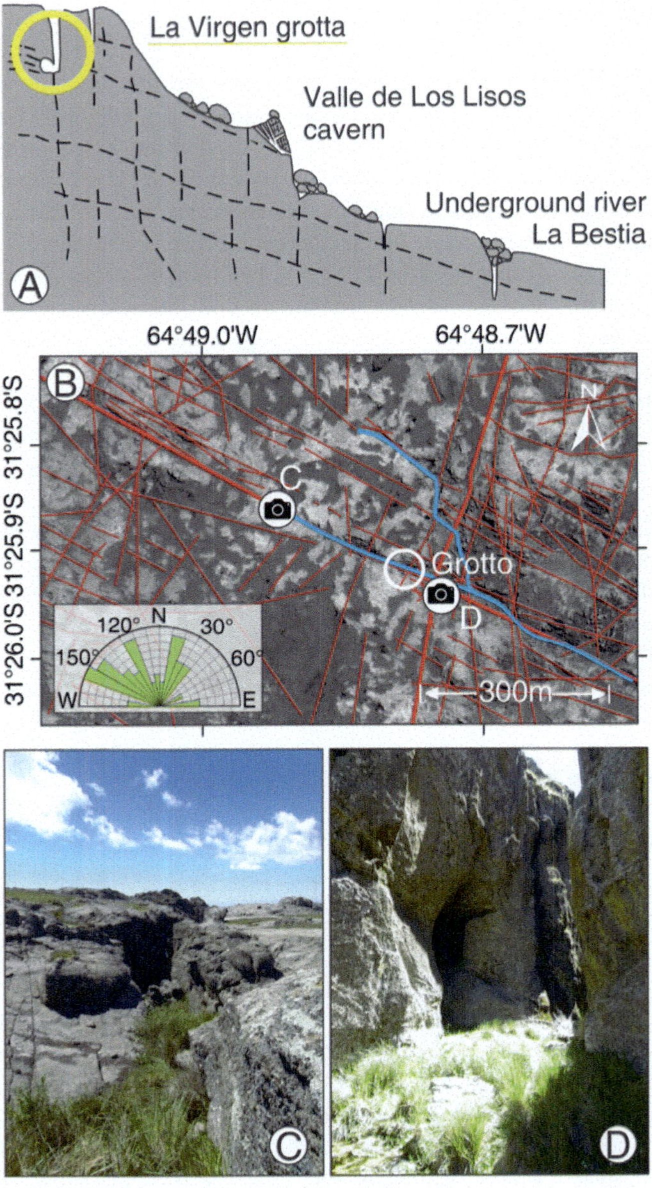


**Fig. 2.** Surroundings of the LV grotto. (A) Schematic profile of the Los Gigantes escarpment and location of the cavern (Modified from Migon 2006). (B) Alignment map (Satellite Image provided by Google Earth). (C) View of the Achala plains in the gully where the cave is located. (D) Entrance opening to the cavity.

## 5.2 The Grotto Survey

This cavity consists of one lone chamber with the shape of a domed grotto (Fig. 3), that develops at the southern portion of a gully with vertical walls, with NW-SE orientation. Its dimensions are the following: 7 m high, 8 m deep and 6 m wide.

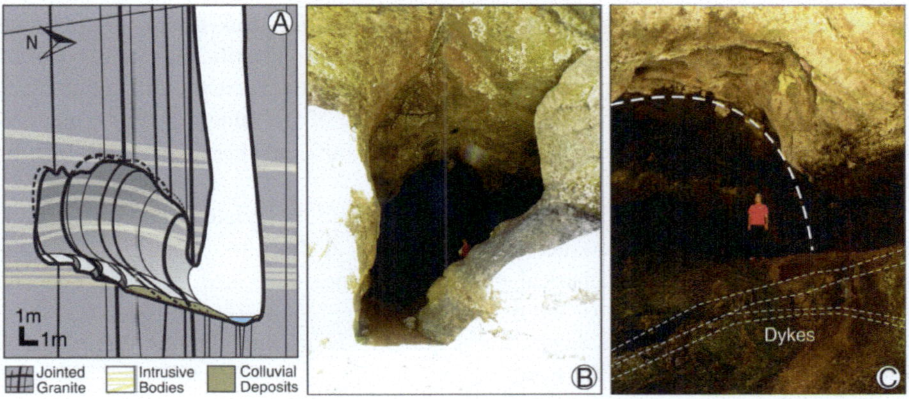


**Fig. 3.** Survey of the LV grotto. (A) Schematic section of the cave. (B) Entrance opening. (C) Inside the cavern. Vaulted ceiling and floor conformed by dykes.

The floor of the cavern is made up by three steps, which ascend towards the inside of the cave. Subhorizontal pegmatitic dykes that are more resistant to erosion (Fig. 3C) bound these steps. Similar dykes have also been observed over the roof of the cavity, protruding at different heights of the rock massif.

All this grouping (dykes and rock massif) is cut across by subvertical discontinuities that present two main directions, with values up to 115° and 130°, creating blocks like sharp wedges, distinguishable both at the roof and on the walls. The steadiness of the wedges and the absence of fallen blocks allow the interpretation of geomechanic conditions of static balance. The floor of the cavern is covered by middle to coarse sands as a result of the weathering and erosion of the massif, disregarding an origin or growth of the underground space due to a gravitational collapse from the roof or the sides. In this matter, the distinctive geometry of the cavern (Fig. 3C) evoques those that take place from dissolving in soluble rocks, where the resistance to the outward pull of the massif controls the circumference radius that forms the roof, and the water flow determines its growth.

Water is present in almost all the mentioned structures (Fig. 3C), mainly on the southeastern walls. The circulation concentrates in the intersection of the discontinuities, forming convex grooves or hollows of different sizes with smooth and polished surfaces.

### 5.3 Hypothesis About Its Origin

The shape of the cavity and the absence of fallen blocks inside it allow the classification of this cavern as a “tafone” and explaining its origin from the elastic model of forming cavities proposed by Vidal Romaní (1985).

According to this model, the stress concentration in endogenous conditions on specific points of a discontinuity plane (joint plane or fracture plane) would produce the breakage of the massif surrounding these points, generating weaker areas, more prone to the weathering conditions, called lacunar precincts or plastification zones (Vidal Romaní 1989). Then, under the influence of exogenous processes, the differential alteration of the massif would produce the withdrawal of the mineral grains contained in the enclosure and would generate the cavities with different shapes and sizes (Mayor Rodríguez 2011).

Therefore, the origin of the cave (Fig. 4) would be related to a differential weathering process channeled across the subvertical fracture planes- that would develop at higher speed in spherical or elliptical zones generated by stress concentration around specific points of the massif.

The presence of several subhorizontal pegmatitic dykes would control the progress of the erosion processes and the resulting morphology.

The evolution scheme proposed in Fig. 4 suggests that the emplacement of the dykes has been previous than the stress concentration that the massif has been put through. This conjecture is based on two main observations:

1. The forming process of the cave affects both intrusives as rock-bearing units, particularly in the middle sector, where the cavity presents its larger dimensions.
2. No residual blocks of the dykes are observed. Although some of these have shown more resistance to erosion, they could have been affected only partially by the stress concentration.

## 6 Valle de Los Lisos Cavern

### 6.1 Description of Its Surroundings

The Valle de los Lisos cavern is located in the Los Lisos valley, in the middle section of the Los Gigantes scarpment (Fig. 5).

Los Lisos valley is a hydrological basin with an elongated geometry in N–S orientation, with an approximate surface of 133 ha. This basin has the peculiarity of not having a drainage mouth at the surface. Its hydrographic closure takes place from the infiltration of the main bed between the blocks and sediments that form the bottom of the small valley that precedes the cavern (Fig. 5B).

In this sector, the sedimentary cover (Fig. 5E) is formed by a sequence of dark coloured clay-slime materials, interwoven with lenticular deposits of middle- and fine-grained sands of low density that have been interpreted as facies from marsh or lake environments interbedded with eolian accumulation events that suggest flooding periods in the area (García Aráoz et al. 2018).

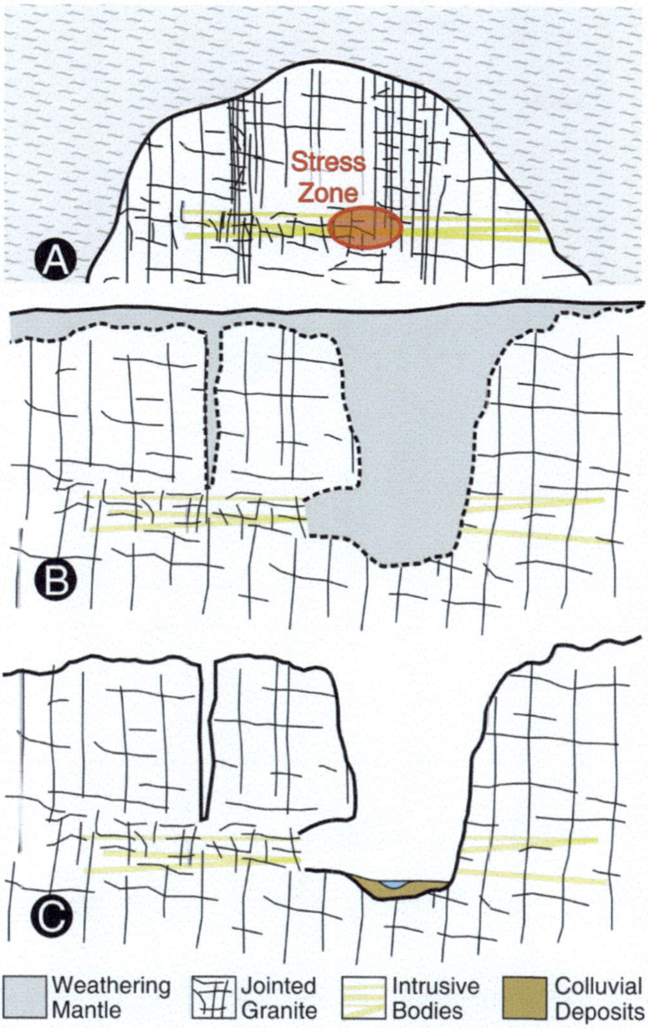


**Fig. 4.** Hypothesis about the origin of the LV grotto. (A) Generation of a stress concentration zone that forms the lacunar precincts during the intrusive stage of the massif. (B) Massif exhumation and differential advance of the weathering front. (C) Cavern formation by erosive processes.

The size of the blocks observed in the surroundings is varied and, as a whole, they have sizes that excede the stream loading capacity (with diameters from 0.5 m to 5 m). They show a prismatic geometry (Fig. 5C) developed by the intersection of three families of major discontinuities (9°/56° W, 16°/48° E, 112°/87° N). These same fracturing patterns have also been recognized in the inside of the cavern (Fig. 5D), therefore showing a strong structural control upon the development of its galleries. The consistent orientation of the major blocks shows that the massif is essentially in place,

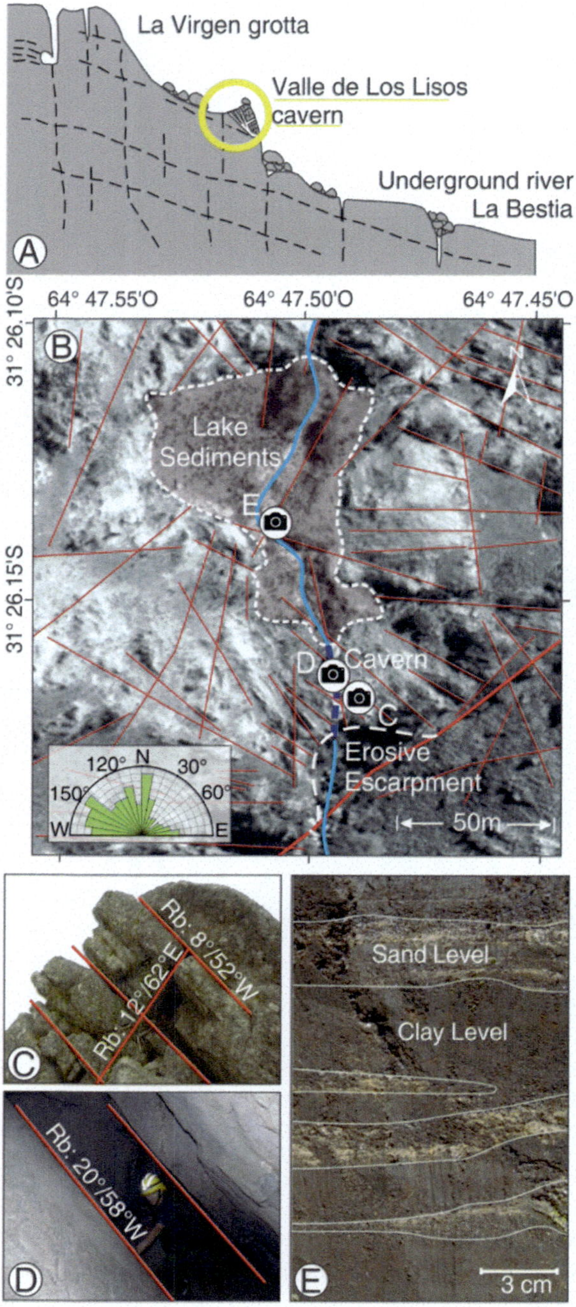


**Fig. 5.** Surroundings of the LL cavern. (A) Schematic profile of the Los Gigantes escarpment and location of the cavern (Modified from Migon 2006). (B) Alignments map and landscape characteristics. (C) Fractures around the cavern related to the outcropping granitic massif. (D) Fractures inside the cavern. (E) Lake sediments.

therefore disregarding a possible origin due to a gravitational collapse (García Aráoz et al. 2018).

## 6.2 The Cavern Survey

This cavern consists of a series of intricate corridors developed in a granitic lithology that present 20 m of horizontal advance with N–S orientation and 25 m of total elevation difference (Fig. 6).

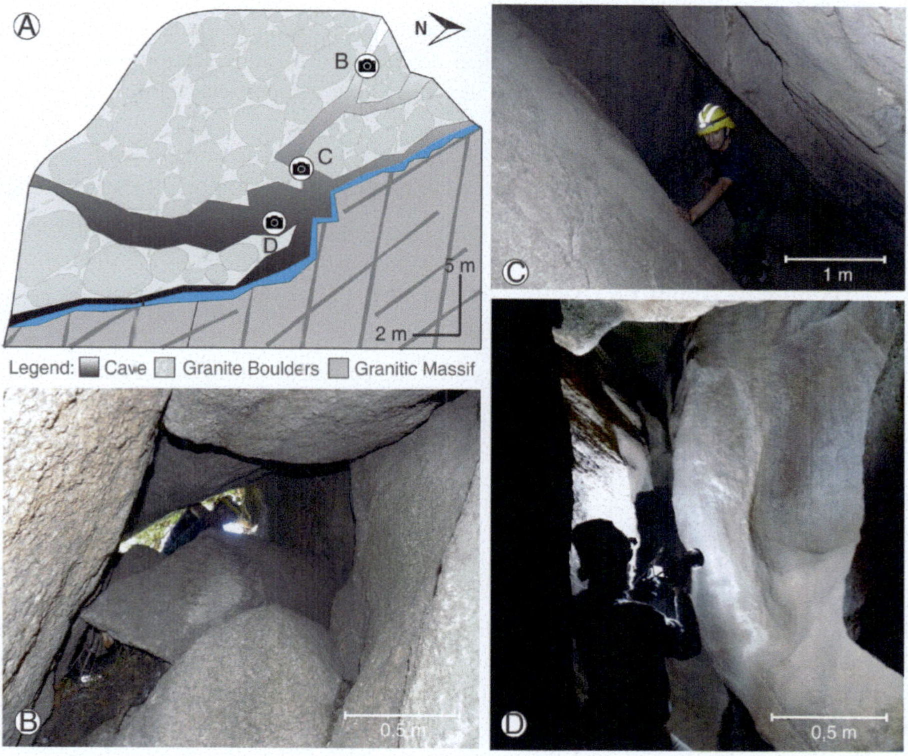


**Fig. 6.** Survey of the LL cavern. (A) Vertical schematic section of the cave. (B) Upper entrance of the cavern. (C) Inside the cavern in the northern sector. (D) Inside the cavern in the southern sector.

Three entrance openings and two exit holes have been identified in different topographic levels (Fig. 6A), with signs of water erosion. The disposition of these entrances and the identification of several erosion surfaces at the base of major blocks in the surroundings indicate the evolution of the cave in successive erosion phases, from the higher to the lower levels.

At present, this stream flows through a deeper level, connecting the entrance and the lower exit through insurmountable small pipes.

Since the access galleries are individual paths that join at one point, the architecture of the underground clearance is solved in two sections with different characteristics: a northern section, upstream of the junction, and a southern section, below this point.

The northern section features narrow galleries (Fig. 6B and C), in general with tabular shape between 0.5 to 1 m wide, defined between the fractures planes housing minor blocks. In some areas, the galleries are so narrow that it is difficult to enter through them.

Towards the southern section (Fig. 6D), on the contrary, when the different entrance paths come together the space is much wider, even defining chambers that reach 2 to 3 m wide with a similar vertical development. In this section, the signs of abrasive water erosion are outstanding; all rocks show a superficial varnish and sink-holes and holes-pot are common features.

The exits of the cave bound with an abrupt slope scarp. The numerous blocks that can be seen at the base of this scarp show the process of erosion recoil that is taking place on it.

## 6.3 Hypothesis About Its Origin

The schematic model that summarizes the evolution and development of the cavern is shown in Fig. 7. It comprises three stages after the emplacement of the intrusive body:

The first stage (Fig. 7A) corresponds to a period of tectonic and climatic stability, where a deep weathering profile develops. It determines the formation of an etchplain with a preponderance of rounded shapes. The second stage (Fig. 7B) is produced with the generation of the topographic projection that can be observed at the exit of the cave. It is a mixed scarp, a product of colluvial and tectonic processes.

This new landscape configuration generated the hydrographic basin of the Los Lisos valley. In the initial stages of this drainage system, the most depressed zone of the basin (where the cavern is presently located) would have operated as a pond during the wetter season. The recurring flooding would increase the infiltration into the discontinuities of the massif, producing a higher alteration of the walls and increasing the material available for erosion.

The third stage (Fig. 7C) of the proposed model foresees the construction of the underground clearance in successive erosive phases from the upper to the lower levels.

Each underground pipe has been formed by the erosive action of the water that removes the previously altered material. Each time a new underground way is released, and the material available for erosion is removed, there occurs another gravitational re-arrangement of the overlapping blocks, blocking each other and leaving different empty spaces between them. Finally, the cavern is formed by the fusion of these empty spaces.

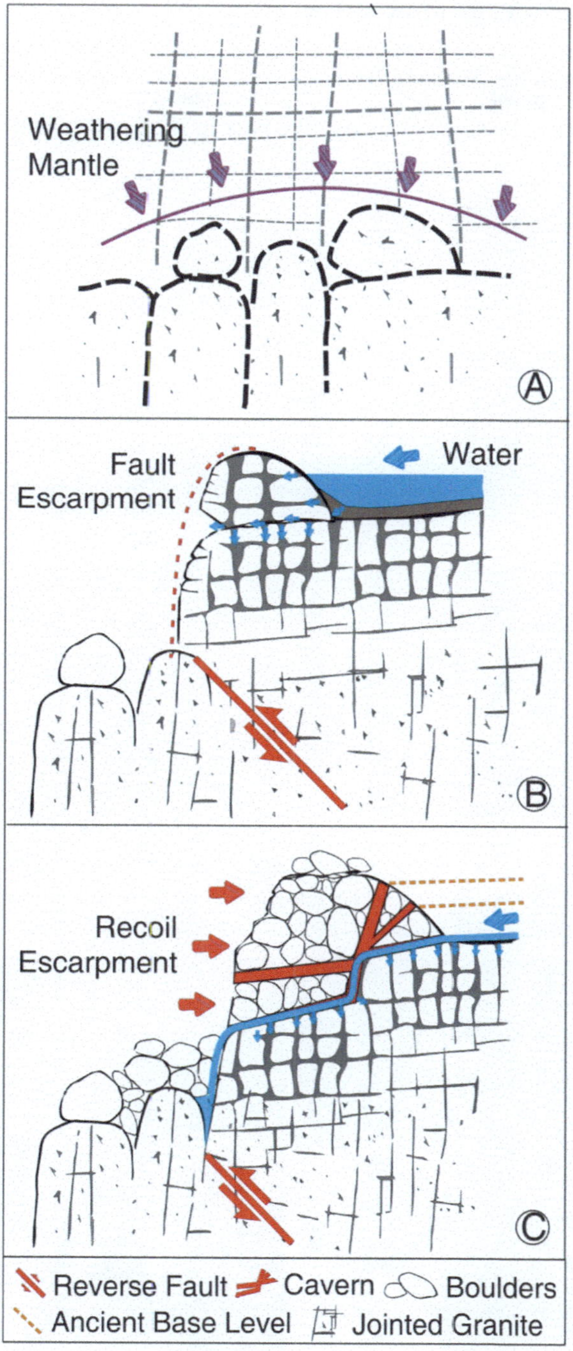


**Fig. 7.** Genetic-evolutionary model of the LL cavern. (A) A deep weathering profile development. It determines the formation of a paleo-relief with a predominance of rounded morphology. (B) Hydrographic basin formation from a mixed scarp (erosive and estructural). Deposition of lake sediments. (C) Opening of the main galleries in successive erosive phases that removed previously altered material.

# 7 La Bestia Underground River

## 7.1 Description of Its Surroundings

La Bestia underground river is located at the base of the Los Gigantes scarp on the El Cajón river bed (Fig. 8).

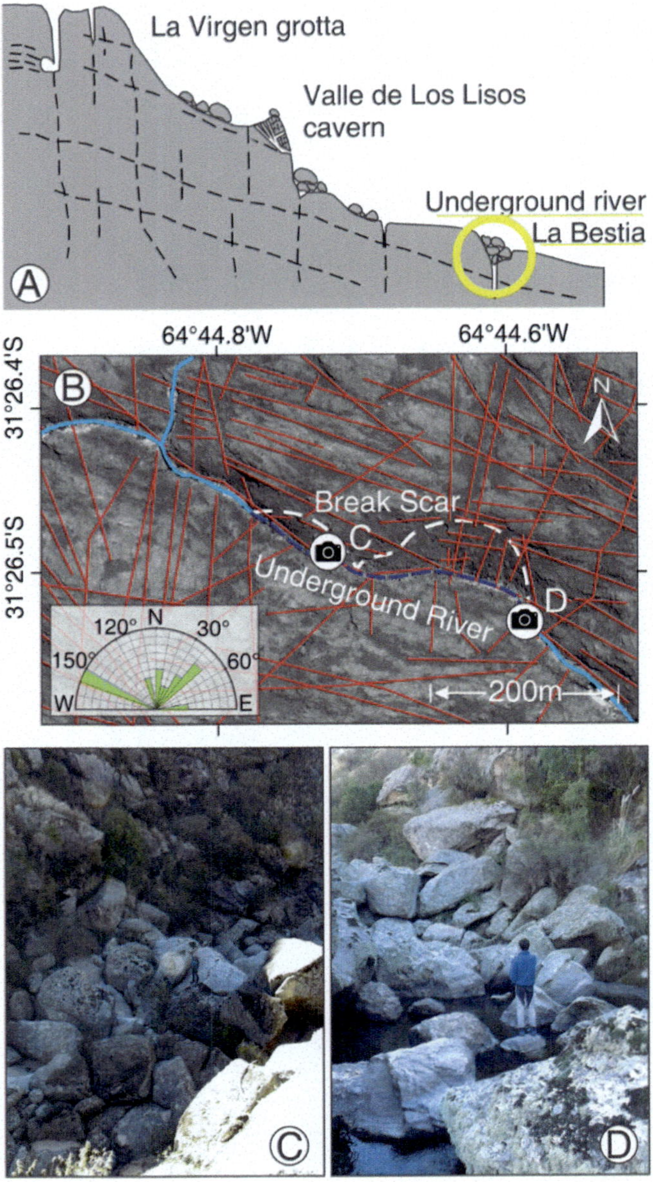


**Fig. 8.** Surroundings of the LB underground river. (A) Schematic profile of the Los Gigantes escarpment and location of the cavern (Modified from Migon 2006). (B) Geomorphological environment and structural mapping. Notice the curved scars of the fracture plane on the stable side of the slip. (C) and (D) Collapsed blocks that form the roof of the cavern.

The surrounding area outstands for the presence of hills with rounded and wide tops, together with convex to straight-convex slopes, and average inclinations between 15° to 30°. The valleys show a flat bottom and are filled with alluvial and colluvial sediments. The length of the slopes, from the top to the centre of the valley, varies between 500 m to 750 m.

In the cavern area, El Cajón river runs along a gully of asymmetric section, with an abrupt northern side (with average slopes higher than 50°) and a smoother southern side (with average slopes lower than 30°). Two curved scars can be recognized in the satellite image at the northern side (Fig. 8B). Scattered blocks of different sizes on the riverbed also have been recognized (Fig. 8C and D). Both provide evidence for mass removal.

The local granite rock shows intense fracturing and jointing with a predominance of three sets from vertical to subvertical discontinuities with orientations 105°/80° N, 110°/75° S and 30°/70° SE.

## 7.2 The Underground River Survey

La Bestia is a cavity formed under large, collapsed, differentially tilted granitic blocks that form the roof of a gully of vertical walls in an area of the El Cajón river (Fig. 9). This way, the underground transit is fully done following the river bed.

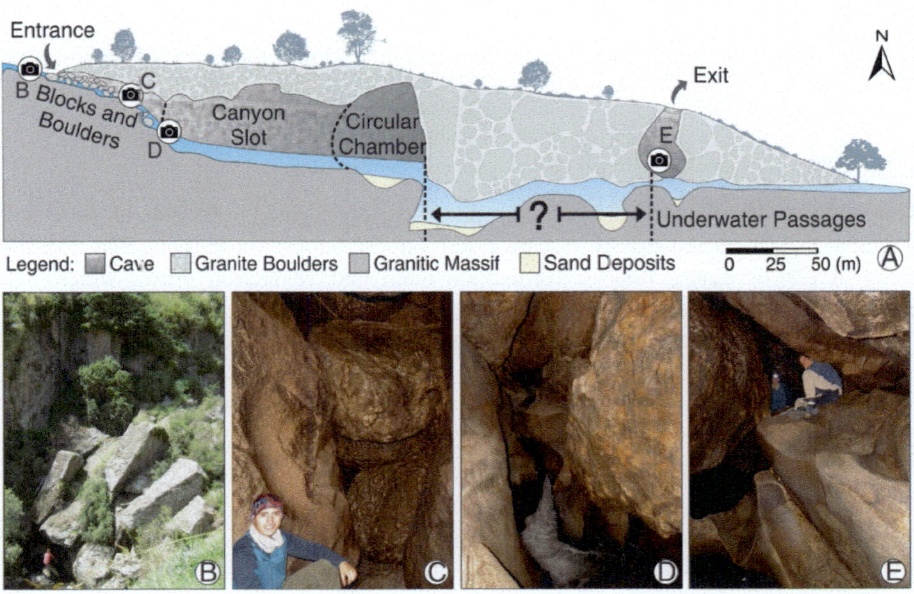


**Fig. 9.** Survey of the LB underground river. (A) Schematic section of the cavern. (B) Entrance formed by differentially tilted tabular blocks in vertical way. (C) Inside the cavern: blocks and boulders. (D) Inside the cavern: access to canyon slot. (E) Inside the cavern: exit hole.

The mouth of the entrance shows metric dimensions and it is composed of tabular blocks limited by three sets of discontinuities (Fig. 9B): 110°/90°, 30°/90° and 45°/30° NE.

The movement of these blocks has been mainly in a downward direction, since its main planes still keep their original orientations.

In the inside of the cave there can be seen different sections with individual characteristics (Fig. 9A). As from the entrance and until the first 50 m, water flows within narrow corridors formed between blocks of different sizes (Fig. 9C). These have been carved and polished by the force of the riverflow. Short waterfalls shorter than 1 m high are common in this section.

This first section connects with a narrow, smooth walled canyon (Fig. 9D) which is approximately 1 m wide, 30 m long and 12 m high. This canyon was formed between two fracture planes. It must be crossed by swimming and it connects an inner chamber of circular shape - approximately 4 m diameter- which keeps a constant water level between 0.5 m and 1.5 m depending on the season. The floor of this chamber is covered with sand and as from there, the river turns into an absolutely underground position, being approachable only with diving equipment.

Following this underwater bed by diving, the cavern continues into other chambers of different sizes and depths regarding the water surface until a final chamber is reached. This final chamber (Fig. 9E) has 10 m to 12 m high and it shows an open roof that can be used as an exit or entrance using rappel techniques.

It has been observed that during flood events, this latter chamber acts as a spillway, filling up impetuously.

The aforementioned circuit is, approximately, 200 m long and 30 m of total internal relief.

## 7.3 Hypothesis About Its Origin

The geoforms locally known as underground rivers are formed in the sets of joints and fractures that affect the massif and are progressively buried under rocks that take the place of a roof. Later, floor is formed on them and vegetation grows which will eventually start the dissolvent processes.

La Bestia underground river has been formed from the process of mass removal that buried part of the river bed. Figure 10 shows this situation in a diagram: at the beginning, (Fig. 10A) weathering conditions alter the areas where the massif is more fractured in a differential way. This allows the installation of a hydrological system that favors those planes of weakness. This way, the rivers deepen their beds, generating vertical walled gullies that can reach tens of meters in height (Fig. 10B). When the foot of these gullies are undermined, or when trembling or earthquakes occur, these walls loose stability and collapse due to gravity, generating the described caverns (Fig. 10C).

Different morphological and structural characteristics have been recognized within the slide (Fig. 10C). Its plan section is triangular, with the base of the triangle limited by the opposite wall of the gully. In the side vertex, the blocks show a predominance of the vertical component of the convergence. It generally keeps its original orientation. Towards the centre, the horizontal slidings are larger, which produces a major rupture

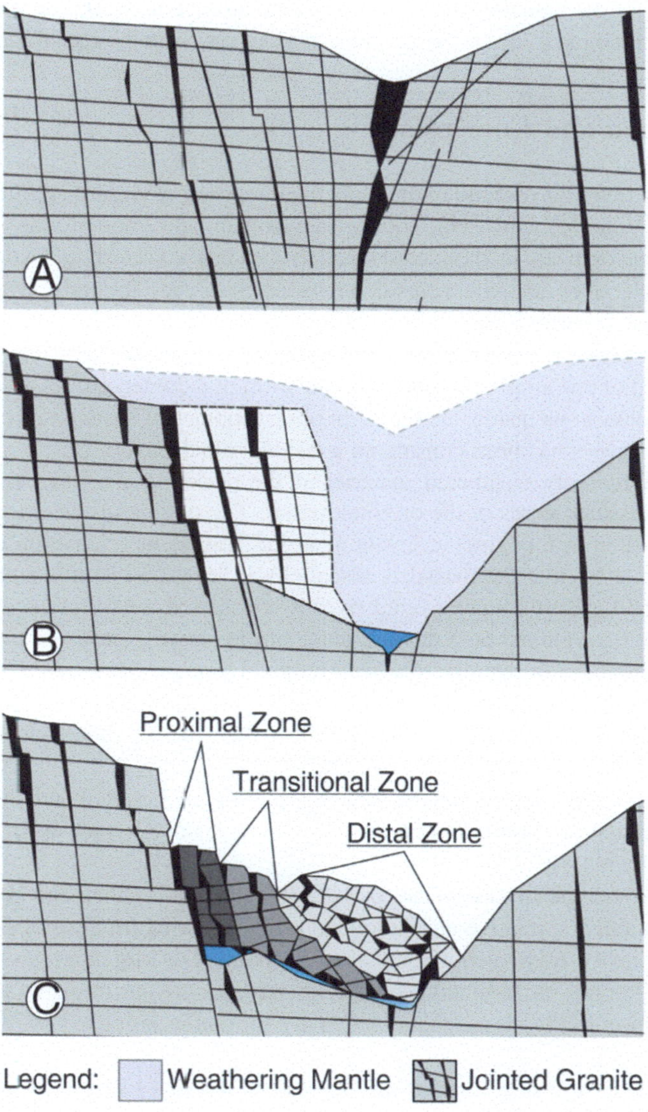


**Fig. 10.** Hypothesis about the origin of the LB underground river (modified from Margielewski and Urban 2017). (A) Development of a deep weathering profile. (B) The stream cuts a steep valley into the rock. (C) Part of the valley wall collpases or falls into the valley. The stream continues to cut downwards, forming a space between the fallen rock and the stream. More collapse adds additional roof material to the cave and weathering forms soil on the fallen rocks. Trees, grass, and other plants begin to grow in soil above the cave. Proximal zone: differentially tilted tabular blocks in vertical way. Transitional zone: tabular blocks and rounded blocks. Distal zone: rounded blocks and irregular blocks with chaotic arrangement.

and scattering of the material, with a more chaotic arrangement. On the stable slope, is is possible to recognize the breakage plane that leaves a curve scar in the massif.

## 8 Discussion and Conclusions

The described caves correspond to the three main types of cavities that different authors (Twidale and Vidal Romaní 1998; 2005; Vidal Romaní and Vaqueiro 2007; Song et al. 2011; 2015) recognized in their classification systems - according to their morphological characteristics and the genetic processes involved- fissural cavities, residual blocks cavities and tafoni.

La Virgen grotto has been considered a tafoni. Its special characteristic morphology, composed of one single chamber with the form of a domed grotto, and the absence of remnant blocks in its inside, are the main observations that sustain this classification.

Los Lisos cavern has been considered a residual block cavern case that was formed while the differentially weathered material of the discontinuity was being removed, leaving only resistant cores of the original massif. The outline of the entrance and exit openings in different topographic levels, plus the consistent orientation of the major blocks (which show that the massif is essentially in its place) is the main evidence.

Finally, the La Bestia underground river is considered a fissural cavity filled with blocks that have developed on a fracture plane of the massif, where the river carved its bed and then has been buried due to a mass removal process that roofed the cavity. The evidence arises from the own characteristics of the underground course described, and from the analysis of the satellite images, where the scars of this detachment can be seen.

The three caves are located on the same scarp of the rock massif and, therefore, they share a geological and geomorphological common history, which is shown in their evolutionary hypothesis.

Based on the type of cavity and certain observations of their environments it is possible to establish a relative chronological order in their formative ages.

The first cavity originated would have been the La Virgen grotto, since being considered a tafone, its formation can be traced back to the intrusive stages of the massif, according to the elastic theory of cavity formation proposed by Vidal Romaní (1985).

The second cavern originated would have been the Los Lisos cavern, because the lake sediments found in their environment have shown the existence of long flooding periods in that place, prior to the formation of the underground space.

Finally, the La Bestia underground river would be the youngest cavity, since it has been formed mainly by erosion processes and mass movement processes, which can be considered instantaneous in terms of geological time.

Taking this into consideration and its common context, it can be seen that these underground environments (as well as other geographical features and granitic geoforms) are not chaotic at all, but - on the contrary- they are structured according to geological and geomorphological processes along time that respond to a common evolutive frame for all the rock massif.

## 9 Final Remarks

- In the development of each cavern three historical evolutionary stages of the granitic landscapes of the Achala batholith can be identified:
  1. Development of a deep weathering profile that conditions the formation of a paleo-relief with predominance of rounded morphologies.
  2. Generation of the geomorphological surroundings due to tectonic or erosive processes in each of them.
  3. Main galleries formation through gravitational and erosion processes.
- According to the observations in all the caverns, in order for the cavern process to take place and be effective in this type of environments, three essential factors must coincide: water availability, a jointed massif and a favorable topographical situation.
- It is significant to point out that the study of the underground cavities must be addressed bearing in mind the historical perspective of the massif. Only in this way hypotheses about their particular origins can be established, thus achieving a full interpretation of the granitic environment.
- Likewise, more detailed studies of these environments can –at the same time-become new tools to clarify the history of the landscape where these landforms can be found, since not all cavities need the same time to be formed and it is possible to locate them in a specific moment of the geomorphological evolution of the massif.

**Acknowledgments.** The authors acknowledge the support of IEYDAT CORP SRL for providing us with access to the study sites. In addition, we extend our sincere thanks to the reviewers Dr. Emilia Aguilera and Dr. Jorge Rabassa, whose comments enriched this work. Also, we thank to the Argentine Consejo Nacional de Investigaciones Científicas y Técnicas (CONICET) and the Facultad de Ciencias Exactas, Físicas y Naturales of the Universidad Nacional de Córdoba (FCEFyN-UNC) for financial support.

## References

Bates, R.L., Jackson, J.A.: Glossary of Geology. p. 788, American Geological Institute, Virginia (1987)

Bense, F.A., Löbens, S., Dunkl, I., Wemmer, K., Siegesmund, S.: Is the exhumation of the Sierras Pampeanas only related to Neogene flat-slab subduction? Implications from a multi-thermochronological approach. J. S. Am. Earth Sciences **48**, 123–144 (2013)

Campbell, E.M.: Granite landforms. J. Roy. Soc. W. Aust. **80**(3), 101–112 (1997)

Campbell, E.M., Twidale, C.R.: The various origins of minor granite landforms. Cadernos do Laboratorio Xeolóxico de Laxe, vol. 20, pp. 281–306. A Coruña (1995)

Capitanelli, R.G.: Clima. In: Vásquez, J.B., Miatello, R.A., Roqué, M.E. (eds.) Geografía Física de Córdoba, pp. 45–138, Córdoba, Boldt (1979)

Carignano, C., Cioccale, M., Rabassa, J.: Landscape antiquity of the Central-Eastern Sierras Pampeanas (Argentina): geomorphological evolution since Gondwanic times. Zeitschrift fur Geomorphologie, Supplementband, pp. 245–268 (1999)

Carignano, C., Kröhling, D., Degiovanni, S., Cioccale, M: Geomorfología. In: Martino, R.D., Guereschi, A. (eds.), Relatorio XIX Congreso Geológico Argentino: Geología y Recursos Naturales de la Provincia de Córdoba, pp. 747–822, Asociación Geológica Argentina, Córdoba (2014)

Cerutti, N.: Geoformas graníticas en el Batolito de Achala, Sierras Grandes de la Provincia de Córdoba. Unpublished Thesis for Bachelor degree, Universidad Nacional de Córdoba, 114 p. (2002)

Crespi, M.F.: Morfogénesis del paisaje granítico del batolito de Achala, provincia de Córdoba: análisis del control geológico. Unpublished Thesis for Bachelor degree, Universidad Nacional de Córdoba, 199 p. (2015)

Dorais, M.J., Lira, R., Chen, Y., Tingey, D.: Origin of biotite-apatite-rich enclaves, Achala batholith, Argentina. Contrib. Miner. Petrol. **130**(1), 31–46 (1997)

Finlayson, B.: Underground streams on acid igneous rocks in Victoria. Helictite **19**(1), 5–14 (1981)

Gaal, L., Bella, P.: Granites and granite caves in the Western Carpathians. Cadernos do Laboratorio Xeolóxico de Laxe, vol. 33, pp. 11–18, A Coruña (2008)

García Aráoz, E., Cioccale, M.: Geoformas graníticas: distribución espacial y procesos formadores. Batolito de Achala, Provincia de Córdoba, Actas VI Congreso Argentino de Cuaternario y Geomorfologia, Ushuaia, Editorial Utopías, pp. 274–275 (2015)

García Aráoz, E., Madelón, N., Cioccale, M.: Caverna del Valle de Los Lisos: una hipótesis sobre su origen en el contexto geomorfológico de las Sierras de Córdoba. Argent. Rev. Mex. Cienc. Geol. **35**(1), 18–27 (2018)

Gordillo, C.E., Lencinas, A.N.: Sierras Pampeanas de Córdoba y San Luis. In: Leanza, A.F. (ed.) Geología Regional Argentina, pp. 577–650, Academia Nacional de Ciencias, Córdoba (1979)

Krapovickas, J.M., Tauber, A.A.: Estratigrafía de las áreas cumbrales de las Sierras Pampeanas de Córdoba: geocronología, modelo regional, paleoambiente y paleoclima en una región poco conocida de Argentina. Rev. Mex. Cienc. Geol. **33**(1), 105–121 (2016)

Lazaridis, G., Pipera, K.: Preliminary report on granite caves in Greece. Cadernos do Laboratorio Xeolóxico de Laxe, vol. 33, pp. 101–113, A Coruña (2008)

Lazarte, S.: Descripción topográfica de la cueva de Los Lisos, Los Gigantes, Provincia de Córdoba. vol. 11, pp. 32–35, Salamanca (2002)

Löbens, S., Bense, F.A., Wemmer, K., Dunkl, I., Costa, C.H., Layer, P., Siegesmund, S.: Exhumation and uplift of the Sierras Pampeanas: preliminary implications from K-Ar fault gouge dating and low-T thermochronology in the Sierra de Comechingones (Argentina). Int. J. Earth Sci. **100**(2), 671–694 (2011)

Manzur, A.: Aspectos genéticos de los suelos presentes en Atum Pampa y Athos Pampa, Sierras Grandes, Córdoba, Argentina. Unpublished Ph.D. Thesis, Facultad de Ciencias Exactas, Físicas y Naturales, Universidad Nacional de Córdoba, 114 p. (1995)

Margielewski, W., Urban, J.: Gravitationally induced non-karst caves: tectonic and morphological constrains, classification, and dating; Polish Flysch Carpathians case study. Geomorphology **296**, 160–181 (2017)

Martino, R.D., Guereschi, A.B., Carignano, C.C.: Influencia de la tectónica Preandina sobre la tectónica Andina: El caso de la falla de la Sierra Chica, Sierras Pampeanas de Córdoba. Rev. Asoc. Geol. Argent. **69**(2), 207–221 (2012a)

Martino, R.D., Toledo, A.E., Guereschi, A.B.: Lineamientos transtensionales en el margen noroccidental del batolito de Achala y su relación con la exhumación de rocas de alto grado y la extensión cretácica en las Sierras de Córdoba. 15° Reunión de Tectónica, Libro de Resúmenes, Facultad de Ciencias Exactas Físicas y Naturales, Universidad Nacional de San Juan, 90, San Juan (2012b)

Martino, R.D., Guereschi, A.B., Caro Montero, A., Abril, E.G.: El control de la tectónica frágil Preandina en la estructura y morfología actual del Batolito de Achala, Sierras Pampeanas de Córdoba. Actas XX Congreso Geológico Argentino, Tucumán, Simposio de Tectónica Pre-Andina, pp. 105–107 (2017)

Mayor Rodríguez, J.A.: Génesis de cavidades graníticas en ambientes endógenos y exógenos, Unpublished Ph.D. Thesis, Instituto Universitario de Geología Isidro Parga Pondal, Universidad da Coruña, 396 p., Spain (2011)

Migon, P.: (ed.): Granite Landscapes of the World. 417 p., Oxford University Press, Oxford (2006)

Montes, A.: Cronología de nuestra prehistoria. Revista de la Facultad de Ciencias Exactas, Físicas y Naturales, Universidad Nacional de Córdoba, vol. 18, no. 3–4, pp. 1–41, Córdoba (1956)

Ollier, C.D., Smith, J.M.B.: Crombie's cave. In: A Granite Cave in New England, vol. 32, no. 1, pp. 17–20, NSW, Helictite (1994)

Rabassa, J., Ollier, C.: (eds.): Gondwana landscapes in Southern South America: Argentina, Uruguay and Southern Brazil. 561 p., Springer, Dordrecht (2014)

Rabassa, J., Zárate, M., Cioccale, M., Carignano, C., Partridge, T.C., Maud, R.: Paisajes relictuales de edad Gondwánica en áreas cratónicas de Argentina. Actas 13° Congreso Geológico Argentino and III Congreso de Exploración de Hidrocarburos, vol. 4, p. 219. Asociación Geológica Argentina, Buenos Aires (1996)

Ramos, V., Cristallini, E.O., Pérez, D.J.: The Pampean flat-slab of the central Andes. J. S. Am. Earth Sci. **15**, 59–78 (2002)

Rapela, C.W., Baldo, E.G.A., Pankhurst, R.J., Fanning, C.M.: The Devonian Achala batholith of the Sierras Pampeanas: F-rich, aluminous A-type granites. In: Actas 6° South American Symposium on Isotope Geology, 8 p., San Carlos de Bariloche (2008)

Roqué, C., Linares, R., Rodríguez, R., Zarroca, M.: Granite caves in the North-East of the Iberian Peninsula: artificial hypogea versus tafoni. Zeitschrift für Geomorphologie **55**(3), 341–364 (2011)

Scheuerer, M., Lundberg, J., Sjöberg, R.: Gobholo cave: a long granite cave in Swaziland (Southern Africa). In: Proceedings of the 16th International Congress of Speleology, vol. 3, pp. 305–307. Brno (2013)

Sjöberg, R.: A proposal for a classification system for Granitic Caves. Abh **2**(5), 25–28 (1969)

Song, Z., Zhou, J., Li, H., Liu, X.: Discovery of especially large bubble-like miarolitic granites in Laoshan and its geological and geomorphic significance. In: 2011 International Conference on Remote Sensing, Environment and Transportation Engineering (RSETE). pp. 3068–3071. IEEE (2011)

Song, Z.J., Tang, W.J., Liu, X.Q., Wang, L.B., Xiang, S.B., Li, J.P., Yang, S.S.: Genesis and geological significance of granite caves in Laoshan of China. Chem. Eng. Trans. **46**, 763–768 (2015)

Tauber, A.A., Goya, F.: Estratigrafía y fósiles del Pleistoceno-Holoceno de las pampas de altura en el departamento Cruz del Eje, Córdoba, Argentina. Actas III Congreso Argentino de Cuaternario y Geomorfología, Córdoba **2**, 717–726 (2006)

Twidale, C.R.: Granite Landforms, p. 372. Elsevier Publishing Company, Amsterdam (1982)

Twidale, C.R.: La iniciación subsuperficial de las formas graníticas y sus implicaciones en las teorías generales de evolución del paisaje. Cadernos do Laboratorio Xeolóxico de Laxe, O Castro, Sada, vol. 13, pp. 49–68, A Coruña (1989)

Twidale, C.R.: Granite landforms, p. 371. Elsevier, Amsterdam (2012)

Twidale, C.R., Bourne, J.A.: Caves in granitic rocks: types, terminology and origins. Cadernos Lab. Xeolóxico de Laxe, vol. 33, pp. 35–57, A Coruña (2008)

Twidale, C.R., Vidal Romaní, J.R.: Landforms and Geology of Granite Terrains, p. 352. Balkema, Amsterdam (2005)

Vaqueiro Rodríguez, M.: The cave-complex "A Touba do Brión–Cobreiras" cavernous weathering related to mass wasting downward sheeting planes, Cadernos do Laboratorio Xeolóxico de Laxe, vol. 30, pp. 193–212, A Coruña (2005)

Vaqueiro Rodríguez, M., Barriero Ben, B., Costas Vázquez, R., Suárez Pérez, R., Groba González, X.: Relation between structure and morphology in the development of the granite cave of 'O Folón' (Vigo, Galicia - Spain). Cadernos do Laboratorio Xeolóxico de Laxe, vol. 31, pp. 87–103, A Coruña (2006)

Vidal Romaní, J.R.: El cuaternario de la provincia de La Coruña. Modelos elásticos de formación de cavidades. Ph.D. Thesis, Servicio de Publicaciones, Universidad Complutense de Madrid, Serie Tesis Doctorales, 283 p., Madrid (1985)

Vidal Romaní, J.R.: Granite geomorphology in Galicia (NW España). Cadernos do Laboratorio Xeolóxico de Laxe, vol. 13, pp. 89–163 (1989)

Vidal Romaní, J.R., Twidale, C.R.: Formas y Paisajes Graníticos. Servicio de Publicacións da Universidade da Coruña, Serie Monografías, vol. 55, 411 p., A Coruña (1998)

Vidal Romaní, J.R., Yepes, J.: Historia de la morfogénesis granítica. Cadernos do Laboratorio Xeolóxico de Laxe, vol. 29, pp. 331–360, A Coruña (2004)

Vidal Romaní, J.R., Vaqueiro, M.: Types of granite cavitites and associated speleothems: Genesis and evolution. Nat. Conserv. **63**(6), 41–46 (2007)

Willems, L., Compère, P., Hatert, F., Pouclet, A., Vicat, J.P., Ek, C., Boulvain, F.: Karst in granitic rocks, South Cameroon: cave genesis and silica and taranakite speleothems. Terra Nova **14**(5), 355–362 (2002)

Yañez, G., Ranero, G., Von Huene, R., Díaz, J.: Magnetic anomaly interpretation across a segment of the Southern Central Andes (32°–34°S): implications on the role of the Juan Fernández Ridge in the tectonic evolution of the margin during upper Tertiary. J. Geophys. Res. **106**, 6325–6345 (2001)

# Scenic and Geotouristic Potential of Volcanic Landscapes in Southern Patagonia: Necks, Buttes and Vents, Santa Cruz Province, Argentina

Juan Pablo Velázquez[1], Elizabeth Mazzoni[2](✉), and Jorge Rabassa[3]

[1] Carrera de Ingeniería en Recursos Naturales Renovables, Unidad Académica Río Gallegos, Universidad Nacional de la Patagonia Austral, Río Gallegos, Argentina
juanpva88@gmail.com

[2] Unidad Académica Río Gallegos, Universidad Nacional de la Patagonia Austral, Río Gallegos, Argentina
elimazzoni@yahoo.com.ar

[3] Laboratorio de Geomorfología y Cuaternario, CADIC-CONICET, Ushuaia, Argentina
jrabassa@gmail.com

**Abstract.** This paper analyzes the spatial distribution of volcanic landforms that compose natural geomorphological elements, which occur as positive features in the Southern Patagonia tablelands. In the province of Santa Cruz (Southern Patagonia, Argentina) an inventory of 106 relevant landforms has been completed: 52 necks, 28 buttes and 26 vents or isolated eruptive centers, which provide a quite high degree of variability in the visual components of the landscape. A database has been prepared, containing information about their geographical position and localization with respect to the main circulation routes, cities and towns, elevation, relative relief and geological age. Likewise, a database of satellite imagery and aerial photographs was compiled and thematic cartography was prepared. This information allowed the identification of the most appropriate sectors in which detailed case studies could be performed, with the aim of evaluating their scenic possibilities and their potential to develop activities associated with Geodiversity and Geotourism.

**Keywords:** Touristic resources · Geodiversity · Geotourism · Volcanic landscapes · Patagonia tablelands · Santa cruz · Patagonia · Argentina

## 1 Introduction

Endogenous processes, such as orogenesis and volcanism, are responsible for the genesis of positive topography in most landscapes. Starting with these initial primary landforms, weathering and the action of the different erosion agents excavate, carve and model different landforms, generating a varied assortment of landform features and geomorphological landscapes.


M. M. Collantes et al. (Eds.): *Advances in Geomorphology and Quaternary Studies in Argentina*, SPRINGEREARTH, pp. 233–261, 2020.
https://doi.org/10.1007/978-3-030-22621-3_12

As a result of these secondary processes, the Patagonian landscape exposes erosion remnants of volcanic landforms, particularly "volcanic necks", "buttes" and "table mountains", which are clearly distinguishable in the Patagonian lowlands, as sharp, abrupt landforms, easily identifiable above the surroundings. Together with these landforms, aligned eruptive centers, either isolated or grouped, also provide substantial Geodiversity to the landscape, which is dominantly composed of extremely flat, endless tablelands, most of them composed of Late Cenozoic basaltic lavas and piedmont deposits (Feruglio 1950).

The study area is located in the southernmost portion of Argentine Patagonia, in the province of Santa Cruz. This remote and secluded region extends between latitudes 46° and 52° S, with a superficial extension of over 249,000 $km^2$ (Fig. 1) and a population density of just slightly above 1.0 inhabitants per $km^2$ (INDEC 2010), perhaps one of the lowest for temperate areas in the entire world. Its physiognomy comprises two large landscape units:

(1) the Cordilleran mountain environment, with prevailing cooler and wetter climate and rugged mountain landscape in the western portion, partially ice covered by remnants of the Late Pleistocene mountain ice-sheets, and
(2) the tableland environments (locally known as "mesetas"), which extend from the Andean piedmont towards the Atlantic Ocean coast in its eastern side, with semiarid climate and ample and bounteous surfaces which reach the Atlantic coastline.

In the tableland environments an almost horizontal topography prevails, only occasionally interrupted by large fluvial valleys and wadis. These grassland surfaces may extend for hundreds of kilometers. These topographic conditions coexist with a homogeneous plant cover, composed of shrubby steppes (Fig. 2). In these environments, the presence of natural elements, and occasionally, of anthropogenic origin, interrupt the landscape dullness and they appear as positive features for the observer, providing variability and diversity to the visual constituents.

Three major geological provinces may be identified from a geological point of view (Ramos 1999a): (1) the Southern Patagonian Andes along the western side, (2) the Patagonian tablelands that occupy a small latitudinal stripe in the northern part of the studied sector and most of the central southern area and, (3) the Deseado Massif, a large, outsized ancient cratonic area. This unit is part of the Patagonian basement, with an uneven, irregular landscape, with low hills and ample ranges of very small local relief (100–150 m), interrupted by sedimentary and basaltic high plains of varying altitude. This unit occupies most of the central and northern portions of the province and it presents abundant landforms as those included in this study (Fig. 1).

The objective of this paper is to analyze the spatial distribution of the landforms of volcanic origin that represent the main natural components exposed as paramount positive landforms above the flat tableland landscape of the province of Santa Cruz. These landforms are such as necks (lava and pyroclastic infilling of volcanic chimneys), buttes (mountains with small and flat lying remnants of lava flows covering their tops) and volcanic vents or eruption centers. Geomorphological and morphometric analysis were performed to identify those landforms with prospective higher potential for Geodiversity landmarks, such as Geosites and Geotourism resources.

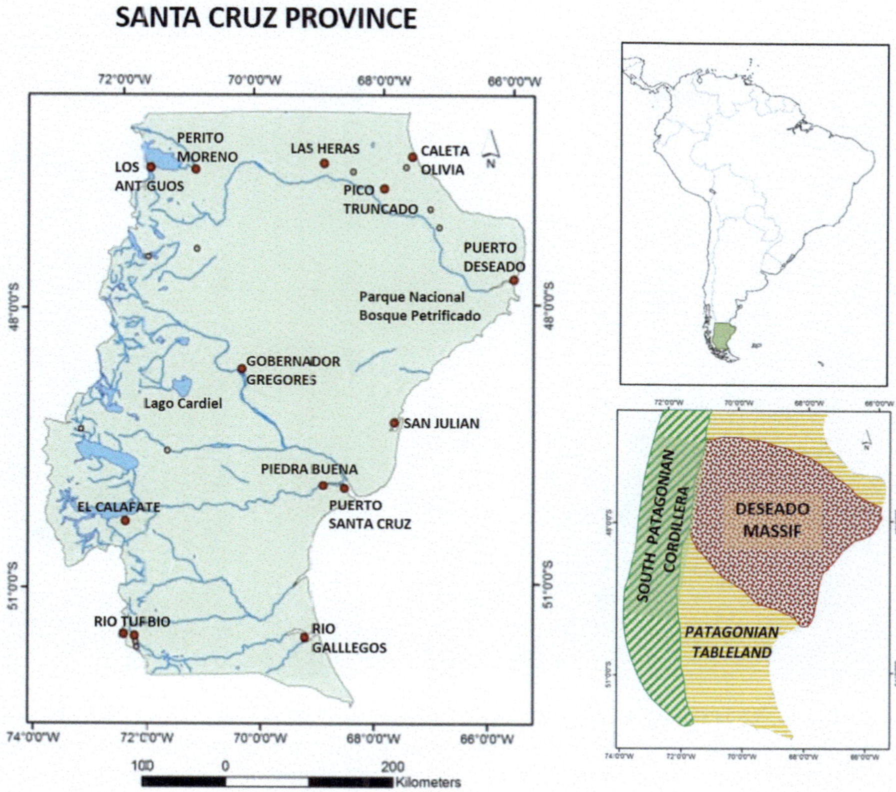


**Fig. 1.** Location of the study area and of the main cities and towns mentioned in the text. Right, below: schematic map of the Geological Provinces of the region.

## 2 Methodology

The methodology used in this study is based upon the detailed visual interpretation of high spatial resolution images available at the Google Earth Pro web site, following systematic itineraries and circuits in east-west and north-south directions. This visual interpretation of the available imagery was checked by subsequent field observations.

The inventory included volcanic necks, buttes and isolated volcanic centers, whose remnants are clearly visible in the surroundings due to its local relief and unique geomorphological features. Once these features were identified, they were recorded in files with specific ID information for each type of landform. Counting upon the available database within the available software tools, the following data were obtained: geographical location, elevation, and local relief compared with the surrounding areas. The local relief parameter was calculated as the average of up to four measurements, which were obtained along different orientations around the volcanic feature. Additionally, their location with respect to main roads, dirt roads and trails and their position in connection to the nearest towns were identified. Simultaneously, screen views were

**Fig. 2.** Landscape of the Patagonian "meseta". The extensive homogeneous surfaces are broken only by the occurrence of a few, higher topographic features. The photograph shows several guanacos (*Lama guanicoe*), an autochthonous, wild camelid, typical of southern South America. Photograph by E. Mazzoni.

captured which allowed to build a digital base of the various outcrops found, together with oblique land photographs shot during fieldwork. Moreover, using available geological information, and geological and topographic sheets at different scales, the age of the befalling volcanic formations was recorded.

The information obtained was organized in a GIS framework and statistically analyzed. For the Elevation and Local Relief variables, basic descriptive techniques were applied, obtaining their maximum, minimum, mean and standard deviation values, which have been presented in graphs and tables. The spatial information (absolute and relative position of the studied volcanic outcrops) was processed by means of techniques available in the Spatial Analysis module of the ArcGis 10.1 software. The Euclidian Distance command was applied, which made available to locate in the maps those landforms of interest in relation to cities and towns, as well as roads and trails.

This set of information permitted to make a preliminary selection of the landforms and geographical areas more suitable to develop future detailed studies to evaluate their scenic qualities and their Geotourism potential, including trekking, mountain climbing, bird watching, particularly condors and eagles, landscape photography, etc.

## 3 Volcanic Landforms in the Province of Santa Cruz

The volcanic landforms in Patagonia are distributed along different active zones of the Andean arc and in the extra-Cordilleran area (Clapperton 1993 and Kay 2002).

There are few volcanic cones and structures along the Cordilleran axis in Santa Cruz province. Six strato-volcanoes of small dimensions are found quite disseminated in the western portion of the Andean ranges, included in the so-called "Austral Volcanic Zone" (AVZ), which is coincident with the segment of the Antarctic oceanic plate that extends under the South American continental shelf (Stern 1989). Geographically, these volcanoes are situated along the international boundary or within Chilean territory. Some of them, as the Lautaro and Viedma volcanoes, are placed inside the Patagonian Ice Field. The Lautaro volcano is the most active one, with historical records that confirm its activity (Martinic 1988).

Contrarily, in extra-Andean Patagonia, the volcanic landforms are quite abundant, concentrated in a wide belt extending up to 300 km E of the Andean ranges. Its origin is associated to multiple volcanic episodes related to extensional processes developed in the back-arc and intra-plate environments (Ardolino et al. 1999; Ramos 1999b and Kay 2002). These processes originated systems of fissures, cracks and veins from which alkaline basalts erupted during different volcanic episodes, being the more important those eruptions which took place during the Eocene and the Late Miocene-Pleistocene, from the point of view of their volume and regional extent (Ramos 2002). Morphologically, they are characterized by the presence of basaltic flows forming plateaus as the more representative volcanic activity. However, monogenetic cinder cones may also be recognized, frequently superimposed to the larger volcanic cones and even to maars (Lapido and Pereyra 1999; Corbella 2002; Mazzoni and Rabassa 2010 and 2018a).

Some of the more important sequences have been dated by means of radiometric techniques. The results obtained show that the ages of the stratigraphically older basaltic flows exceed 80 Ma and that the youngest superficial flows are younger than 1.0 Ma. These dates also suggest that four volcanic activity cycles took place, the oldest in the Late Cretaceous (81–77 Ma), followed by others in the Early Tertiary (57–36 Ma), Middle Tertiary (35–34 Ma, 29–25 Ma) and Late Tertiary to Holocene (16 Ma to 10,000 years ago) (Ramos et al. 1982; Aguilera et al. 2018). The regional distribution shows a progressive migration of the volcanic activity towards the south, exposing the youngest outcrops in the Pali Aike Volcanic Field (PAVF), located in southernmost Santa Cruz and extending across the border in Chile.

In Fig. 3, the localization of the different volcanic landscapes of Santa Cruz is taken into consideration. These landscapes are grouped following the age of the lava flows (Mazzoni and Rabassa 2010, 2018a). The oldest ones, corresponding to the Late Mesozoic and beginnings of the Tertiary, are scarce. They are mostly located in the central zone of the Deseado Massif and form small tablelands, highly dissected by various erosion processes, such as fluvial, wind and mass movement processes, with locally abundant debris cover.

The Eocene basaltic cycle is widely represented in the central-northern portion of the Deseado Massif (Cerro del Doce Basalt, Panza 1982) and in the area comprised between the Posadas and Pueyrredón lakes (Posadas Basalt, Riggi 1957), extending also in the Lago Cardiel area and the basal sector of the Meseta Lago Buenos Aires (Ramos 1982). In general, they integrate highly dissected, extensive lava flows and volcanic necks. Concerning the radiometric ages (K-Ar), these basalts were erupted during the 60–40 Ma period, with a maximum volcanic activity between 45 and 48 Ma (Middle Eocene, Ramos et al. 1982).

After this magmatic cycle, other similar events took place during the Oligocene and the Early to Middle Miocene, but with smaller extent than the previous volcanic period. Their radiometric ages are clustered between 29–25 Ma and also, between 13–11 Ma, respectively (Ramos et al. 1982; Nullo et al. 1993 and Gorring et al. 1997). Outcrops of the Oligocene volcanic cycle are mostly found in the Deseado Massif, named as the Alma Gaucha Basalt (Panza 1982, 1995), which includes a set of volcanic rocks that occur mainly as "mesetas" or their relicts, necks and dykes. Due to its higher resistance to the erosion agents, these outcrops are clearly noted in the landscape, forming some of the most important landforms at the local level, including the Cerro Cojudo Blanco (1335 m a.s.l.) and the "meseta" of the same name, which are the highest points all over the Deseado Massif (Panza and Franchi 2002).

The main outcrops of the Early to Middle Miocene cycle are located in the central portion of the province. Among the more representative examples, the Gregores Basalt (Marín 1982), placed northeast of the town of this name, the El Pedrero Basalt (Panza 1998), whose lava flows comprise a high tableland 60 km NNW from Gobernador Moyano, and the Cóndor Cliff Basalt (Cobos and Panza 2001), in the northern margin of the Río Santa Cruz valley, should be mentioned.

Starting in the Late Miocene, the collision of the South American Plate with the dorsal separating the Nazca and Antarctica plates took place. This process is associated to the eruption of huge volumes of plateau lavas, estimated in 1,000 $km^3$ (Ramos 2002). The set of volcanic outcrops found in the northwestern part of the province pertain to this volcanic cycle. These rocks are generally named as the Strobel Basalt (Ramos 1978, 1982), which include the "mesetas" located in the region of the Cardiel, Strobel, Belgrano and Posadas lakes, and the high "mesetas" of de la Muerte and Lago Buenos Aires, just naming those most archetypal. Likewise, in this volcanic period the lava flows that cover the high plains located north, south and southwest of Gobernador Gregores and east of the Río Chico valley are included.

Following this event, another one may be mentioned, which has radiometric ages between 4.0 and 5.5 Ma, approximately (Early Pliocene cycle), which would have taken place after the landscape was modified by the first glaciation recorded for this area, sometime between 7.0 and 5.0 Ma, approximately (Mercer and Sutter 1982 and Rabassa 2008). The basalts found in the western portion of the province correspond to this event, in a southernmost position compared to the other basalts, forming ample "mesetas" from south of the Lago Cardiel up to the north of the town of Río Turbio. These rocks are found superimposed to Miocene basalts in the Meseta Lago Buenos Aires and in the tablelands close to the town of Gobernador Gregores. Northeast of this city, the Cerro Tejedor (848 m a.s.l.) should be noted as one of the more characteristic basaltic volcanoes of this period.

The following eruption cycle has been temporarily placed between the Late Pliocene – Early Pleistocene, with radiometric ages between 3.5 and 1.0 Ma, approximately. The outcrops of this cycle are widely distributed in the province, including extensive lava plains located within the Deseado Massif, in the higher portions of the Chalía - Santa Cruz watershed, south of the valley of the latter in Camusú Aike and the southern portion of the province, forming the Pali Aike Volcanic Field. These rocks are also found in the Meseta Lago Buenos Aires, where they have been named as El Sello Basalt (Lapido 1979), where the frequent eruption vents integrate the highest sites of the local relief.

Holocene volcanism is found only partially composing the Pali Aike Volcanic Field.

The widely dominant landforms in the province are the basaltic plateaus, which cover around 24,500 $km^2$, that is, 14% of the total provincial surface (Mazzoni and Rabassa 2010, 2018a). The geomorphological evolution of these landscapes allowed the formation of volcanic necks, as relict forms of the disintegration of the eruption centers, and buttes, the result of the dissection and degradation of the basaltic "mesetas", frequently accompanied by belts of coarse, slumped blocks (Mazzoni and Rabassa 2018b and Mazzoni and Diez 2017).

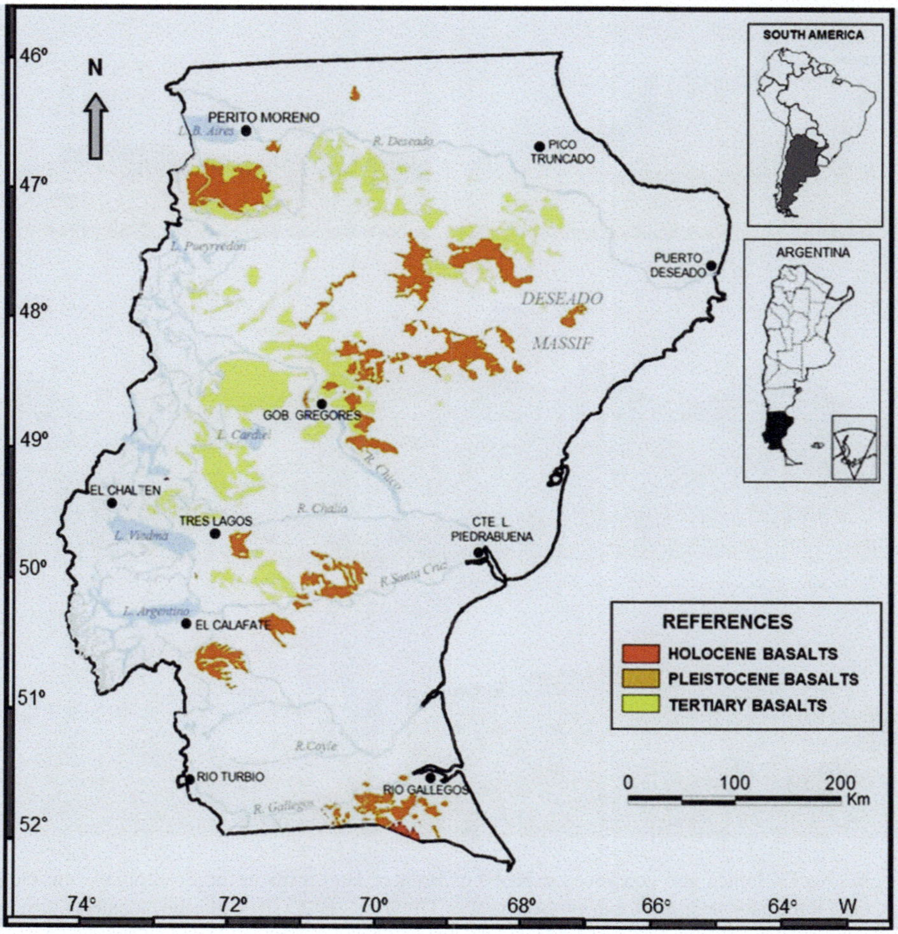


**Fig. 3.** Distribution of the basaltic lava flows in Santa Cruz, grouped following their radiometric absolute age. Source: Mazzoni and Rabassa, 2010, 2018a.

The landforms analyzed in this paper, which have not been represented in Fig. 3 due to scale reasons, are the following:

(a) Volcanic necks (Ollier 1988): landforms composed of magma which became solidified inside the volcanic chimneys, which was not sent off to exogenous environments due to the halt of eruption conditions and which was later on exposed due to denudation of the volcanic edifice (Fig. 4a, b);
(b) Buttes or small remnants of volcanic mesas (Ollier 1988): relicts of the erosion of volcanic flows. These features are topographic heights crowned by the lava mantle with a wide and flat summit, which visually resembles a table or shelf (Fig. 4c);
(c) Volcanoes: these are structures resulting of the eruption processes and the accumulation of lavas and pyroclastic materials around the vent. Considering the kind of volcanic environment of the study area, the volcanoes are mostly of the

**Fig. 4.** Necks, buttes and volcanoes exposed as isolated landforms in the Patagonian "meseta". A: Cerro Redondo, near the town of Gobernador Gregores; B: Cerro Colmillo, located north of Lago Pueyrredón; C: Sequence of buttes in the Parque Nacional de los Bosques Petrificados, in the Deseado Massif; D and E: Volcanoes found in the Pali Aike Volcanic Field. Photographs: E. Mazzoni.

monogenetic type, composed of pyroclastic sediments and small fields of associated basaltic lavas. Most of them are found superimposed to the larger basaltic "mesetas". However, in the Pali Aike Volcanic Field and some other sectors of the large provincial space, they are exposed as independent landforms of the landscape, following in some cases structural lineaments (Fig. 4d, e).

## 4 Morphometry and Spatial Distribution of Necks, Buttes and Volcanoes in the Province of Santa Cruz

Within the provincial space, 106 volcanic landforms of interest have been identified. Of this number, 52 are necks, 28 are buttes and 26 are volcanoes and other vents that appear as isolated features of the landscape. Thus, the abundant eruption vents that overlie the basaltic plateaus have not been included in the inventory, as for instance those occurring on top of the Meseta Lago Buenos Aires (47° S–71° W). Their spatial distribution may be observed in Fig. 5. The erosional remnants are mostly located in the central-northern portion of the Deseado Massif, associated with the location of the oldest lava flows, those erupted during the Mesozoic and Early Tertiary, usually of Eocene age.

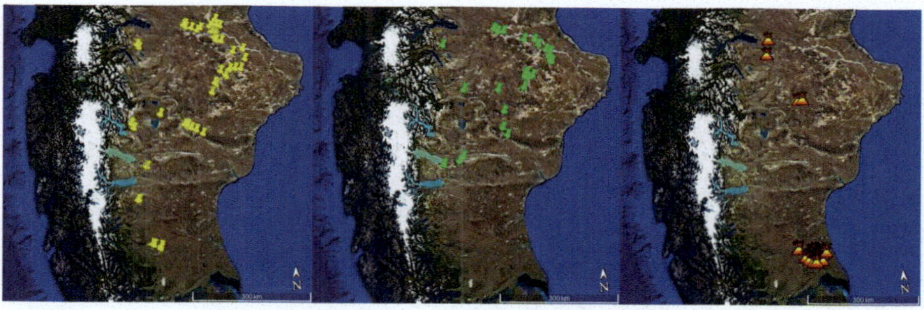

**Fig. 5.** Distribution of necks (left), buttes (center) and volcanoes (right) in the province of Santa Cruz, observed on the satellite image mosaic of Google Earth.

The volcanoes, contrarily, are concentrated in the Pali Aike Volcanic Field, where they are distributed following several structural lineaments with predominant direction SE – NW (Fig. 6) and they occur, in isolation, in the western sector of the Deseado Massif. These are small vents, mostly scoria and cinder cones, with just one single lava flow, whose genesis has been referred to the Quaternary (Corbella 2002, Mejía et al. 2004). Many of them show horse-shoe craters, with portions of the cone eroded and displaced by the outpouring lava flows. A good example is the Diablo Negro volcano, located in the vicinity of the city of Río Gallegos, where the nature of its most recent, Holocene lava flow is observed (Fig. 7).

**Fig. 6.** Volcanoes aligned in the Pali Aike Volcanic Field. Photography: E. Mazzoni.

**Fig. 7.** Located in the volcanic lineament known as "Diablo Negro", in the Pali Aike Volcanic Field, these eruption vents show different morphological characteristics. A magnificent horseshoe crater is observed in the central portion of the image. The image shows also the texture of the Holocene basaltic flows. Image: Google Earth.

The absolute elevation of the set of volcanic landforms varies from W to E following the general topography of the terrain, with a wide range of altitude between 1,381 and 125 m a.s.l. The maximum value recorded is an extreme value, compared with the surrounding features. It is a neck, identified as N-10 in the inventory (49° 02′23″S–72° 06′48″ W), whose location is very close to the Southern Andean ranges.

The comparison of the elevations between the three groups of studied landforms (Table 1, Fig. 8) shows that the buttes and necks occupy the highest topographic sites, whereas the volcanoes have lower altitudes. Necks show the largest dispersal of the data, with a mean of 551 m a.s.l. and standard deviation of 262 m, with the most frequent values in the interval of 300–400 m a.s.l. Buttes, contrarily, are more homogeneous concerning these variables. They are placed in the range comprised between 325 and 947 m a.s.l., with an average value of 612 m and standard deviation of 189 m. High frequency values appear in the 400–500, 700–800 and 800–900 m a.s. l. intervals.

Volcanoes are distributed in two groups. The first one concentrates 81% of the vents, and it has elevations between 100 and 300 m a.s.l. The second group is located between elevations of 600 and 1000 m a.s.l. The mean general elevation is 313 m a.s.l. with a standard deviation of 264 m. However, these values seem to be non-representative, because of the differences that each group presents. The first group has a mean elevation of 190 m with a standard deviation of 42 m, whereas the second group has a mean of 831 m, with a standard deviation of 117 m. This altitude distribution is exclusively related to the geographical location of the vents: the first group corresponds to the Pali Aike Volcanic Field, where base level is around 120 m, whereas the second group is located in more northwestern positions, above the Deseado Massif topographic level, with elevations higher than 600 m a.s.l.

**Table 1.** Basic statistical parameters of necks, buttes and volcanoes

| | Elevation (m a.s.l.) | | | Local Relief | | |
|---|---|---|---|---|---|---|
| | NECKS | BUTTES | VOLCANOES | NECKS | BUTTES | VOLCANOES |
| Maximum | 1381 | 947 | 938 | 558 | 439 | 123 |
| Minimum | 146 | 325 | 125 | 30 | 81 | 15 |
| Mean | 551 | 612 | 313 | 151 | 190 | 72 |
| Standard Deviation | 262 | 189 | 264 | 99 | 98 | 33 |

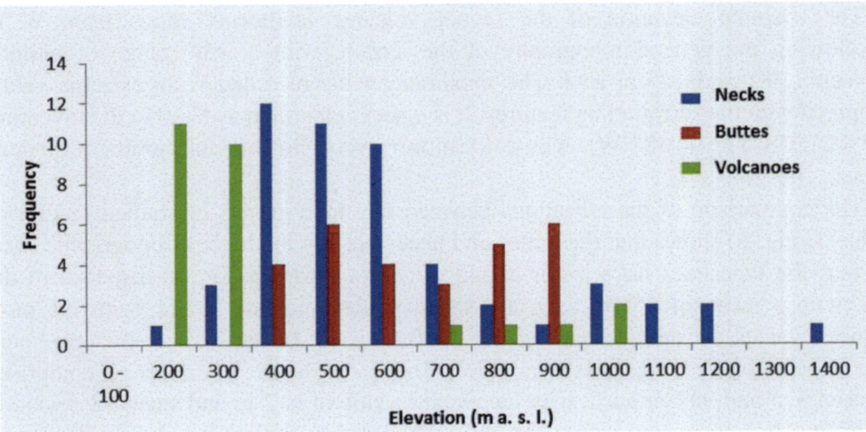


**Fig. 8.** Elevation of necks, buttes and volcanoes.

The local relief or relative elevation of these landforms is one of the more significant morphometric parameters in relation to the objectives of this work, since this parameter intervenes both in the visual perception as in the type and difficulty of the related activities that may take place on them. In Fig. 9, the frequency distribution for this variable is depicted. The local relief of this set of landforms varies between 15 and 558 m. Buttes are, considering the mean values, higher elevation landforms, with 190 m and a standard deviation of 98 m. Almost 50% of these landforms have a local relief higher than 150 m, 30% is located in the 100–150 m interval and the remaining 20% has between 50 and 100 m of altitude difference. The maximum height for these landforms is 439 m.

Necks present a larger dispersion, with a range that extends in between 30 and 558 m in elevation; however, 96% of them possess an altitude lower than 300 m. The mean value is 151 m, with a standard deviation of 99 m. The most frequent values, with 37% of the cases, are found in the 100–150 m interval, whereas 33% have elevations above this modal interval. This type of landforms has the maximum value recorded (Cerro Kach Aike, 49° 9′ S–72° 2′ W, Fig. 10).

Volcanoes appear as the landforms with smaller local relief. They are distributed in an altitudinal range between 15 and 123 m, with a mean relative height of 72 m and a standard deviation of 33 m. The elevation interval with highest frequency is 50–100 m, with 54% of the observed cases. Those volcanoes with smaller relative height, which are almost undistinguishable in the "meseta" landscape, are composed of very recent, unweathered lava flows, of darker colors and tones, which still show their original volcanic textures. This condition enhances their geotouristic interest and importance, since younger lava flows preserve better the original volcanic features. One of these volcanic features, the "Laguna Azul" volcano, also possesses a water body in the inner part of the cone, which gives the name to the volcano. This feature increases even more the scenic value of this landscape (Fig. 11).

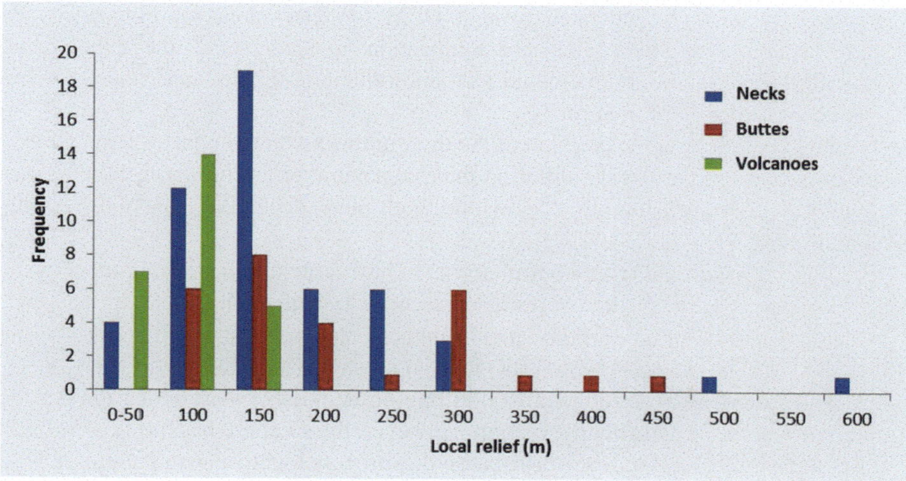


**Fig. 9.** Local relief of necks, buttes and volcanoes.

**Fig. 10.** Cerro Kach Aike, located in the glaciolacustrine plain between the San Martín and Tar lakes; this is the landform of highest local relief in this region, with 558 m. Below: Oblique view seen from the southeast. Photograph: J. Rabassa; Satellite image: Google Earth.

**Fig. 11.** Aerial views of the Laguna Azul volcano. The water body, whose depth reaches 60 m (Zolitschka et al. 2006), is placed within three nested craters. The upper photograph, oriented towards the north, shows the Holocene basaltic flow that still exposes its original volcanic features (Corbella 2002). The lower photography allows the observation of the eruption vents of the surroundings, located towards the SW. Photographs: E. Mazzoni.

The spatial distribution of the studied volcanic landforms was analyzed according to their relative localization with respect to the towns and villages of the provincial territory, as well as to the main roads and trails. In Fig. 12, the provincial towns and the location of necks, buttes and volcanoes are shown according to distance range with 20 km intervals. The results obtained permitted the observation that the cities of Las Heras, Gobernador Gregores and Río Gallegos are those with a larger number of landforms of geotouristic value within a 100 km radius. In the discussion of the values quoted below, it should be taken into consideration that the distances between towns and villages of the Santa Cruz Province are enormous, unconceivable in the European continent, for instance. This province has a total surface extent close to that of Italy or Great Britain, but instead with a population density of roughly less than one inhabitant per square kilometer. Moreover, there are areas of the provincial territory as large as some smaller European countries where there are no towns at all, totally uninhabited except by the rural population working in huge ranches devoted to extensive husbandry and farming, locally known as "estancias", in gold and silver mines, or in oil and gas wells and camps.

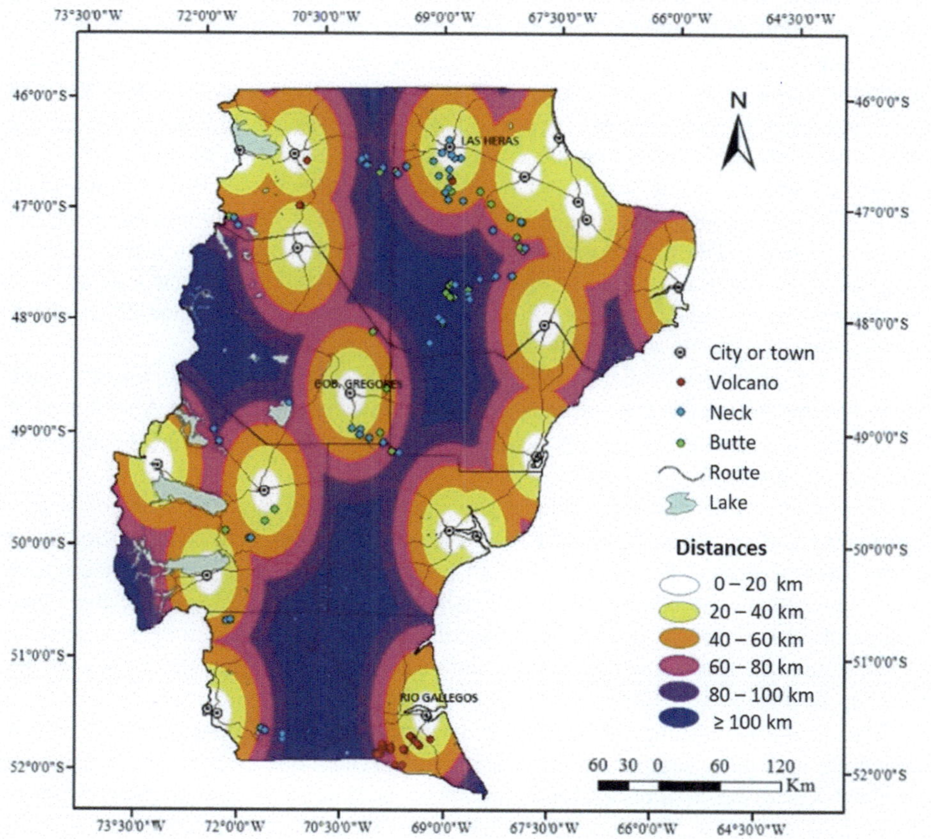


Proyection: Transverse Mercator; Datum: WGS 84; Central Meridian: 69 W

**Fig. 12.** Location of the necks, buttes and volcanoes with respect to the main cities and towns of Santa Cruz province.

The city of Las Heras (46° 33′ S–68° 55′ W) presents the largest abundance of erosion remnants inventoried (such as necks and buttes) within its surroundings. Six pertinent, relevant features are found within a distance smaller than 20 km, 17 of them are so at less than 60 km and up to 28 are situated within a 100 km range. Figure 13 allows the observation of this location in detail. It is a town situated in the northern provincial region, whose economy is based upon mining and oil and gas exploitation, presently with almost null tourism offer. However, this region is traversed by a paved provincial road (RP 43) communicating the coastal zone with the Cordilleran environment. This situation could propose an alternative touristic itinerary based upon these geotouristic resources. This zone possesses also abundant and valuable paleontological resources (Suárez et al. 2014 and Casal et al. 2015), thus allowing the convergence of all these resources and attributes into a regional itinerary.

Gobernador Gregores is a nice, quite small city that provides services to the rural surroundings. It is located in the northern margin of the Río Chico, an allochthonous stream that has plentiful riverbed vegetation and implanted grass, which give green tones to the regional semiarid environment. Within this valley, six necks and two small buttes are found, erosional remnants of extensive basaltic "mesetas" found in the area. These landforms expose a great geodiversity of lithological types and geological structures. Some examples are shown in Figs. 14 and 15, such as the Cerro Ventana (i.e., "mountain of the window", that is a particularly valued attraction for the local inhabitants to a point that its profile is included in the town coat of arms. This neck exposes columnar jointing basalts of Late Miocene age (the Strobel Basalt; Panza et al. 2002, 2005) and a large opening, a rocky "window", near its summit and being the origin of its name. These landforms are located between 40 and 80 km, approximately, away from the town in a SE direction. To these attractive features, and at equivalent distances, a neck named as Cerro Puntudo (49°09′ S–71°05′ W) should be added, oriented towards the W, and a butte occurs in the opposite direction.

From the city of Río Gallegos, the capital of the Santa Cruz Province, located in the SE end of the provincial territory, most of the volcanic cones of the Pali Aike Volcanic Field may be easily observed. The closest of these volcanic cones are located at a distance of only 30 km, in a SW direction.

Nearby other towns or small villages, such as Perito Moreno, Bajo Caracoles, Tres Lagos, El Calafate and Río Turbio, a few volcanic landforms were mapped. Southeast of the town of Perito Moreno, less than 15 km away from it, one of the few eruptive vents which does not belong to the Pali Aike Volcanic Field, is located: the Cerro Volcán (46°39′ S–70°47′ W), whose lavas, of known Pleistocene age (Ton That et al. 1999), extended in a 6 km radius from the central vent, reaching the Valley of the Deseado River.

The accessibility to these landforms is determined by the existence of roads, which sometimes are totally absent, and/or the available services, due to the colossal distances involved. The distance of necks, buttes and other volcanic features to the main routes is represented in Fig. 16. As it may be noted, most of them are in a radius of less than 20 km from the communication ways, being the provincial routes 12, 27 and 43 the nearest ones to the majority of these landforms.

Route 12 crosses the Deseado Massif with an approximate direction NE – SW, linking the towns of Pico Truncado, to the north, with Gobernador Gregores, in the

**Fig. 13.** Detailed view of the location of erosion remnants near the town of Las Heras and the Provincial Route 43. The lower image allows the observation of the high density of oil and gas production sites in the region, seen as whitish dots. The photographs expose some of the necks and buttes seen from this town. The dominant steppe conditions are also noted. Photographs: E. Mazzoni; Satellite image: Google Earth.

**Fig. 14.** Oblique views of a group of erosion remnants located near the town of Gobernador Gregores and the Provincial Route 27. Below, the mountains known as Ventana (foreground), Cuadrado (left) and Redondo (right) are seen. Images: Google Earth.

center of the province. Along its middle section, numerous landforms of interest are situated, some of them easily identifiable due to their nearness to the road as, for instance, an unnamed volcano (V-26, 48° 12′ S–69° 43′ W; Fig. 17) and various necks and buttes on both sides of the route.

The main difficulty of this itinerary is that it is a dirt road, poorly managed, not frequently graded, with a total absence of service centers between the two towns, separated by a large distance of 380 km. Moreover, many of the ranches of this area were abandoned by their owners after the huge cinder, eruptive explosions of the Hudson Volcano in 1991–1995, which drastically increased their isolation.

On Provincial Route 27, in the section between the towns of Comandante Luis Piedra Buena and Gobernador Gregores, and actually being closer to the latter, a

**Fig. 15.** Terrain photographs of the necks known as Cerro Ventana, Cerro Cuadrado and Cerro Redondo. In the upper right photograph a detail of the summit of the first one is shown, with its well-known "ventana" (i.e., Spanish for "window"), modeled by erosion agents and frequent columnar jointing structures. Photographs: E. Mazzoni.

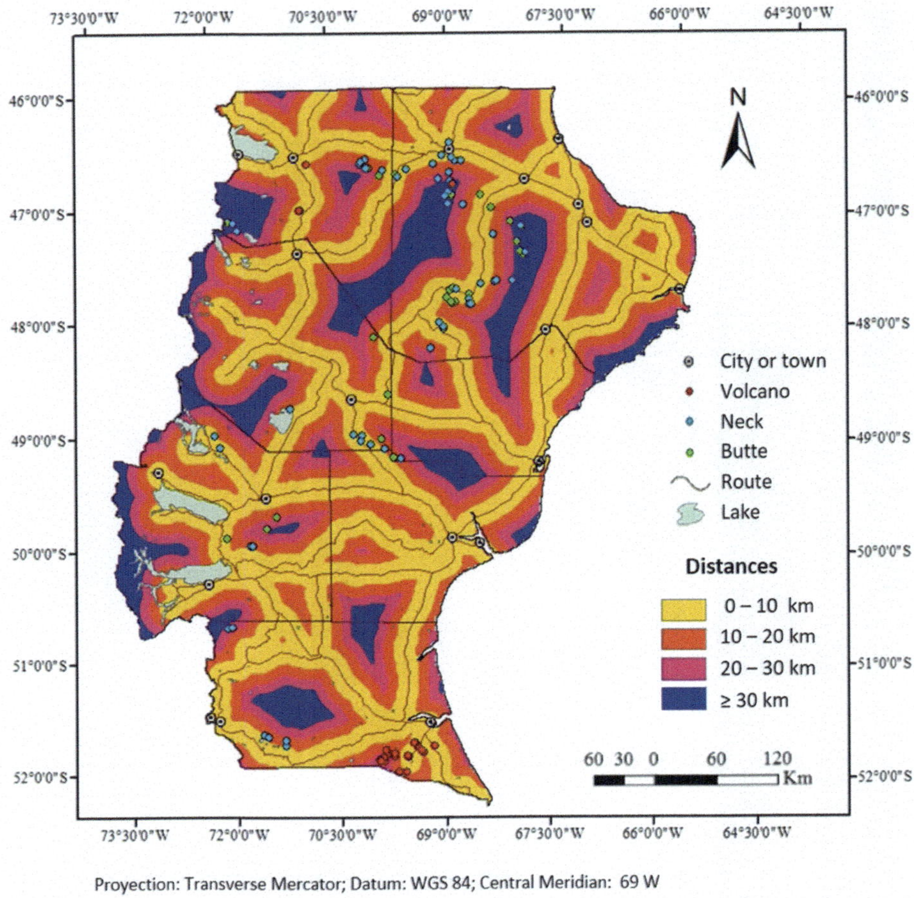


**Fig. 16.** Location of necks, buttes and volcanoes with respect to the main national and provincial highways.

sequence of eight erosional remnants along the southern margin of the Río Chico are observed. These remnants include the aforementioned Cerro Ventana and the Cuadrado, Redondo, Puntudo, Pistola and Ventura hills, which offer beautiful views along this road (Figs. 14 and 15). Basaltic columnar jointing is a frequent, paramount feature in these mountains.

Provincial Route 43 communicates the towns at the northern sector of the province, such as Pico Truncado, Las Heras and Perito Moreno. Nearby the road, a large group of volcanic landforms may be noted, mostly necks and buttes (Fig. 13). Some of them are easily accessible from the town of Las Heras, but most of them are located in the southern margin of the Río Deseado valley, where access is scarce and very difficult, since only secondary roads or trails may be used, or even trekking is the only possibility.

**Fig. 17.** Eruption vent located close to the Provincial Route 12. Photography: E. Mazzoni. Image: Google Earth.

Along other roads of the province, the amount of inventoried landforms is smaller; notwithstanding, due to their geographical location some of them are of high significance, as for instance some necks placed near National Route 40 in different sections.

Route 40, with a total length of more than 5,000 km, traverses Argentina from N to S, following the Andean eastern front, from the border with Bolivia to the Magellan Straits. Due to the variety of landscapes that this route crosses, it is permanently visited by tourists from all over the world, who follow it as individuals or in groups, either in cars, trucks, 4x4 vehicles, motorbikes and even bicycles (Ferrari et al. 2015; Mazzoni et al. 2016 and Mazzoni 2017). Amongst the landforms that may be observed from this road, the following should be noted:

- The Philippi (385 m a.s.l.), Domeyko (363 m a.s.l.), Gay (180 m a.s.l.) and Cuadrado (231 m a.s.l.) mountains (Fig. 18a), in the southern sector, located between Río Gallegos and the town named as "28 de Noviembre". These hills are huge necks of an average of more than 500 m in diameter, which occur breaking the smooth glacial topography of the surroundings (Mazzoni 2017), formed by lavas erupted during the Late Miocene, with ages close to 8.0 Ma (Meglioli 1992, Panza and Franchi 2002, D′Orazio et al. 2004, Corbella et al. 2014).
- Cerro Puntudo (Fig. 18b), a mountain in the northern margin of Lago Cardiel, not far from the town of Gobernador Gregores.
- Two volcanoes identified in the inventory as V-22 and V-23 (Fig. 18c), located in the Estancia Casa de Piedra ("estancia", as it has been mentioned before, is the term used in Argentina for a large ranch, an extensive rural property), at the eastern margin of the Meseta Lago Buenos Aires.
- Cerro Volcán (Fig. 18d), an already mentioned volcanic peak, occurring in the neighborhood of the town of Perito Moreno.

Along National Road 3 (RN3), that crosses from N to S the eastern portion of the province of Santa Cruz, the eruption vents of the Pali Aike Volcanic Field are clearly exposed in the southernmost portion, between the city of Río Gallegos and the international border with the Republic of Chile. Numerous volcanoes appear aligned, following geological structures, and some of them are located in distances less than 1 km in between them. Near the international border pass, National Route 3 dodges between the most recent lava flows. The complex of Holocene lava flows, known as "Diablo Negro", should be noted with several lined-up craters that expose a great variety of volcanic geomorphology features (see Fig. 7). The most important cone rises almost 90 m above the regional surface, reaching an altitude of 261 m a.s.l. Moreover, the "Laguna Azul" volcano is found in this sector, as it was previously mentioned, which is the best known of all volcanic craters and therefore more frequently visited by tourists going to and from Río Gallegos and local dwellers in holidays or recreation days, making use of the advantage of access through a good, paved road.

Towards the northern portion of the province, in the Deseado Massif, National Route 3 offers access to a protected area, the "Parque Nacional Bosque Petrificado" (National Park of the Petrified Forest), where an arid landscape deeply eroded by fluvial and eolian erosion processes is exposed. The fluvial erosion in arid environments has originated plentiful buttes and necks, among which the "Madre e Hija" (i.e. "Mother and Daughter") volcanoes dominate (47° 42′ S–68° 06′ W - Fig. 19). The name of the volcanoes refer to the existence of a main volcanic cone (the "Mother") with an adventitious cone (the "Daughter") on its flanks. The extreme aridity of the surroundings and the varied and contrasting coloring of the rocks, such as darker Cenozoic basalts and highly weathered, reddish sedimentary and pyroclastic Mesozoic layers, make this a highly attractive landscape. This landscape and the bearing of huge fossil trees and dinosaur bones, and the impressive, untouched local flora and fauna, add positively to a paramount impressive volcanic and arid land geomorphology.

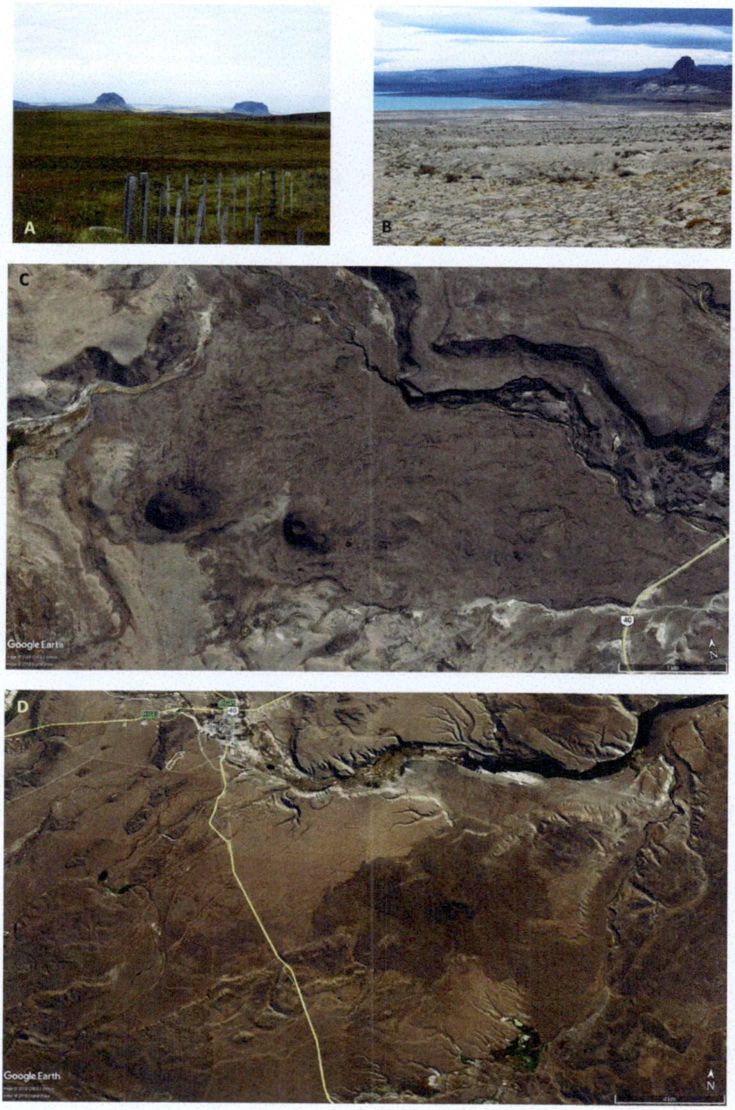


**Fig. 18.** Different volcanic outcrops located close to the National Route 40. A: Philippi and Domeyko necks in the southern portion of this road; B: Cerro Puntudo, in the center of the province (northern margin of Lago Cardiel). At the foreground, a field of coastal dunes is observed; C and D: Quaternary volcanoes located in the northern sector of Route 40 in the province of Santa Cruz. Figure 18C shows two eruption vents at Estancia Casa de Piedra, near the Meseta Lago Buenos Aires, whereas 18D corresponds to Cerro Volcán. In this last satellite image the town of Perito Moreno and the upper reach of the Río Deseado are also observed. The volcanic flow emerging from this vent reached the bottom of the Río Deseado valley. Photographs: E. Mazzoni. Images: Google Earth.

**Fig. 19.** View of the Cerro Madre e Hija (i.e., “Mother and Daughter”), within the “Parque Nacional Bosque Petrificado”, in the Deseado Massif. This is a huge, main volcanic cone (the “Mother”) with an associated, adventitious volcanic cone (the “Daughter”). Photograph: E. Mazzoni.

## 5 Conclusions

The province of Santa Cruz possesses a large number of volcanic landforms that expose high potential for Geotourism due to their topographic, geological and geomorphological attributes. The inventory performed in this paper allowed to build up a list of 106 geological/geomorphological landforms of high interest, which glowingly appear in the landscape due to their high topography and relative isolation from the surroundings, both factors that contribute to their easy visualization.

Most of the necks present in their upper, almost vertical portion, steep rocky walls with typical columnar jointing, as a product of the cooling of basaltic lavas after their eruption, whereas their bases are usually covered by debris of varied nature. These lithological conditions may be appropriate for rock climbing, as well as for nature photography and bird watching. Many of these birds, and particularly Andean condors (*Vultur gryphus*), and also eagles and falcons, use these outcrops as nesting places or observation sites (Mazzoni and Rabassa 2018c). Some necks and buttes have also special meaning as cultural resources, because they have been used by the aboriginal populations as burial sites or “chenques” or for rock art of varied types (Ortiz Troncoso 1973, Borrero and Charlín 2010).

As a follow up of the present work, several areas of interest have been identified, which have been defined both for the abundance of necks, buttes and/or volcanoes, as well as for their accessibility to roads, towns and services.

Firstly, the Río Chico valley should be noted because of the sequence of necks and buttes that appear along the Provincial Route 27, with the town of Gobernador Gregores as the main service center, because it provides a good offer of lodgment (hotels and cabins), restaurants, markets and service stations. It is herein proposed the informal denomination of "Valley of the Necks" to this portion of the route, becoming one of the sectors suggested for developing more detailed studies that may permit in the future their inclusion in touristic itineraries for this area.

Related to this same locality, the Cerro Puntudo, a huge neck with neat columnar jointing, is located at the northern margin of Lago Cardiel. This lake is one of the few water bodies found in the "meseta" environment. It is highly noted for its large dimensions (approximately 370 $km^2$) and greenish blue, turquoise color, the latter being highly contrasting with the darker tones of the basaltic flows that partially bound it and the brighter colors of the sedimentary strata, heavily eroded, that occur along its eastern margin. This area may be reached by car/truck from Route 40 and it is frequently visited by trout fishermen. There are ranches in the area that already offer limited touristic services. From the Cerro Puntudo impressive views may be obtained and it may be suitable for rock climbing, but with careful, previous evaluation of the quality of the exposed rocks.

Besides, it is proposed to use the access from Route 3 to the "Parque Nacional del Bosque Petrificado" to provide appropriate mural illustration and precise geological information to the tourists. Thus, the tourists and visitors may perceive the intensity and dynamics of the erosion processes in the region, and also understand the magnificent examples of necks and buttes exposed there. This would strongly complement the information associated to the presence and distribution of enormous fossil trees, embedded in Jurassic pyroclastic rocks.

It is also suggested that the Pali Aike Volcanic Field may be developed for touristic purposes with the design of interpretation trails that would allow the observation of the large Geodiversity of the region, mostly associated with the Pleistocene/Holocene volcanism. Presently, the "Laguna Azul" crater has been declared as "Provincial Geological Reserve" but it has not been provided with a management plan or educational posters. This volcanic field is one of the youngest ones in Argentina and it presents a unique morphology where aligned eruption vents, maars, lava plains and other complex landforms, such as nested volcanic cones or craters in maar structures, are combined (Corbella 2002). In this sense, it should be noted that there already exists a Chilean National Park covering the portion of this volcanic field in the Republic of Chile, which already actively promotes the tourism in the Chilean sector.

This paper proposes to carefully evaluate the geomorphological characteristics and the visual scenarios of the necks located nearby the town of Las Heras, with the aim of providing recreation alternatives to the inhabitants of the different towns of the oil and gas basin. This purpose should also apply for those tourists that drive through them towards the Atlantic Ocean coast or, inversely, to the towns of Perito Moreno and Los Antiguos in the Cordilleran piedmont environment. Many environmental problems are common in this region, being related to the present mining activities and the cattle and sheep raising and husbandry (Mazzoni and Vázquez 2009). It would be necessary to offer several activities based upon the proper knowledge of nature that would allow the understanding and valuation of the environment, as well as enriching the life quality of their inhabitants, providing jobs and promoting the preservation of the landscapes and Geodiversity.

Finally, the Cerro Kach Aike should be of special interest, not only for being the tallest of all studied necks, but also for its spectacular surroundings and close location to the Lago San Martín. Its isolated presence in the glaciolacustrine plain between the San Martín and Tar lakes, its long geomorphological history that has permitted its survival to perhaps more than twenty Pleistocene glaciations and its impressive local relief, exceeding 550 m, indicate that it is a site of highest exceptionality (for the use of this term, see Leopold 1969 and Rabassa 2018), visible from very long distances and almost unique in this portion of Patagonia. However, its access today presents many difficulties due to the poor state of conservation of the Provincial Route 31, which links the town of Tres Lagos with this dominant feature, located 71 km towards the NW.

The study of the scenic and geotouristic potential of the landforms and landscapes herein described, as well as the evaluation of alternatives for its rational use, may promote a greater diversity in the economic activities for the inhabitants of the province of Santa Cruz and contribute to the valuation and preservation of the natural environment.

Likewise, the offer of these landscape resources as a valid objective of non-conventional tourism, may favor the economic diversification of the little towns of the region. In this sense, those necks, buttes and volcanoes surveyed in this work might be visited by small special groups based upon specific interests, such as science, education, bird watching, etc. Besides, other practices of specific activities, such as rock climbing, trekking, landscape observation, nature photography and flora and fauna watching would find in this region excellent scenarios for their performing.

**Acknowledgments.** The authors are greatly indebted to the Universidad Nacional de la Patagonia Austral (UNPA), Unidad Académica Río Gallegos, and CADIC-CONICET, Ushuaia, Tierra del Fuego for their support. Funding for field work was provided by the cited institutions. Valuable comments and suggestions by an anonymous reviewer were extremely useful to improve the quality of this manuscript.

## References

Aguilera, E., Mazzoni, E., Rabassa, J.: Patagonian cenozoic magmatic activity. In: Mazzoni, E., Rabassa, J. (eds.) Volcanic Landscapes and Associated Wetlands of Lowland Patagonia. Springer Book Series on Latin American Studies, pp. 31–68 (2018)

Ardolino, A.M., Franchi, M., Remesal, C., Salani, F.: El volcanismo en la Patagonia Extraandina. In: Geología Argentina, Anales, vol. 29, pp. 579–612. SEGEMAR, Buenos Aires (1999)

Borrero, L., Charlin, J.: Arqueología de Pali Aike y Cabo Vírgenes: Santa Cruz, Argentina. Buenos Aires, Editorial Dunken, 152 p. (2010)

Casal, G., Allard, J., Foix, N.: Análisis estratigráfico y paleontológico del Cretácico Superior en la Cuenca del Golfo de San Jorge: nueva unidad litoestratigráfica para el Grupo Chubut. Rev. Asoc. Geológica Argent. **72**(1), 77–95 (2015)

Clapperton, C.: Quaternary Geology and Geomorphology of South America. Elsevier, Amsterdam (1993)

Cobos, J., Panza, J.L.: Informe preliminar de la hoja geológica 4752 – IV, Tres Lagos, Provincia de Santa Cruz. SEGEMAR, Buenos Aires (2001)

Corbella, H.: El campo volcanotectónico de Pali Aike. In: Haller, M. (ed.) Geología y Recursos Naturales de Santa Cruz. 15° Congreso Geológico Argentino, Relatorio, Buenos Aires, vol. 1, no. 18, pp. 258–301 (2002)

Corbella, H., Ercolano, B., Coronato, A., Tiberi, P.: Tectónica distensiva, magmatismo y glaciaciones en la cuenca superior y media del Río Gallegos, Patagonia austral. Actas XIX Congreso Geológico Argentino, Córdoba (2014)

D'Orazio, M., Innocentia, F., Manetti, P., Haller, M.J.: The Cenozoic back-arc magmatism of the southern extra-Andean Patagonia (44.5–52°S): a review of geochemical data and geodynamic interpretations. Rev. Asoc. Geológica Argent. **59**, 525–538 (2004)

Ferrari, S., Albrieu, C., Navarro, V., Mazzoni, E., Espinosa, S., Biott, J.M., Almendras, A.: Inventario y jerarquización de atractivos y recursos naturales/culturales en el tramo Austral de la RN40 (Santa Cruz). In: Torre, M.G. (ed.), XII Jornadas Nacionales de Investigación Acción en Turismo. EDUCO, Universidad Nacional del Comahue, 1–12 Neuquén (2015). http://170.210.83.98:8080/jspui/handle/123456789/678

Feruglio, E.: Descripción geológica de la Patagonia. Tomo 3. Yacimientos Petrolíferos Fiscales, Buenos Aires (1950)

Gorring, M., Mahlburg Kay, S., Zeitler, P., Ramos, V., Rubiolo, D., Fernández, M., Panza, J.L.: Neogene Patagonian plateau lavas: continental magmas associated with ridge collision at the Chile Triple Junction. Tectonics **16**(1), 1–17 (1997)

INDEC: Censo Nacional de Población y Vivienda 2010. Instituto Nacional de Estadística y Censo, Buenos Aires (2010)

Kay, S.M.: Magmatic sources, tectonic setting and causes of Tertiary to Recent Patagonian plateau magmatism (36° to 52° S Latitude). XV Congreso Geológico Argentino, Actas. El Calafate, Santa Cruz (2002)

Lapido, O.: Descripción Geológica de la Hoja 51 a Los Antiguos, Provincia de Santa Cruz. Servicio Geológico Nacional, unpublished report, Buenos Aires (1979)

Lapido, O., Pereyra, F.: Cuaternario en la Patagonia Extrandina. In: Caminos, A. (ed.) Geología Argentina, pp. 704–709. SEGEMAR, Buenos Aires (1999)

Leopold, L.B.: Landscape esthetics: How to quantify the scenics of a river valley. Natural History **36–45**, 454–467 (1969)

Marín, G.: Descripción geológica de la hoja 55c, Gobernador Gregores, Provincia de Santa Cruz. SEGEMAR, unpublished report, Buenos Aires (1982)

Martinic, M.:. Actividad volcánica histórica en la región de Magallanes. Rev. Geológica Chile **16**(2), 181–136 (1988)

Mazzoni, E.: Landscapes of Southern Patagonia, Argentina, their scenic characteristics and their importance for Geotourism. In: Rabassa, J. (ed.) Advances in Geomorphology and Quaternary Studies in Argentina. Proceedings of the Sixth Argentine Geomorphology and Quaternary Studies Congress. Springer Series on Earth System Sciences. Springer, pp. 140–179 (2017)

Mazzoni, E., Diez, P.: Potencialidad del geoturismo en Santa Cruz, Argentina. Sitios de interés geológico en proximidad de la RN 40, tramo Río Turbio – Lago Viedma. In: XX Congreso Geológico Argentino. San Miguel de Tucumán, Agosto 2017. Acta Simposio de Patrimonio Geológico, Desarrollo Sostenible y Estilos de Vida Saludables, pp. 35–38 (2017). http://congresogeologico.org.ar/gestor/wp-content/uploads/2017/08/TU-S8.pdf

Mazzoni, E., Ferrari, S., Navarro, V.: Paisajes y modalidades turísticas en el Sur de la Patagonia continental argentina. In: Vieira, A., Bento Gonçalves, A., Costa, F., Cunha, L., Lima Troleis, A. (eds.) A Geografia Física e a Gestão de Territórios Resilientes e Sustentáveis, pp. 743–754. Universidade do Minho. UMDGEO - Departamento de Geografia (2016)

Mazzoni, E., Rabassa, J.: Inventario y clasificación de escoriales basálticos de Patagonia en base a utilización de imágenes satelitales y SIG., Provincia de Santa Cruz, Argentina, Buenos Aires. Rev. Asoc. Geológica Argent. **66**(4), 608–618 (2010). http://www.scielo.org.ar/scielo.php?script=sci_arttext&pid=S0004-48222010000300019

Mazzoni, E., Rabassa, J.: Geomorphology of the Patagonian volcanic landscapes: provinces of Neuquén (Northern Patagonia) and Santa Cruz (Southern Patagonia). In: Mazzoni, E., Rabassa, J. (eds.) Volcanic Landscapes and Associated Wetlands of Lowland Patagonia. Springer Book Series on Latin American Studies, pp. 69–122 (2018a)

Mazzoni, E., Rabassa, J.: An evolutionary model of volcanic landscapes. In: Mazzoni, E., Rabassa, J. (eds.) Volcanic Landscapes and Associated Wetlands of Lowland Patagonia. Springer Book Series on Latin American Studies, pp. 155–176 (2018b)

Mazzoni, E., Rabassa, J.: The environmental value of volcanic landscapes and wetlands of Lowland Patagonia and their state of conservation. In: Mazzoni, E., Rabassa, J. (eds.) Volcanic Landscapes and Associated Wetlands of Lowland Patagonia. Springer Book Series on Latin American Studies, pp. 271–299 (2018c)

Mazzoni, E., Vázquez, M.: Desertification in Patagonia. In: Latrubesse, E. (ed.) Geomorphology of Natural and Human-Induced Disasters in South America. Developments in Earth Surface Processes. Elsevier, Amsterdam, pp. 351–377 (2009)

Meglioli, A.: Glacial Geology of Southernmost Patagonia, the Strait of Magellan and Northern Tierra del Fuego. Unpublished Ph.D. dissertation, Lehigh University, Bethlehem, Pennsylvania, U.S.A. (1992)

Mejia, V., Opdyke, N.D., Vaas, J.F.: Plio-Pleistocene time-averaged field in southern Patagonia recorded in lava flows. Geochem. Geophys. Geosyst. **5**, 1–15 (2004)

Mercer, J.H., Sutter, J.F.: Late Miocene-earliest Pliocene glaciation in southern Argentina: implications for global ice sheet history. Paleogeography Paleoclimatology Paleoecol. **38**, 185–206 (1982)

Nullo, F., Haller, M., Panza, J.L., Marín, G., Pardo, M.: Basaltos alcalinos eocenos y miocenos de algunas localidades de la Patagonia (Chubut y Santa Cruz). Rev. Asoc. Geológica Argent. **48**(1), 33–40 (1993)

Ollier, C.: Volcanoes, p. 228. Blackwell, Oxford (1988)

Ortiz Troncoso, O.: Artefactos de sílex de una tumba de Morro Philippi, valle medio del Río Gallegos. An. Del Inst. Patagon. **4**(2), 131–139 (1973)

Panza, J.L.: Descripción geológica de las Hojas 53e, Gobernador Moyano y 54e, Cerro Vanguardia, Provincia de Santa Cruz. Servicio Geológico Nacional, 197 p. Buenos Aires (1982)

Panza, J.L.: Hoja Geológica 4969-II, Tres Cerros, Provincia de Santa Cruz. Servicio Geológico Argentino, Boletín **213**, 1–103 (1995)

Panza, J.L.: Hoja Geológica 4769-IV, Monumento Natural Bosque Petrificado, Provincia de Santa Cruz. SEGEMAR, Boletín 257, Buenos Aires (1998)

Panza, J.L.: Franchi, M.: Magmatismo basáltico cenozoico extrandino. In: Haller, M. (ed.) Geología y Recursos Naturales de Santa Cruz. Relatorio del XV Congreso Geológico Argentino. El Calafate, vol. 1, no. 14, pp. 201–236. Buenos Aires (2002)

Panza, J.L., Saccomani, L., Cobos, J.: Mapa Geológico de Santa Cruz. Updated version, SEGEMAR, Buenos Aires (2002)

Panza, J.L., Saccomani, L., Cobos, J.: Hoja geológica 4969-III Laguna Grande. Boletín 362. SEGEMAR, Buenos Aires (2005)

Rabassa, J.: Late Cenozoic Glaciations in Patagonia and Tierra del Fuego. In: Rabassa, J. (ed.) The Late Cenozoic of Patagonia and Tierra del Fuego. Developments in Quaternary Sciences, vol. 11, pp. 151–204. Elsevier, Amsterdam (2008)

Rabassa, J.: Natural Heritage: quantitative evaluation of landscape scenic values. In: Lopes da Cunha, F., dos Santos, M., Rabassa, J., (eds.) Latin American Heritage. Interdisciplinary dialogues on Brazilian and Argentinian Case Studies. Springer Latin American Studies Book Series, pp. 111–120 (2018)

Ramos, V.A.: Descripción geológica de la Hoja 55b, Meseta de la Muerte. Servicio Geológico Nacional. Unpublished report. Buenos Aires (1978)

Ramos, V.A.: Geología de la región del lago Cardiel, provincia de Santa Cruz. Rev. Asoc. Geológica Argent. **37**, 23–49 (1982)

Ramos, V.A.: Las provincias geológicas del territorio argentino. Geología Argentina, Anales **29** (3), 41–96 (1999a)

Ramos, V.A.: Evolución tectónica de la Argentina. Geología Argentina. Anales 29, SEGEMAR, Buenos Aires, 715–759 (1999b)

Ramos, V.A.: El magmatismo neógeno de la Cordillera Patagónica. In: Haller, M. (ed.) Geología y Recursos Naturales de Santa Cruz. Relatorio XV Congreso Geológico Argentino, El Calafate, Buenos Aires, I-13, pp. 187–199 (2002)

Ramos, V.A., Niemeyer, H., Skarmeta, J., Muñoz, J.: Magmatic evolution of the Austral Patagonian Andes. Earth Sci. Rev. **18**, 411–443 (1982)

Riggi, J.C.: Resumen geológico de la zona de los lagos Pueyrredón y Posadas, Buenos Aires, Provincia de Santa Cruz. Rev. Asoc. Geológica Argent. **12**(2), 65–97 (1957)

Stern, C.H.: Pliocene to Present migration of the volcanic front, Andean southern volcanic zone. University of Colorado, Department of Geological Sciences, USA (1989)

Suárez, M., Márquez, M., De La Cruz, R., Navarrete, C., Fanning, M.: Cenomanian-? Early Turonian minimum age of the Chubut Group, Argentina: SHRIMP U-Pb geochronology. J. S. Am. Earth Sci. **50**, 67–74 (2014)

Ton That, T., Singer, B., Mörner, N., Rabassa, J.: Datación de lavas basálticas por $^{40}Ar/^{39}Ar$ y geología glacial de la región del Lago Buenos Aires, provincia de Santa Cruz, Buenos Aires, Argentina. Rev. Asoc. Geológica Argent. **54**(4), 333–352 (1999)

Zolitschka, B., Schäbitz, F., Lücke, A., et al.: Crater lakes of the Pali Aike volcanic field as key sites for paleoclimatic and paleoecological reconstructions in southern Patagonia, Argentina. J. S. Am. Earth Sci. **21**, 294–309 (2006)

# Early Human Occupations in the Valleys of Northwestern Argentina: Contributions to Dating by the Varnish Micro-Laminations Technique

J. P. Carbonelli[1(✉)] and M. M. Collantes[2]

[1] CONICET, IDECU, UBA- FFYL, Museo Etnográfico, Moreno 350, Buenos Aires CP 1091, Argentina
juanp.carbonelli@gmail.com
[2] Instituto de Geociencias y Medio Ambiente, Facultad de Ciencias Naturales e Instituto Miguel Lillo, Universidad Nacional de Tucumán, Miguel Lillo 205, 4000 San Miguel de Tucumán, Argentina

**Abstract.** In this paper, we present the results of dating by micro-laminations made on artifacts from a quarry-workshop located in the Yocavil Valley, Northern Argentina. The obtained data indicate that the minimum exposure age of the surface where these artifacts were recorded is 9,400 yr BP. This allows us to start the discussion of the peopling of this mesothermal valley region in northern Argentina throughout the Holocene. Secondly, dating from varnish micro-laminations results a contribution to the knowledge of the regional paleoclimates, based on the identification of at least ten wetter events along the entire sequence.

**Keywords:** Varnish dating technique · Holocene · Northwestern Argentina · Lithic quarry

## 1 Introduction

At present, there is a gap in the knowledge regarding hunter-gatherer occupations in the lowlands during the Early and Middle Holocene in South America due to a shortage of sites and the absence of information from multiple research lines, including paleoclimate, during such period (Aldenderfer and Flores Blanco 2011). This problem becomes particularly important in investigations based on surface sites, where the absence of materials for radiocarbon dating prevents the record to be interpreted properly.

In this paper, it will be intended to reassess the chronology of a quarry-workshop site located in the Yocavil Valley, in Northwestern Argentina (Fig. 1). This information is of great importance, given the scarcity of data on prehistoric occupations during the Early Holocene, outside the higher altitude areas. At present, strictly considering Northwestern Argentina the earliest data are concentrated in the Puna region. The occupation of hunter-gatherers was placed in the Late Pleistocene, after the last glacial advance of the Tauca Phase (Yacobaccio and Morales 2011). However, new data was


M. M. Collantes et al. (Eds.): *Advances in Geomorphology and Quaternary Studies in Argentina*, SPRINGEREARTH, pp. 262–282, 2020.
https://doi.org/10.1007/978-3-030-22621-3_13

published from outside the Puna macro-region; these data refer to the findings from the Quebrada de Los Corrales in the El Infiernillo (Fig. 2), Tucumán (Martínez et al. 2013), where a lithic workshop with plenty of surface material (Workshop PV1) lays. The excavations have yielded evidence of a primary context, with manufacturing as well as maintenance of artifact areas; artifact diversity, high density of findings and the stratigraphic context allows thinking of this site as a residential base (Martínez et al. 2013). The materials from the earliest excavated layer have provided a radiocarbon date of 7420 ± 25 years AP, which is the oldest radiocarbon record outside the Puna region (Martínez et al. 2013).

**Fig. 1.** Location of the Ampajango site in the Yocavil Valley, Northwestern Argentina.

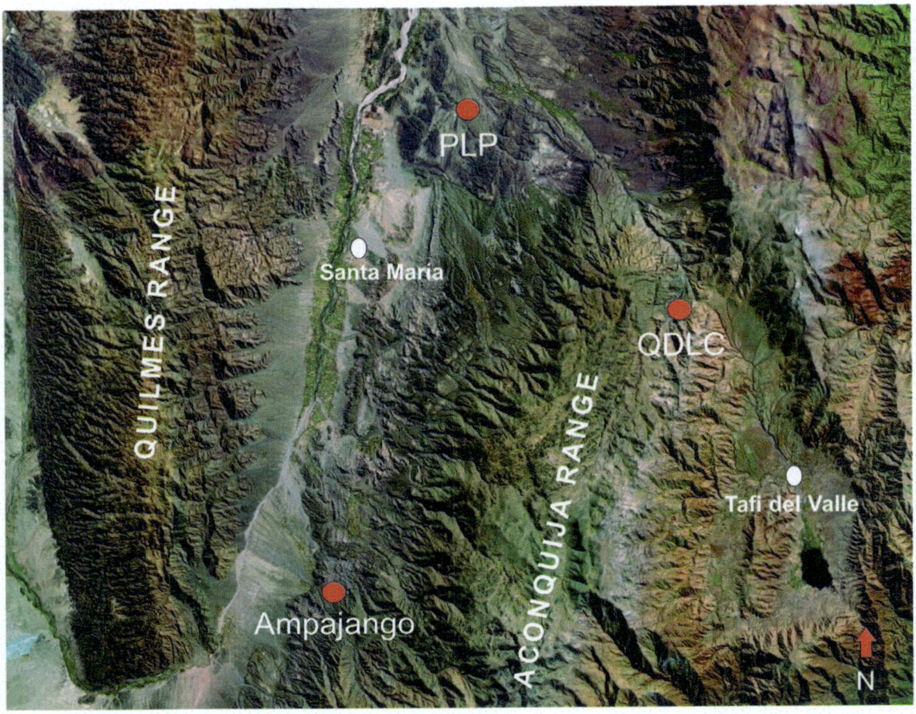


**Fig. 2.** Specific sector of the Argentine Northwest with the main sites mentioned in the text. References: PLP, Planchada La Puntilla; QDLC, Quebrada de los Corrales.

Another set of evidence comes from open-air sites, which also present "Ampajanguense" assemblages. These sites are located nearby our study area, more specifically in the Quebrada of Amaicha del Valle, provinces of Catamarca and Tucumán, in areas considered as effective sources for the acquisition of lithic raw material (Baied and Somonte 2011). In these sites, the reuse of formatized artifacts was detected, with a remarkable difference between the original discard and subsequent returns, inferred through the presence of fresh negative flakes over the edges of artifacts covered by rock varnish. From the confirmation of reclaiming cases in Planchada La Puntilla (Fig. 2) it is feasible to consider La Puntilla as a tertiary source for subsequent populations (Somonte and Baied 2013). Baied and Somonte (2012) dated the negatives of flaked lithic artifacts covered by varnish in the aforementioned site using the VML (Varnish Micro Laminations) technique. After several years of research (Dorn 2007, Liu and Dorn 1996) that covered the understanding of the varnish genesis and its applicability as a dating technique, a body of growing evidence reaffirms that the micro-stratigraphy of the varnish may be used as a climatic record. The investigations of Liu and Dorn (2004) reveal that the varnish is formed by superimposed layers that can be differentiated. The chemical composition of the layers was generated by local and regional climatic variations (Liu and Broecker 2007). Since these variations are regionally contemporary, the VML technique can be used as a correlative dating tool. The obtained results by Somonte and Baied (2013) constitute the first dates on materials corresponding to the "Ampajanguense" Industry, establishing the occupation of this area prior to the 6500–5900 radiocarbon years BP lapse.

This information provides an expectation that may be extended to the Ampajango quarry – workshop due to the morphological similarity among the materials involved (Baied and Somonte 2012, Somonte and Baied 2013), as well as a set of new analytical and methodological paths for approaching the chronology of the Ampajango quarry - workshop and reassessing its functionality within the landscape. The main objective of this work is to offer a relative chronology to a key "industry" in the history of early human settlement in the Argentine northwestern region. The information obtained in open-air sites, as Borrero (2011) points out, belongs to a type of data that could change the current perception of the Pleistocene-Holocene transition within the discussions of early Andean human settlement.

## 2 The Ampajango Site

The Ampajango site is located in the Yocavil Valley (Province of Catamarca), a tectonic depression with a north-south orientation, bounded to the east by the mountain ranges of Aconquija and Cumbres Calchaquíes and to the west by the Sierra del Cajón (Figs. 1 and 2). It presents a semiarid to arid climate, with an annual rainfall regime (which is highly seasonal) between 200 and 300 mm, which partly determines a vegetation from the mountain geographic province at the bottom of the valley and the foothills, represented by the shrub steppe, and pre-Puna vegetation on the slopes (Collantes and González 2012).

The history of research regarding quarries-workshops in this valley is associated with the search for the oldest early human occupations in this region in particular, and the American continent in general. This was the main concern of the late Professor Dr. Eduardo Cigliano, who initiated archaeological research from the Universidad Nacional del Litoral (Santa Fe, Argentina) in the 1960's. In the locality of Ampajango, just below a level of Cenozoic terraces, Cigliano et al. (1962) recorded a large amount of preforms, nodules, cores with small extractions and instruments laying over a large glacis, which allowed the classification of such site as a quarry-workshop.

Cigliano (1961) used the term "Ampajanguense" to refer to a hand ax industry mainly composed of bifaces manufactured by big strokes, hypothesizing a large, ample spread over Northwestern Argentina. Cigliano et al. (1962: 102) based on the technique of artifact manufacture, morphology and stratigraphic position of the materials, granted the "Ampajanguense" industry an estimated antiquity of more than 10,000 years. An interesting fact is that Cigliano (1961) and Cigliano et al. (1962) reported that a strong patina, or rock varnish, covered every piece, but he did not realize then that this was a solid argument undoubtedly proving their old age.

These statements were made within a particular socio-historical context and in the framework of research regarding the peopling of the Americas. By that time, the sites that presented a technology of artifacts on flakes and thick bifaces without projectile points, were grouped into the "Pre-Projectile points" stage (Krieger 1964), characterizing the human populations occupying these spaces as generalized hunter-gatherers.

A subsequent review, led by those who defended the Paleo-Indian paradigm and the notion of specialized hunter-gatherers, discussed the geological and technological linkage proposed by Cigliano (1961). In this sense, Durando et al. (1986) determined that the artifactual assemblage without projectile points would have worked only as "quarry sites" and thus they could not be associated with the hunter-gatherer groups of the biface Pre-ceramic Horizon (Durando et al. 1986; Nami 1988). This site (the Ampajango quarry workshop) belongs to the archaeological locality of Ampajango, where other sites corresponding to the Formative, Late, and Inca periods are known (Lorandi 1966; Tarragó and González 2005).

The Ampajango quarry-workshop lays inside a depression in the shape of an amphitheater, between remains of erosion piedmont surfaces. The material found in the quarry-workshop corresponds to the Caspinchango Formation (mainly appearing as an outcrop in the third piedmont level), chronologically assigned to the Quaternary by Ruiz Huidobro (1972). Bossi et al. (2001) referred precisely to the Ampajango Cenoglomerate, which presents flow deposits of debris coming from alluvial fans containing abundant volcaniclastic pebbles and clasts as by-products from volcanic sources. Among the different pebbles, we found volcanic clasts, mainly andesites and basalts, from the Los Corrales Formation. The size of such clasts, following Spiegelman (1971) may vary between fine gravel and blocks; among the latter some specimens reach a diameter of up to 1 m. Finer sediments are distributed in flood areas of the region main rivers and in the lower areas of the valleys of Santa María and Tafí.

Taking this information as a starting point it may be argued that the glacis where the Ampajango quarry-workshop is located conformed a potential source of raw material. The predominant rocks found throughout the landform are andesites, presenting a grayish-brownish color, characteristic of the presence of desert varnish. Complementarily, a low frequency of white quartz was recorded. Regarding the quality of these raw materials, the andesite rocks present conchoidal fracture, with a mainly aphanitic texture, but in most cases presenting phenocrysts and inclusions in the matrix. In this sense, these rocks may be considered a raw material of good to regular quality. On the other hand, a defining feature of quartz is its tenacity, with difficult predictability and directionality of the strike during the manufacturing process. Therefore, it may be considered as a bad quality raw material.

Regarding the distribution of these rocks it may be inferred that, following Nami (1992), lithic resources are concentrated (more than ten rock pieces by square meter). Commonly, they appear in the form of big blocks and/or nodules, whose diameter is larger than 50 cm. The largest concentration of artifacts is located at the base of the depression, decreasing towards the glacis. Large blocks and nodules with evidence of extraction, a large amount of shaped instruments discarded on surface as well as knapping debitage, all of them covered by desert varnish, were recorded throughout the quarry-workshop.

From fieldwork and survey, we defined that the total area of the Ampajango site (2500 m a.s.l., 26°55'29, 9" S, 66°4'6,1" W) reaches up to 15.6 hectares. It constituted a large and relevant space for the procurement of lithic resources in the Yocavil Valley, since andesite blocks and nodules of finer texture and good quality are concentrated there (Fig. 3). After determining the total area of this site during the first fieldwork season (Carbonelli 2013), a smaller area was selected as the universe of study; this area was chosen due the higher concentrations of lithic artifacts (Fig. 4) and the more adequate visibility and accessibility (Gallardo and Cornejo 1986). There, a simple probabilistic sampling was performed, applied by making transects. As a result, we realized that, in this sector, flakes predominates over cores and scarcely formatized artifacts. However, this record is not observed everywhere in the site, since in two specific areas of the quarry-workshop a concentration of artifacts with bifacial reduction and thinning, as bifaces, blanks of bifacial pieces and bifacial edges with sinuous ridges was recorded (Carbonelli 2013). By analyzing different production sequences as well as recycling processes and superimposed edges in some artifacts, the present authors agree with Durando et al. (1986) who argued that Ampajango functioned as a quarry-workshop, where andesite was exploited as a secondary source (Carbonelli 2013).

**Fig. 3.** Blocks with rock varnish and rock art in the quarry-workshop.

**Fig. 4.** Detailed figure of the archaeological site of Ampajango. References: 1. Ampajango River. 2. Area of Ampajango quarry-worlshop. The red rectangle marks the collection sector. 3. Bad land, end of the Ampajango quarry-workshop. 4. Ampajango II, Inca site.

As previously described, the Ampajango quarry-workshop probably configured itself, through time, as a space that could have functioned as an important reserve of raw material. Therefore, this site has been interpreted as a transit zone and subject of continuous procurement excursions. Consequently, one of the effects in the use of landscape was the generation of a palimpsest of accumulated artifacts that resulted from a complex interrelation of natural and cultural processes differentially marking the landscape (Barton et al. 2002). It was then started from a position of "archaeological record transformation" by adhering to the posture that emphasizes on the spatial distribution of the record, highlighting the action of the processes transforming or distorting the materials formally, spatially, quantitatively and even in relational terms (Schiffer 1983).

In the first survey performed, 233 artifacts were analyzed. The most important erosion agents detected during the first prospection were the action of wind and rain (Carbonelli 2013). The sediments conforming the matrix of the analyzed sample consist of silt and sand fractions. Regarding this composition, the action of the wind could have caused the displacement of smaller flakes. In addition, the action of rain may have provoked a sliding of archaeological materials. The observation of deep grooves on the sediment indicates an intense hydrological activity developed in the site, that caused the transportation and reordering of artifacts. The effect of fluvial transportation on such artifacts was evidenced through two indicators as well: damaged edges and surface abrasion (Barton et al. 2002). The former is defined as the removal of material (irregularly and randomly) along the artifact edges, while the latter is manifested by the presence of striations, signals of flattening and rounded edges. Only 50% of collected artifacts (N = 233) presented micro-chipping or micro-flaking, although these attributes were not recorded on flakes or cores. On the other hand, shaped artifacts exhibiting striations correspond to the sectors with highest artifactual density, located away from abrupt slopes and the action of water runoff. In sum, it is herein considered that this runoff acted on both the distribution and density of findings, but were limited to a specific area of the sector sampled, therefore not constituting a relevant bias in the sample.

Besides the lithic artifacts, other materialities were recorded in this quarry-workshop. A few ceramic fragments corresponding to later centuries before the Inca conquest were documented in the depression. At the same time, over one of the contiguous terraces, a ceramic urn was found which may be stylistically ascribed to around 1000 yr AD. Regarding rock art representations, several blocks and nodules covered with desert varnish present carvings that can be linked with the Aguada iconography, a religious-political phenomenon with large dispersion in Northwestern Argentina (Núñez Regueiro and Tartusi 2002; Cruz 2007; Gordillo 2010). For the case of rock art in Ampajango quarry workshop, Lorandi (1966) and Fiore (2007) agreed in the presence of motifs similar to the ones registered in other sites of the Argentine Northwest during the periods known as "Early" and "Middle" ones, this is to say, approximately during the entire first millennium AD.

This dataset allow us to point out the persistence of the Ampajango site as a node, a recurring reference point in the landscape. In order to understand the role of this site over time, the Ampajango quarry workshop is proposed as a site built upon the base of a succession of superimposed landscapes, a quarry-workshop, a transit zone, and a place particularly chosen for symbolic expression through art (Carbonelli 2013). As mentioned before, the use and reoccupation of the landscape by several communities during a prolonged period yield a palimpsest (Holdaway et al. 2004). Considering the quantity of materials observed in the Ampajango quarry workshop, it is feasible to think about successive episodes of material deposition, of activity layers (sensu Bailey 2007), where the differentiation between activities could result a very difficult task. Nevertheless, taking into consideration both the size of the quarry workshop and the distribution of the material on surface, sectors for the recognition of past activities with a good resolution in interpretative terms could be isolated. In this way, the conception of such space would adjust to Bailey (2007) definition of "spatial palimpsests".

Then, we still find ourselves dealing with the problem of chronological control. What temporality each one of the detected activities refer to? Since when this space was used? Unable to get dates by conventional dating, the varnish micro-lamination (VML) appears as a reasonable tool, a suitable technique to approximate the chronology of this site, considering that such patina covers most of the instrumental assemblage found in Ampajango.

## 3 Materials and Methods

Rock varnish is one of many patinas that can affect the surface of stones (Somonte 2009), with the particularity that this is a slow accretion formation process (<1 to 40 μm/ka) (Liu and Broecker 2000). Regarding its composition, manganese (Mn) and iron oxide (Fe) represent 30% of the materials involved, the rest being composed by silica (Si), aluminium (Al), magnesium (Mg), potassium (K) and calcium (Ca) (Liu and Broecker 2000). Manganese and iron oxide concentrations are precisely the ones that provide the colors of this patina; when the manganese content decreases and the iron abundance increases, the color of the varnish turns to orange (Dorn 2007).

Regarding the conditions in which such varnish appears, Somonte and Collantes (2007) argued that it can develop in different morphoclimatic environments (yet more frequently in arid and semiarid environments), although the landforms that give origin to the rock varnish must possess a morphogenetical stability. In this sense, it is important to mention the observation made by Somonte and Baied (2017), who are currently working in other sites similar to Ampajango. They argued that the presence of rock varnish covering lithic artifacts and clasts is an indicator of the stability of the landform where such items were deposited, given that (as it will be described below) the formation and growing of this patina demands periods of even thousands of years.

Currently, the most accepted model regarding the origin of this varnish is the one known as polygenetic. This establishes that "varnish formation is a nanometer-scale marriage between clay minerals and the millennial accumulation and decay of Mn-Fe-encrusted bacterial cast. Nanometer-size Mn-Fe remains of bacteria both feather and cement clay minerals to the rock and to prior varnish" (Dorn 2007: 269), meaning that the chemical composition of the varnish is independent from the rock serving as substrate.

Since the 1980's, a new dating method emerged based on the observation of the composition of varnish. Given its sedimentary origin, sometimes it can present a micro-stratigraphy, observable with a petrographic microscope, after thin sections are polished to reach a thickness between 5 to 10 μm (Liu and Broecker 2008a). The Microprobe Chemical Mapping method reveal layers of dark tonality in ultra-thin sections of varnish, rich in manganese (Mn) and barium (Ba), but poor in silica (Si) and aluminum (Al), while orange and yellow layers are poor in Mn and Ba, but enriched in Si and Al (Liu and Broecker 2008b). Liu and Dorn (2004) interpreted that orange / yellow layers poor in Mn indicate arid climates, whereas rich dark layers in Mn indicate wetter conditions.

It is worth mentioning the fact that these two types of layers appear interleaved, forming a distinctive micro-stratigraphy strongly influenced by regional climatic variations. As these are regionally contemporary, varnish micro-laminations can be used as a correlative dating tool that allows knowing the minimum exposure ages of geomorphic surfaces presenting varnishes (Somonte 2009).

As a correlative age determination technique, VML substantially depends on the calibration of radiometric age of varnish micro-stratigraphy and climatic correlation (Somonte 2009). These studies originated in the arid area of the Great Basin and the Western United States, where the radiometric calibrated micro-stratigraphy of the varnish for the Late Pleistocene was conformed. Samples from different Pleistocene and Holocene landforms with an established chronology, since they have been already dated by radiometric methods, were also obtained in these places (Liu and Broecker 2007). In this case, climatic correlation was obtained through the measurement of detrital carbonate pulses in records of the North Atlantic deep sea sediments, which mark colder events that match with the wetter events recorded in the varnish (Liu and Broecker 2007).

The sequence that covers the entire Holocene, the LU-1 unit, is visualized under the microscope as a yellowish layer, which would signal the interglacial dry climate. This unit is sub-divided into 12 dark layers interspersed with 13 yellow/orange layers (Fig. 5). The calibration indicates that six of the dark layers (WH1–WH6), located in the upper portion of the sequence, may have been deposited in the last 6000 years coinciding with wetter events from the Late Holocene. Particularly, it may be mentioned that the WH1 layer would mark evidence of the Little Ice Age (XVIth. to XIXth. centuries AD) in the Yocavil valley. On the other hand, five of the dark layers (WH8–WH12) were deposited between 7000–1000 years BP, coincidental with the humid events of the Early Holocene, while the WH7 layer would signal the Middle Holocene wetter phase (Liu and Broecker 2008a).

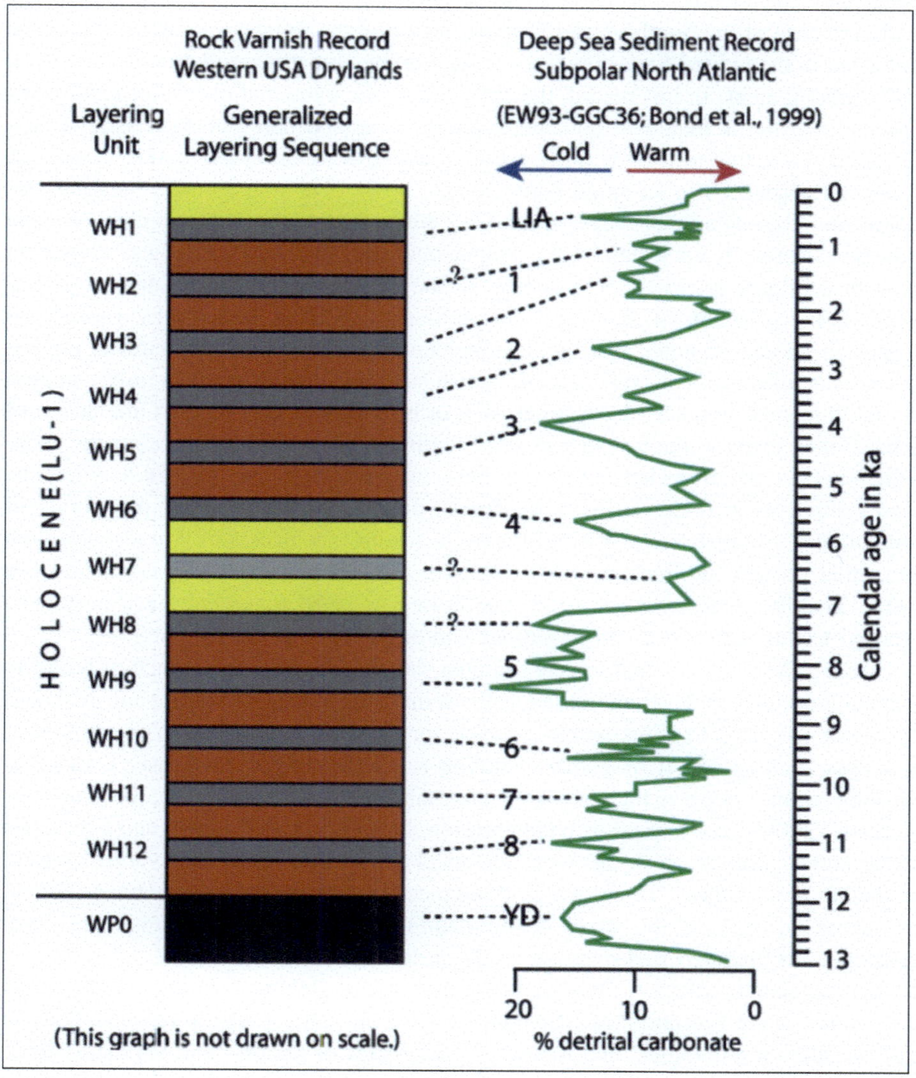


**Fig. 5.** Sequence of varnish layers corresponding to the Late Pleistocene and Holocene in the Western United States. Lu: Layer Unit, LIA: Little Ice Age. Modified from Liu and Broecker (2008a).

In summary, dating by varnish micro-lamination is an age correlative determination technique that can be used to date and correlate various geomorphological features in desert environments (Liu and Broecker 2008a). In the case of the sequence constructed for the Holocene, the temporal resolution has an error margin between 500 and 1000 years. With respect to its archaeological applicability, it must be considered that this technique is used to determine the minimum age of exposure for surfaces where archaeological remains are found; therefore "… is not about dating archaeological events themselves, but as a dating technique of geological processes affecting the archaeological remains…" (Somonte 2009 :89).

In the case of the Southern Hemisphere, adopting the sequence constructed for the Holocene in the United States involves accepting the climatic correlation between two very different areas. In this regard, the present authors argue that only after that micro-lamination patterns match in both hemispheres is established (for each sample submitted) it is feasible to use the VML as an effective correlation and dating technique.

In the absence of a stratified context that allows obtaining suitable materials for dating under conventional ways, and taking advantage of the large number of lithic artifacts covered by rock varnish in Ampajango, the VML technique was chosen in order to configure a tentative chronology of the quarry-workshop exploitation. A key aspect is the selection of the area where artifacts and rocks covered by varnish are located. It should be noted that the development of desert varnish over rocks implies that the rate of erosion since the clock varnish starts must be zero. Even only a micron of rock thickness impacted by wind erosion or biochemical activity (like the action of lichens) may result in the loss of a layer of varnish and, therefore, part of, or the entire sequence (Liu and Broecker 2008b). In this sense, it should be intended to select the most stable rock surfaces in the field, that is, those having the highest probability of containing a complete varnish stratigraphy, a feature not necessarily associated with rocks that present the darkest, more extensive or thicker varnish (Broecker and Liu 2001). Thereby, it is advisable to choose those varnished rocks that have only been exposed to dust deposition, precipitations and/or dew (Dorn 2009).

Among the attributes that the chosen area should present, the presence of boulders or natural nodules over 30 cm buried in the sediment and covered with varnish should be mentioned (Fig. 6). This would account for a surface with no geomorphological processes such as runoff or gullies denoting rock transfer. In this sense, a factor to consider is the slope, which controls the stability of the landform. As stated by Somonte (2009), factors such as wind abrasion, lakeside actions, cracks and alterations of the cortex product of physical and chemical weathering can restart varnish action, producing new surfaces. Therefore, all those pieces with varnish that may have suffered the action of water flow, water accumulation or pieces whose varnish has been developed on fractured rocks should be discarded (Dorn 2009).

In Ampajango, after selecting the area, the present authors proceeded to the characterization of the geomorphological context where the varnish had developed. For this purpose, the morphological and metric characteristics of nodules and varnished lithic artifacts was recorded, assessing their spatial distribution and its disposition on surface in order to obtain the paleoenvironmental and archaeological context for varnish formation. The survey and sampling was performed using a collection unit of 4.0 $m^2$, subdivided into four sub-units following clockwise direction.

**Fig. 6.** Sample collection area for VML analysis.

During this process, the position of each piece before collecting it was noted, and to register this a piece of paper tape was attached on the side exposed towards the viewer (Fig. 6). This kind of procedure is important for determining which faces presented rock varnish. This collection methodology also enables the determination of another indicator of surface stability: 50% of collected artifacts (n = 19) present rock varnish on both faces, whereas 31.5% (n = 12) only present varnish in the face exposed to the operator. Rock varnish is one of the slowest accretional processes registered (Liu and Broecker 2000). If there are artifacts presenting varnish on both faces, that means that the stability conditions required for its formation were maintained through a long period, without geomorphological processes interfering with its development (Somonte and Baied 2017).

Every item, both natural rocks and artifacts with presence of varnish (n = 38), was collected and subsequently analyzed in the laboratory, under 60x binocular lens, in order to observe the micro-depressions where varnish is found and dispersed (such depressions measure between 1–2 mm). When observed under binocular microscope, this varnish should look shiny, uniform and thick, as an assurance that these features will appear once again when observed under the microscope. After observing some artifacts with lenses, some of them with angular shapes were discarded, since this morphology could be the result of erosion processes after the original deposition of the clasts.

Two debitage flakes were selected for further analysis, an angular flake and a chipped edge of the core (Fig. 7). The locations in the selected pieces for dating and in which ultra-thin sections were effected (with a thickness between 5–10 microns), were choices of the VML Dating Lab.

## 4 Results and Discussion

Four dates were obtained from two archaeological samples from the Ampajango quarry workshop (Carbonelli 2014), which were climatically correlated and radiometrically calibrated following the general sequence of layers used for the Late Pleistocene and Early Holocene in the western United States. The report provided by Dr. T. Liu states that "a similar climatic correlation and age scale may likely exist for the Argentine varnish…" (Dr. Thanzuo Liu, personal communication). Therefore, although age estimates are still speculative, these data add to the first dates of surface sites in the mesothermal valleys of Northwestern Argentina (Baied and Somonte 2011, 2012).

The results of the four VML dates are presented in Table 1. These ages, estimated by VML, provide minimum exposure ages of the landforms where lithic artifacts were recorded. Based upon these results, it is interpreted that the oldest layer of varnish was deposited at least around 9400 years BP (Carbonelli 2014). It should be noted that these estimated ages are speculative, at least until the construction of a VML scale of calibrates ages for Argentina (cf. Baied and Somonte 2012). In other words, this date marks the minimum age since varnish started growing. The presence of ten layers signals the geomorphological stability of that space, allowing a successive development without alterations.

**Fig. 7.** Flakes used for dating varnishes. A, B, C and D are the sectors where ultra-thin cuts were performed.

**Table 1.** VML results of the analysis of rock varnish samples from Valle de Yocavil

| Rock varnish sample (Fig. 5) | Number of ultra-thin sections | Oldest layering pattern observed | Image of oldest layering pattern (Fig. 6) | VML age estimate (in cal. yr. BP) |
|---|---|---|---|---|
| Muestra-A (D11) | | | | |
| D 11-A | 1 | LU-1 (WH 10) | A | 9400 |
| D 11-B | 1 | LU-1 (WH 9) | B | 8100 |
| Muestra-B (C10) | | | | |
| C 11-C | 1 | LU-1 (WH 9) | C | 8100 |
| C 11-D | 1 | LU-1 (WH 9) | D | 8100 |

This interpretation of an archaeological level allowed the present authors to argue that these artifacts have a greater antiquity than the one presented here, although it cannot be specified the exact time of its production. As shown in Fig. 7, the places chosen by Dr. Liu to make ultra-thin sections are located on the pieces' negative flakes, which allowed the authors to sustain that manufacturing took place before the development of varnish.

In the case of Sample A, two different dates were obtained: 9400 years BP and 8100 years BP. It should be insisted that one of the disadvantages of VML dating is that there are actions such as wind abrasion and other alterations of the cortex that can restart the varnish clock. Therefore, it is conceivable that either the varnish clock was restarted in sector B of Sample A, or that this patina presented a differential development in different depressions of the flake. Nevertheless, an archaeological explanation can also be proposed: these two dates could be representing temporarily decoupled human interventions over the same artifact (Carbonelli 2014). This does not mean that a systemic context is under observation (sensu Schiffer 1991) oblivious to post-depositional alterations. However, considering the reclamation process inferred through the VML analysis, new ways of occupying the space and exploiting the lithic resources in the past can be detected and inquired. This process of using quarry-workshops as raw material sources (tertiary sources sensu Church 1995) by several human populations over thousands of years, was an activity already proved in different open-air sites in the southern portion of the Valles Calchaquíes (Baied and Somonte 2012; Somonte and Baied 2013, 2017).

In technological terms, based on the investigations performed in the Yocavil Valley (Carbonelli 2012, 2013), it is argued that the instrumental assemblage of the Ampajango quarry-workshop matches the expectations proposed by Andrefsky (1994) where, within a regional structure of abundant good quality rocks, both complex instruments and expeditious cutting tools are manufactured. From these statements, it is possible to argue that these sites were used as a continuous supply of local raw materials of good quality. A future task will involve investigating whether nearby sites also present reclaimed artifacts coming from such a tertiary source.

The question now is: Since when the Ampajango quarry workshop was used? In first place, the data collected through VML dating indicate that the minimum age of the surface where these artifacts were found is 9400 years BP. Therefore, this signals that the act of transformation of these raw materials in debitage occurred before. Then we can argue that, although there is still no elements to refine this coarse-grained chronology, the first record of human populations in Ampajango approximates quite well the age of 10,000 years, long ago proposed by Cigliano et al. (1962), who obtained that age from a comparison of the "Ampajanguense" and "Ayampitín" artifactual assemblages.

The results presented here become a contribution on the study of climate conditions throughout the Holocene for this sector of the Southern Hemisphere, still characterized by a scarcity of data and proxies regarding paleoclimate. In Fig. 8, some pictures of the sample micro-stratigraphy are shown. These images reveal that, using this technique, ten humid events have been detected during the Holocene, dated between 9400 years AP and 300 years AP.

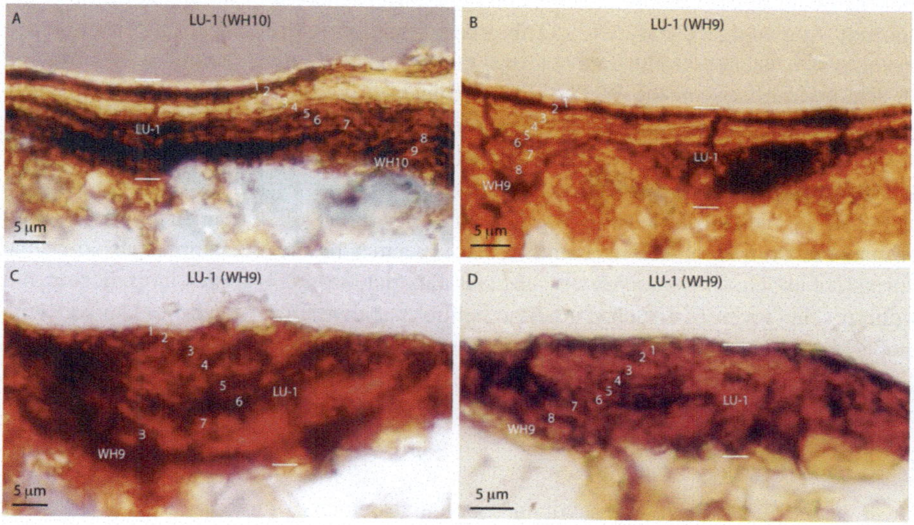


**Fig. 8.** Sections of desert varnish present in ultra-thin cuts from the Ampajango site.

The record of the WH10 and WH9 wetter events indicate higher moisture and lower temperature conditions in the Early Holocene; nevertheless, there are no other data or proxies that can corroborate them in our study region. Tchilinguirian (2011) stated that this trend could be extrapolated to the entire Northwestern Argentina by bringing together studies from different sectors of the region.

It is in the Late Holocene where it is feasible to find enough data to make comparisons with the results obtained through VML analyses. The investigations of Strecker (1987) detected two sand layers rich in organic matter in the terraces of both margins of the Santa María River. The dating of both layers provided an age of 2190 ± 53 and 1470 ± 50 $^{14}$C years BP. This is translatable to a wetter period during such chronological frame. These data are concordant with the pollen analysis performed by Garalla (2003) on a sedimentary sequence of the Abra del Infiernillo (province of Tucumán). In this sequence, among the identified species, the most abundant are from herbaceous steppe, low shrubs and a vegetation characteristic of humid environments. Such vegetation is considered as characteristic of this sector around 2000 ± 60 years BP (Garalla 2003). By crossing these data with the ones obtained through VML, it may be assumed that the mentioned references are close to the WH4 wetter event, but they are better adjusted to the calibrated radiometric age of the WH3 event. It must be kept in mind that the calibration of each layer is still speculative and not definitive for the case of the Southern Hemisphere.

On the other hand, there is scientific consensus regarding the existence of the Medieval Warm Period in the region of the valleys of Northwestern Argentina. The development of dune fields in the pre-Andean valleys of Catamarca and Tucumán would have started between 1000 and 1300 years BP, during an extremely dry period (Sayago et al. 2010). This is corroborated by the analysis of the sedimentary geological deposit found in El Alacrán, a site located at the right margin of the Santa María River (Gómez Augier and Caria 2012). This trend regarding increased aridity during such period is consistent with the data obtained through varnish micro-laminations, signaling an arid period between the WH3 and WH2 layers.

Finally, correlation is found in some local geomorphological investigations with the last humid event that gave origin to varnish layers dated between 650 and 300 years BP.

Collantes and González (2012) have detected for such period the development of a paleosoil in Santa María valley, which is the remains of a superficial soil, whose original material is of the loess-like type, deposited before the humid period of the Formative epoch. These authors argued that this paleosoil constitutes an indicator of a wetter event, given that such climatic conditions allowed the formation of a soil with moderate development. This paleosol was dated via AMS obtaining an age of 435 ± 15 $^{14}$C years BP (Collantes and González 2012).

## 5 Conclusions

One of the contributions of this work points towards establishing paleoclimatic indicators for the Holocene, conforming parameters that are useful for this part of South America (Northwestern Argentina). Currently, in regions such as the Titicaca Basin and Northern Chile, there is evidence of a Holocene period composed of several wetter episodes alternating with more arid phases (Aldenderfer and Flores Blanco 2011). Secondly, this research line allowed the establishment of a tentative chronology for a surface site, characterized for not yielding material record in stratigraphy suitable for radiocarbon dating. More specifically, dating by varnish micro-lamination presented here, account for ten events occurred under wetter conditions, beginning 9400 years BP and interspersed to episodes of aridity.

In general, and being cautious regarding the interpretation of a relative dating technique, the Ampajango quarry-workshop can be considered as a site belonging to the Early Archaic Period −10000 to 8000 years BP (Aldenderfer and Flores Blanco 2011). As described above, this technique provides an estimated timeline for the minimum exposure ages of landforms on which the artifacts were deposited. The obtained results indicate that the oldest varnish layer registered, started its deposition in

Ampajango 9400 radiocarbon years ago, although another layer of varnish, deposited 8100 years BP, was also recorded. Solving the lack of chronological resolution present in these spatial palimpsests is achievable by working with a microscopic approach (Bailey 2007). The first VML analyses performed in the Ampajango quarry-workshop also constitute the first steps for approaching the diverse temporalities crossing the lithic materials. This diversity of temporalities is also observable in this particular space in other materialities, such as pottery and rock art.

To interpret the raw material procurement, a series of data allowed the present authors to consider the Ampajango quarry-workshop as a tertiary source: (a) the large amount of material available; (b) the different tonalities of rock varnish registered in the surface of a large artifact density; (c) the diachrony in the dating registered on a debitage from the quarry-workshop; (d) the presence of other materialities (pottery, rock art) which suggest this was a space transited and re-occupied for thousands of years. This last feature is shared with other nearby "persistent spaces", such as Planchada La Puntilla (Somonte and Baied 2017). In other words, varnish micro-lamination allowed thinking about the Ampajango quarry-workshop as a space constituting a raw material source used by a continuum of human populations throughout the Holocene. The Ampajango quarry workshop worked through millennia as a tertiary source of raw material, where various societies claimed the raw material and artifacts deposited there. In order to achieve a deeper understanding of the processes regarding raw material procurement in Ampajango, to increase the number of VML dates over reclaimed lithic artifacts is needed.

The absence of a stratified context as well as lack of information from other lines of evidence prevents to provide, from the Yocavil Valley, responses to the general Prehistoric human occupation during the Late Pleistocene-Early Holocene transition. However, gathering the information presented above, meaning: dating by varnish micro-lamination performed in both Amaicha del Valle and Ampajango, coupled with radiocarbon dating from Taller de Puesto Viejo, it is understood that there are elements to strengthen the Martínez et al. (2013) hypothesis on human occupation at lower locations and outside the Puna high plains throughout the Holocene.

**Acknowledgements.** The authors would especially like to thank Dr. C. Somonte for her advice and collaboration in the research activities developed in Ampajango. We would like to express our gratitude to Paula Cukjati, Eng., for helping us in the translation of this text. The priceless collaboration of Dr. T. Liu, patiently analyzing the artifacts' samples from Ampajango is deeply acknowledged. Thanks to his knowledge and hard work, we were able to obtain the sequence of varnish layers for the Yocavil Valley. This research was largely developed thanks to all the support and trust of the Yocavil Project, to which the authors belong; especially, we would like to thank Dr. Lanzelotti who read and commented on an early draft of this paper.

## References

Aldenderfer, M.S., Flores Blanco, L.: Reflexiones para avanzar en el estudio del período arcaico en los Andes Centro-Sur. Chungara, Revista de Antropología Chilena **43**, 531–550 (2011)

Andrefsky, W.: Raw-material availability and the organization of technology. Am. Antiq. **59**(1), 21–34 (1994)

Baied, C., Somonte, C.: Geocronología, paleoambiente y dinámica ocupacional durante el Holoceno medio y tardío en la Quebrada de Amaicha del Valle, Tucumán, Argentina. In: Mondini, M., Martínez, J., Muscio, H., Marconetto, M.B. (eds.) Poblaciones humanas y ambientes en el Noroeste Argentino durante el Holoceno Medio, pp. 59–65. Taller de Arqueología, Córdoba (2011)

Baied, C., Somonte, C.: Mid-Holocene geochronology, palaeoenvironments, and occupational dynamics at Quebrada de Amaicha, Tucumán, Argentina. Quatern. Int. **299**, 80–89 (2012)

Bailey, G.: Time perspectives, palimpsests and the archaeology of time. J. Anthropol. Res. **26**, 198–223 (2007)

Barton, M., Bernabeu, J., Aura, J.E., García, O., La Roca, N.: Dynamic landscapes, artifact taphonomy, and land uses modeling in the Western Mediterranean. Geoarchaeol. Int. J. **17**, 55–190 (2002)

Borrero, L.: Cazadores tempranos del mundo andino. Comentarios Finales. Boletín de Arqueología PUCP. **15**, 385–394 (2011)

Bossi, G.E., Georgieff, S.M., Gavrilof, I., Ibañez, L.M., Muruaga, C.M.: Cenozoic evolution of the intramontane Santa María basin, Pampean Ranges, northwestern Argentina. J. South Am. Earth Sci. **14**, 725–734 (2001)

Broecker, W.S., Liu, T.: Rock varnish: recorder of desert wetness? GSA Today **11**(8), 4–10 (2001)

Carbonelli, J.P.: Base regional de recursos líticos y fuentes de materias primas al sur del valle de Yocavil. Estudios Sociales del NOA **12**, 111–137 (2012)

Carbonelli, J.P.: El sitio Ampajango, cincuenta años después. Arqueología **19**(1), 41–64 (2013)

Carbonelli, J.P.: Evidencias paleoambientales y de producción lítica en la cantera taller Ampajango, Valle de Yocavil (Catamarca, Argentina). La Zaranda de Ideas. Revista de Jóvenes Investigadores en Arqueología **11**, 9–26 (2014)

Church, T.: Terms in Lithic Resource Studies. In: Church, T., Francis, J.E., Haury, C.E. (eds.) Lithic Resource Studies: a Sourcebook for Archaeologists, pp. 9–31. University of Tulsa, Oklahoma (1995)

Cigliano, E.: Noticia sobre una nueva industria precerámica en el Valle de Santa María (Catamarca): El Ampajanguense. Anales de Arqueología y Etnología. Tomo **XVI**, 169–179 (1961)

Cigliano, E., Bereterbide, S., Carnevali, B., Lorandi, A.L., Tarragó, M.: El Ampajanguense. Publicación del Instituto de Antropología, vol. 5, pp. 7–104. Facultad de Filosofía y Letras, Universidad Nacional del Litoral, Rosario, Argentina (1962)

Collantes, M.M., González, L.: Mecanismos del proceso de desertificación en el valle de Santa María, provincia de Tucumán (Argentina). Acta Geológica Lilloana **24**(1-2), 108–122 (2012). San Miguel de Tucumán

Cruz, P.: Hombres complejos y señores simples. Reflexiones en torno a los modelos de organización social desde la arqueología del valle del Ambato. In: Nielsen, A., Rivolta, M., Seldes, V., Vazquez, M., Mercolli, P. (eds.) Procesos Sociales Prehispánicos en el Sur Andino. La vivienda, la comunidad y el territorio, pp. 99–122. Editorial Brujas, Córdoba (2007)

Dorn, R.I.: Rock varnish. In: Nash, D.J., Mc Laren, S.J. (eds.) Roca barniz, geoquímicos y los sedimentos de los paisajes, pp. 246–297. Blackwell, London (2007)

Dorn, R.I.: The Rock Varnish Revolution: New Insights from Microlaminations and the contributions of Tanzhuo Liu. Geogr. Compass **3**, 1–20 (2009)

Durando, P., García Salemi, M., Platanía, G.: Estudios geomorfológicos, paleoclimas, dataciones relativas y tipología a partir de paleolitos del Valle de Santa María (provincias de Catamarca y Tucumán, República Argentina). In: Iriarte Brenner, F. (ed.) Actas del VI Congreso Peruano: Hombre y Cultura Andina, pp. 89–109. Facultad de Ciencias Sociales, Perú (1986)

Fiore, D.: Analysis of Ampajango rock art, Santa María, Argentina: an approach to the structure of its designs and the technology of its production. Unpublished Master of Arts Thesis, Institute of Archaeology, University of London (2007)
Gallardo, F., Cornejo, L.: El diseño de la prospección arqueológica: un caso de estudio. Chungara, Revista Chilena de Antropología **16–17**, 409–420 (1986)
Garalla, S.: Análisis polínico de una secuencia sedimentaria del Holoceno Tardío en el Abra del Infiernillo, Tucumán, Argentina. Polen **12**, 53–63 (2003)
Gómez Augier, J.P., Caria, M.: Los paleoambientes y los procesos culturales en el noroeste argentino: una aproximación desde la arqueología de Tucumán. Acta Geológica Lilloana **24**(1–2), 80–97 (2012)
Gordillo, I.: De piedra, tierra y madera. Arquitectura y prácticas sociales en la Iglesia de los Indios (Ambato, Catamarca). El hábitat prehispánico. Arqueología de la arquitectura y de la construcción del espacio organizado, pp. 155–185. EDIUNJU, San Salvador de Jujuy, Argentina (2010)
Holdaway, S., Shiner, J., Fanning, P.: Hunter-Gatherers and the archaeology of discard behavior: an analysis of surface stone artifacts from Sturt National Park, Western New South Wales. Aust. Asian Perspect. **43**(1), 35–71 (2004)
Krieger, K.: Early man in the new world. In: Jennings, J., Norbeck, E. (eds.) Prehistoric Man in the New World, pp. 161–181. University of Chicago Press, Chicago (1964)
Liu, T., Broecker, W.: How fast does rock varnish grow? Geology **28**, 183–186 (2000)
Liu, T., Broecker, W.: Holocene rock varnish microstratigraphy and its chronometric application in the drylands of western USA. Geomorphology **84**, 1–21 (2007)
Liu, T., Broecker, W.: Rock varnish microlamination dating of late quaternary geomorphic features in the drylands of western USA. Geomorphology **93**, 501–523 (2008a)
Liu, T., Broecker, W.: Rock varnish evidence for latest Pleistocene millennial-scale wet events in the drylands of western United States. Geology **38**(3), 403–406 (2008b)
Liu, T., Dorn, R.: Understanding spatial variability in environmental changes in drylands with rock varnish microlaminations. Ann. Am. Geogr. **86**, 187–212 (1996)
Liu, T., Dorn, R.: Understanding the spatial variability of environmental change in drylands with rock varnish microlaminations. Ann. Assoc. Am. Geogr. **86**(2), 187–212 (2004)
Lorandi, A.M.: El arte rupestre del Noroeste Argentino. Área del norte de La Rioja y sur y centro de Catamarca. Dédalo **2**(4), 15–172 (1966)
Martínez, J., Mauri, E., Mercuri, C., Caria, M., Oliszewski, N.: Mid-Holocene human occupations in Tucumán (northwest Argentina). Quatern. Int. **307**, 86–95 (2013)
Nami, H.: Arqueología experimental, tecnología, artefactos bifaciales y modelos. Estado actual del conocimiento en Patagonia y Tierra del Fuego. Anales del Instituto de la Patagonia. Serie Ciencias Sociales **18**, 157–176 (1988)
Nami, H.: El subsistema tecnológico de la confección de instrumentos líticos y la explotación de los recursos del ambiente: una nueva vía de aproximación. Shincal **2**, 33–53 (1992)
Nuñez Regueiro, V., Tartusi, M.: Aguada y el proceso de integración regional. Estudios atacameños **24**, 9–19 (2002)
Ruiz Huidobro, O.: Descripción Geológica de la Hoja 11E Santa María, provincias de Catamarca y Tucumán, Bol. Nº 134. Dirección Nacional de Minería, Buenos Aires, Argentina (1972)
Sayago, J.M., Collantes, M.M., Neder, L., Busnelli, J.: Cambio climático y amenazas geoambientales en el área metropolitana de San Miguel de Tucumán (Argentina). Revista de la Asociación Geológica Argentina **66**(4), 546–566 (2010)
Schiffer, M.: Toward the identification of formation process. Am. Antiq. **48**(4), 675–706 (1983)
Schiffer, M.: Los procesos de formación del registro arqueológico. Boletín de Antropología Americana **23**, 39–46 (1991)

Somonte, C.: Tecnología lítica en espacios persistentes de Amaicha del Valle (Tucumán). Unpublished Doctoral Thesis, Universidad de Buenos Aires. Buenos Aires (2009)

Somonte, C., Baied, C.: Edad mínima de exposición de superficies en canteras-taller: reflexiones en torno a las primeras dataciones mediante microlaminaciones del barniz de rocas (VML) para el noroeste argentino. Chungara **45**(3), 427–445 (2013)

Somonte, C., Baied, C.: El palimpsesto como una puerta de acceso a diferentes temporalidades: el caso de Río Las Salinas 2 (Tucumán, Argentina). Chungara **55**, 35–55 (2017)

Somonte, C., Collantes, M.M.: Barniz de las rocas y espacios persistentes, su abordaje desde los procesos de reclamación artefactual lítica en Amaicha del Valle (Tucumán). Mundo de Antes **5**, 119–137 (2007)

Spiegelman, A.: Revista de la Asociación Argentina de Mineralogía, Petrología y Sedimentología, vol. 1–2, no. (2), pp. 1–18. Sedimentología del grupo de Santa María, Buenos Aires (1971)

Strecker, M.R.: Late Cenozoic Landscape in Santa María Valley, Northwestern Argentina. Unpublished PhD dissertation. Cornell University, Ithaca, USA (1987)

Tarragó, M.N., González, L.R.: Variabilidad en los modos arquitectónicos incaicos. Un caso de estudio en el valle de Yocavil (Noroeste Argentino). Chungara **37**(2), 129–143 (2005)

Tchilinguirian, P.: Paleoambientes sobre el Holoceno Medio (Noroeste Argentino). Estado de situación y problemática. In: Mondini, M., Martínez, J., Muscio, H., Marconetto, M.B. (eds.), Poblaciones Humanas y ambientes en el Noroeste Argentino durante el Holoceno Medio, Taller de Arqueología, Córdoba, pp. 13–21 (2011)

Yacobaccio, H., Morales, M.: Ambientes pleistocénicos y ocupación humana temprana en la Puna Argentina. Boletín de Arqueología PUCP **15**, 337–356 (2011)

# Author Index




M. M. Collantes et al. (Eds.): *Advances in Geomorphology and Quaternary Studies in Argentina*, SPRINGEREARTH, pp. 283–284, 2020.
https://doi.org/10.1007/978-3-030-22621-3

The manufacturer's authorised representative in the EU is Springer Nature Customer Service Centre GmbH, Europaplatz 3, 69115 Heidelberg, Germany. If you have any concerns regarding our products, please contact ProductSafety@springernature.com

Printed and bound by CPI Group (UK) Ltd, Croydon, CR0 4YY
30/09/2026
02225233-0004